Romanche fracture zone
Mid-Atlantic ridge
Cape Verde Islands
Sénégal
GAMBIA A.P.
VEMA FRACTURE ZONE
DOLDRUMS FRACTURE ZONE
SIERRA LEONE RIDGE
Freetown
Monrovia
Abijan
Accra
DEMERARA PLATEAU
AMAZON CONE
ST PAUL FRACTURE ZONE
ROMANCHE FRACTURE ZONE
CHAIN FRACTURE ZONE
Amazon
Tapajos
Xingu
Tocantins
Belém
Fortaleza
Rocas
Recife
Sao Francisco
ASCENSION FRACTURE ZONE
Salvador
ST HELENA FRACTURE ZONE
St Helena I.
HOTSPUR FRACTURE ZONE
MARTIN VAS FRACTURE ZONE
Martin Vaz I.
Trinidade I.
Rio de Janéiro
Sao Paulo
SANTOS PLATEAU
Paraná
Uruguay
RIO GRANDE FRACTURE ZONE
RIO GRANDE PLATEAU
Montevideo
TRISTAN DA CUNHA FRACTURE ZONE
GOUGH FRACTURE ZONE
Schwabenland
ARGENTINE RISE
FALKLAND AGULHAS FRACTURE ZONE
MAURICE EWING BANK
MALVINAS CHASM
SCOTIA RIDGE
South Georgia I.
BOUVET
CONRAD FRACTURE ZONE
South Sandwich Islands
BARKER RIDGE
BRISTOL BOUVET RIDGE
SOUTH SANDWICH
South Orkney Is.

GEOPHYSICS: The Earth's Interior

Inside Front Cover

South America and the ocean floor on both sides. Earthquake epicenters for the period 1960–1980 are shown in red. The shade of red indicates the focal depth (pink: less than 34 km; light red: 34 to 100 km; red: 101 to 300 km; deep red: 301 to 700 km). Only earthquakes of magnitude ≥ 4.5 are shown; those of magnitude ≥ 7.5 are circled in black.

The increasing depth of the foci from west to east under South America is very clear, as is the localization of most oceanic earthquakes near the oceanic ridge crests along fracture zones (see especially the Romanche fracture zone in the South Atlantic). In the East Pacific a series of highly seismic fracture zones extends from the present East Pacific Rise to the south coast of Chile. An abandoned East Pacific Rise is also visible between the present one and South America. The Nazca ridge, in contrast, is a typical aseismic ridge. (From Espinosa, A. F., W. Rinehart, and M. Tharp, 1981. Seismicity of the Earth Map (original scale 1 : 46,460,600 in full color display). Sunshine Canyon, Boulder, Colorado: Gerlein and Howard. (Copyright © 1982 by Rebecca M. Espinosa and Marie Tharp; reprinted by permission.)

Inside Back Cover

Part of Africa, Europe and Asia. Earthquake epicenters for the period 1960–1980 are shown in red. The shade of red indicates the focal depth (pink: less than 34 km; light red: 34 to 100 km; red: 101 to 300 km; deep red: 301 to 700 km). Only earthquakes of magnitude ≥ 4.5 are shown; those of magnitude ≥ 7.5 are circled in black.

Africa is essentially a seismic except for the East African Rift. A broad seismic belt of varying width borders the Alpine–Himalayan chain to the north, but most of the foci are shallow (≤ 100 km). The belt covers most of Iran and is remarkably wide over China. In contrast, the seismic belts below Indonesia and in the Western Pacific are sharply localized and follow subduction zones. (From Espinosa, A. F., W. Rinehart, and M. Tharp, 1981. Seismicity of the Earth Map (original scale 1: 46, 460, 600 in full color display. Sunshine Canyon, Boulder, Colorado: Gerlein and Howard. (Copyright © 1982 by Rebecca M. Espinosa and Marie Tharp: reprinted by permission.)

GEOPHYSICS: The Earth's Interior

Jean-Claude De Bremaecker

Department of Geology
Rice University

John Wiley & Sons

New York Chichester Brisbane Toronto Singapore

Library of Congress Cataloging in Publication Data:

De Bremaecker, Jean–Claude, 1923–
Geophysics, the earth's interior.

Includes index.
1. Geophysics. I. Title.
QC806.D39 1985 551 85-17388
ISBN 0-471-87815-4

Printed in the United States of America

10 9 8 7 6 5 4 3 2 1

PREFACE

The advent of the theory of plate tectonics has made it evident that geology and the physics of the earth's interior are inseparable and that study of both is needed to ensure further progress in the earth sciences. Although there are many excellent texts for students of physics or graduate students of geophysics, and even more books on geophysical prospecting, few if any recent introductory texts present geologists with the more global or "academic" aspects of geophysics.

I have attempted to fill this gap. In writing this book I have assumed that students have had at least one year each of mathematics (calculus) and physics, and that they are familiar with the general concepts of plate tectonics. (I have heard it said that some seniors in geology do not know what vectors or integrals are. If such students exist, this text is not written for them!) Several problems also require the ability to write simple computer programs. I have *not* assumed a knowledge of vector calculus, but I have used partial derivatives where needed. They are briefly explained when introduced. A lengthier, intuitive explanation is given in Appendix E, which also deals with the directional derivative and the gradient.

Finally, I have limited the length of the text to what can be used in a one-semester course, with some material left over for each person's choice. Such a length appears to be reasonable for an introduction to "academic" geophysics aimed at nonspecialists.

Within these constraints I have tried to present the methods used as well as some significant results, for both are important in understanding what the science is. In explaining new and difficult concepts, I have used an inductive approach: start with a simple example and generalize from it. Such an approach appears more natural than a purely deductive one. To reinforce the understanding of new ideas, I have placed problems immediately after the material they require. It is essential that students solve as many problems as possible in order to learn, in Thomas Kuhn's words, "consequential things about nature." (A set of solutions is available on request from the publisher.)

I have followed a more or less traditional division of geophysics into its branches and have attempted to show what each contributes to our understanding of the earth's interior. Seismology provides the most abundant information on our subject (and one of its subdivisions, reflection seismology, also provides employment for the majority of geophysicists!). I have thus devoted considerable attention to it. I then deal with gravity and magnetism, and end with the difficult problems of temperatures and convection in the earth. I have not looked here at those branches of geophysics dealing with water, air, and space; students should be aware, however, that these are also important.

In providing the student with references, I have tried to give either recent ones (for obvious reasons) or very early ones. One important reference for any aspect of geophysics deserves special mention: the Quadrennial Report of the U.S. National Committee for the International Union of Geodesy and Geophysics. The latest two of these reports have been published in *Reviews of Geophysics and Space Physics*. As its name implies, the report is updated every four years.

Some practical notes:

- A vector or tensor is indicated by a letter set in boldface type (e.g., **V**), its magnitude by the same letter set in italics (V), and its components by a subscript added to this letter (V_x). A unit vector is indicated by a circumflex, or "hat" ($\hat{\mathbf{V}}$).
- Lines of type more closely spaced, not right-justified, and set off by horizontal lines indicate subject matter that is not essential to an understanding of the material that follows; it is generally more mathematical in nature.
- Tables of units, symbols, and conversions are given in Appendices A and B.
- Selected data on the earth are given in Appendix C.
- Notations are given in Appendix D.
- Except in gravity, I have used SI units (Système International), which is the "rationalized" meter-kilogram-second-ampere system (see Appendices A and B).

I am grateful to the Department of Electrical Engineering for the free use of its word processing system, and to Donald Schroeder for typing much of the manuscript. Dr. Robert Hood of the Department of Computer Sciences graciously devised a computerized indexing scheme.

The text is the result of several years of teaching, of many comments and suggestions by colleagues, and of much suffering by students. I express my gratitude to all of them. Despite my efforts I am certain that many errors remain, and I will be glad to have them pointed out to me. I would also appreciate hearing about the selection or omission of topics or, indeed, about any aspect of this work.

I hope that this text may help bridge the gap between geology and geophysics and thus bring about a better understanding among scientists interested in these subjects.

Jean-Claude De Bremaecker
Rice University

CONTENTS

PART ONE

Seismology

1

Introduction

A. Plate Tectonics

No single unifying idea has had a greater influence on the earth sciences than plate tectonics for at least a century. Before looking at seismology itself, therefore, we should briefly review the basic concepts of plate tectonics. This review will help place seismology in perspective. Extended reviews can be found in most modern introductory geology textbooks (e.g., Press and Siever 1982); a more comprehensive but slightly dated review is given by Le Pichon et al. (1976).

Before the advent of the theory of plate tectonics, the continents were considered fixed with respect to one another, and land bridges raised by vertical motions of the ocean bottom were invoked to explain the migrations of some species. Although Alfred Wegener suggested in 1912 that continents could move, plowing through the oceanic crust, his ideas were rejected by the majority of earth scientists for several decades.

In the 1960s new evidence—discussed below—forced the conclusion that the continents were not fixed relative to each other. Instead, scientists discovered, the surface of the earth consists of six major plates, each comprising continental and oceanic areas, as well as several minor ones. These plates are essentially rigid

FIG. 1-1 *Two effects of the 1906 earthquake in and near San Francisco. (a) The City Hall of San Francisco. (b) Fault scarp near Point Reyes, California, north of San Francisco. (Courtesy of the Earthquake Information Bulletin, U.S. Geological Survey.)*

but move relative to each other. Geological activity is concentrated at plate boundaries.

From a purely kinematic point of view—that is, considering only the velocities and not the forces—boundaries of three different types must exist.

- Convergent boundaries, along which two plates move toward one another. In most actual cases one plate is oceanic while the other is continental along the boundary; the oceanic plate is generally overridden and subducted into the mantle. When continental plates collide, subduction is hampered because the continental crust is appreciably less dense than the mantle. The collision of India with Asia is the best-known example of such a collision. (See inside back cover.)
- Transform boundaries, along which two plates slide past each other. The best-known transform boundary is the San Andreas Fault in California (Figure 1-1), but transform faults are extremely numerous on the ocean bottom, as a good bathymetric map will show. (See inside front cover.)
- Divergent boundaries, along which partially molten material upwells from the depths. Oceanic ridges are typical divergent boundaries. (See inside front cover.)

A fourth type, intermediate boundaries, may or may not exist.

As the implications of plate tectonics for geology became clear, the forces giving rise to the motions—that is, the dynamics of the system—were investigated. At present we can accurately speak of a theory of plate tectonics, even though many of its details are still unclear or controversial.

In a highly simplified form the dynamic model is as follows (Figure 1-2). Heated by radioactive sources and by heat from the cooling core, the material in the earth's mantle expands and becomes lighter. Despite its very high viscosity, comparable to that of glass, the material rises more or less vertically in some places, especially under oceanic ridges. As it rises, it is subjected to less and less pressure and enters a region in which it is partially molten. Eventually the uppermost material may reach the ocean floor, near or at the ridge, as lava flows or dikes. Small earthquakes occur in such regions (Figure 1-3 and inside front cover). The slow stream of material close to the melting point is now forced to turn and travel in a horizontal direction. As it does so, it gradually cools, first near the surface, then to increasingly greater depths, and forms the oceanic lithosphere. Eventually, after traveling several thousand kilometers, it may plunge (mostly under island arcs and continents) as a "cold slab," helping to drag the lithosphere with it and causing major earthquakes. (See inside front and back covers.) The motions of the slab and the mantle under the continents push the latter one way or another. Sometimes they collide, forming mountain chains (e.g., the Himalayas); sometimes they separate, forming first rifts (East Africa) and then new oceans, and eventually perhaps repeating the cycle. In sketching this picture, we have implicitly defined the lithosphere as being bounded underneath by an isotherm (a surface of constant temperature). Because the lithosphere

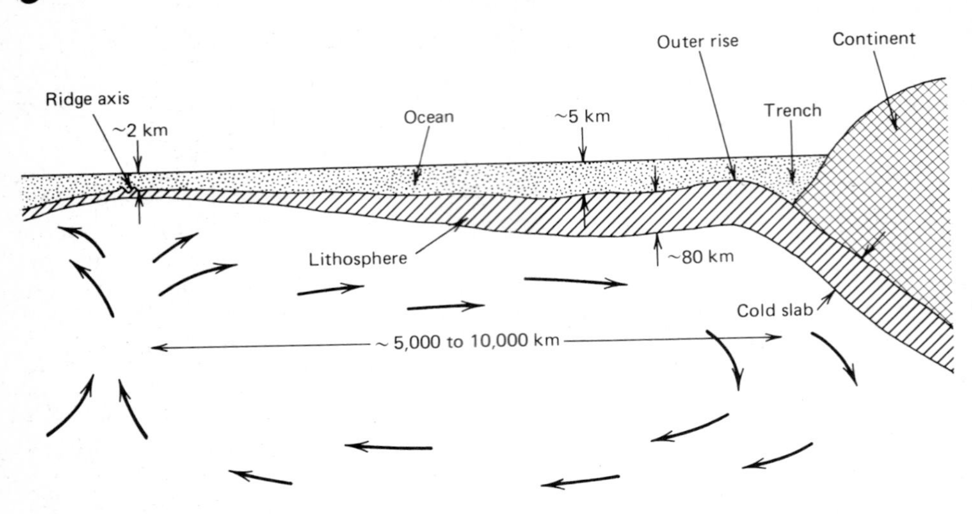

FIG. 1-2 *Schematic diagram, not drawn to scale, of a simple model of the upper mantle. The plate forms at a midocean ridge and subducts under a continent. The principal features are the ridge axis, the lithosphere thickening as it travels, the increasing ocean depth, the outer rise, the trench, and the cold slab. Hypothetical directions of flow are shown by arrows.*

gradually loses heat as it travels away from the ridge, this isotherm gradually increases in depth and the lithosphere gradually thickens. For simplicity's sake, we may consider the continental lithosphere to be bounded by this same isotherm (see also Chapter 5, Section B, and Chapter 12, Section E).

Although this picture as a whole is satisfactory, many of its details remain obscure. We will attempt to evaluate the evidence pertaining to these details as we progress. Because of the uncertainty surrounding many of these points, you will often encounter such expressions as "most scientists believe that. . . ." In such cases, it should be clearly understood, we are relying only on present scientific judgments, which are subject to review as new evidence becomes available.

B. Seismology and Earthquakes

Before turning to seismology, we should note that the great majority of geophysicists in the United States are employed by the oil industry to interpret seismograms obtained by means of artificial sources. In some respects (e.g., seismic stratigraphy), this work is close to geology; in others (e.g., multichannel filtering), it is close to electrical engineering. Numerous books cover various aspects of this subject in detail. From our standpoint, seismology deals with three aspects of earthquakes: the source, the transmission of seismic waves, and the receiver that makes these waves observable. For each aspect many different topics arise, some

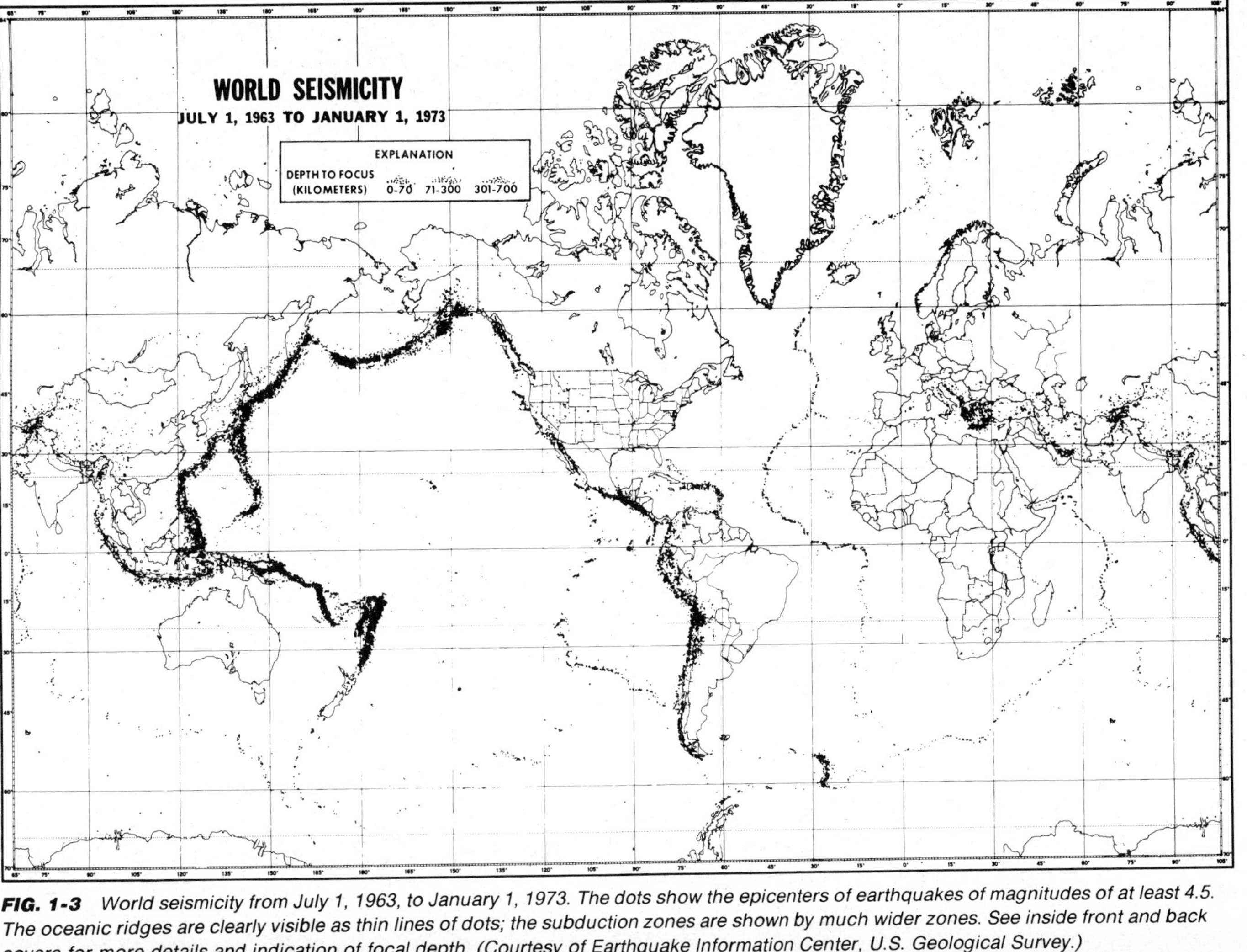

FIG. 1-3 *World seismicity from July 1, 1963, to January 1, 1973. The dots show the epicenters of earthquakes of magnitudes of at least 4.5. The oceanic ridges are clearly visible as thin lines of dots; the subduction zones are shown by much wider zones. See inside front and back covers for more details and indication of focal depth. (Courtesy of Earthquake Information Center, U.S. Geological Survey.)*

of which are discussed in the following sections. Together with other topics, they will also be discussed in subsequent chapters.

Earthquake Sources

The nature of an earthquake source yields information about the state of stress in the source area. The spatial distribution of some sources—the so-called Benioff-Wadati zones—provided the first indication of the existence of a plunging lithospheric slab. The restriction of focal depths to less than 670 km is an important but poorly understood constraint on the dynamics of plate tectonics. The temporal distribution of sources, and its application to earthquake prediction, is the subject of research in several countries. This research may perhaps save many lives in the future.

From a physical standpoint, earthquake sources consist of the abrupt release of the potential elastic energy stored in rocks over a period ranging from a few years to thousands of years. At the earthquake source the laws of elasticity do not apply, but prior to the earthquake or away from the source, these laws govern. Elasticity is thus basic to an understanding of the focal mechanism of earthquakes.

Seismic Waves

A small part of the energy (see Chapter 7, Section B) released by earthquakes is used to heat the volume near the source, but most of the energy is radiated away as elastic waves, also called seismic waves, of various types. While traveling through the earth, these waves are influenced by the characteristics of the media through which they pass. Elasticity enables us to understand and quantify this influence, and thus to determine the earth's structure—the crust, the mantle, the outer core, and the inner core—as well as the lateral changes near the surface, which, for instance, mirror the differences between continents and oceans, and the gradual cooling and thickening of the oceanic lithosphere. Elastic waves may hold the solution to such unsolved questions as the depth to which continents extend, the existence of deep mantle plumes, and the nature of some of the phase changes in the earth.

The Receiver

The primary function of a seismograph, or an array of seismographs is, of course, to record seismic waves. In addition, however, seismographic records can provide information about the earth's structure immediately below the seismographic station.

Other Aspects

Many aspects of seismology will not be addressed in this book, because their dimensions extend beyond the realm of geophysics. Included in this category are, for example,

- The destruction wrought by a great earthquake, which involves not simple statistics but human tragedy.
- The extreme difficulty or impossibility of building earthquake-resistant and inexpensive houses in developing countries where lumber is scarce—a socio-economic as well as an engineering question.
- The problems connected with issuing or not issuing an earthquake prediction, especially for a foreign country, which is nearly a moral question.

These matters, along with many similar ones, are valid and important aspects of seismology, but they fall outside the scope of this text.

2

Elasticity

As pointed out in Chapter 1, earth scientists now realize that the plates on the earth's surface have moved and are moving with respect to one another. This discovery has given rise to a number of significant questions. It is important to determine, for example, how the plates have moved, and why. Scientists are also reasonably sure that some mountain chains are still rising, and they would like to reconstruct the past and understand the forces underlying this phenomenon. Another important question involves folds and faults. These are common geologic features, but under what circumstances do rocks fold rather than break, and what determines the different types of folds? Although many aspects of these questions exceed the scope of this book, this chapter will help set the framework for later investigations.

Specifically, we will see that the concept of deformation—of strain, to be precise—differs appreciably from that of a vector, and that the forces inside a body are not simple vectors. These concepts will enable us to understand how one can, for example, measure the "stress field" near a major fault such as the San Andreas; such measurements may, perhaps, eventually enable scientists to predict a major earthquake.

In this part of the text, we will assume that rocks essentially are elastic. (This idea will be clarified later.) This assumption is warranted because we are here concerned mostly with phenomena of much less than an hour in duration. In Part Four we will examine the consequences that follow from assuming that rocks can flow like a viscous liquid, because the phenomena discussed there involve a time scale of 1 million years and more.

A final note: it is essential that you know exactly what a vector, a normal, and a force are before attempting to deal with the concepts that follow. Although these concepts may be rather hard to grasp at first, the examples should help make them clear.

A. The Stress Tensor

Consider, first, a simple question: What are the resultants of the forces acting on a horizontal plane h_1 through the base of a tree? The equality of action and reaction shows that the force $\mathbf{F}_a$ exerted by the weight of the top of the tree is balanced by an equal and opposite force $\mathbf{F}'_a$ exerted by the base of the tree. Both $\mathbf{F}_a$ and $\mathbf{F}'_a$ are applied at the same point P, which is near the center of the tree.

Now consider the resultants of the forces acting on a vertical plane h_2 passing through the same point P and of the same area as h_1. The resultant $\mathbf{F}_b$, exerted by the left part, is balanced by $\mathbf{F}'_b$, exerted by the right part. But $\mathbf{F}_b$ is, in general, much smaller than $\mathbf{F}_a$. We conclude that $\mathbf{F}_a$ and $\mathbf{F}_b$ are *not* vectors, because vectors possess only magnitude and direction, whereas $\mathbf{F}_a$ and $\mathbf{F}_b$ also depend on the orientation of the plane on which they act. $\mathbf{F}_a$ and $\mathbf{F}_b$ thus represent a new type of quantity: a vector associated with a particular plane. We call this quantity a tensor.

The same conclusion can be reached if we consider, say, the face of a cliff: the resultant of the forces acting on a horizontal area near a point is much greater than and has a different orientation from that acting on a vertical area of the same size near the same point. We must conclude, therefore, that the forces inside a body are not vectors.

To be precise, the quantity discussed above must be called a second-rank tensor. A zero-rank tensor is a scalar, which has magnitude but no direction. A first-rank tensor is a vector, which has both magnitude and direction. A second-rank tensor is a vector associated with a direction—here, that of the normal to the plane.

To understand the nature of the forces inside a body, we must turn to the general case. Consider a body of limited dimensions. Various forces are acting on its surface, deforming it from its state of rest. Gravity is also acting on every particle. Take a point P inside this body (Figure 2-1) and consider an imaginary plane cut of limited size passing through P and having an arbitrary but definite "direction" or "attitude." Geologists might define this "direction" by its dip and its strike. We shall define it by the direction of its normal, $\hat{\mathbf{n}}$, and call it "cut $\hat{\mathbf{n}}$."

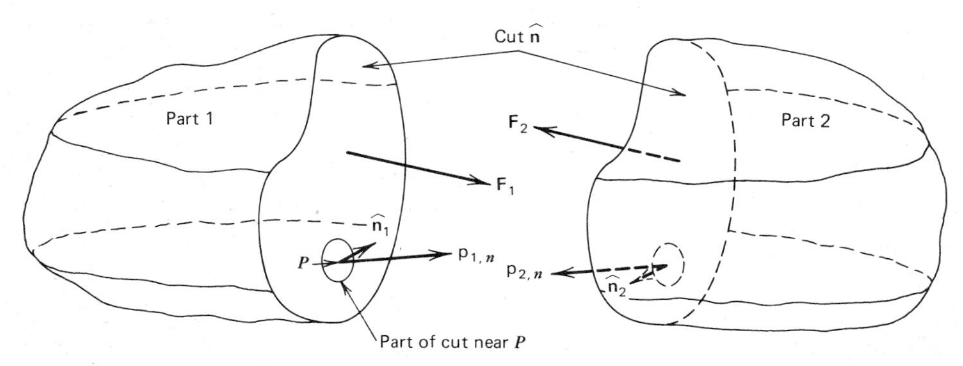

FIG. 2-1 *The two parts of a body divided by an imaginary cut. The parts are shown as separate and the cut is shown as extending through the whole body for the sake of clarity only.*

Because $\hat{\mathbf{n}}$ is a unit vector (i.e., it has direction and unit magnitude), we need only two numbers to identify it. Both $\hat{\mathbf{n}}_1$ and $\hat{\mathbf{n}}_2$ are normal to the cut (see below). In Figure 2-1, for the sake of clarity, the cut is shown as extending through the whole body, but this is not necessarily the case. For the same reason, the two parts of the body are shown separated, whereas in reality the body is joined along the cut, which is purely imaginary. Consider first the part of the body below (or to the left) of the cut, labeled part 1. The resultant of the forces on this cut, $\mathbf{F}_1$, is in general not at P and not normal to the cut. Next, consider the part of the body above (or to the right) of the cut, labeled part 2, and the resultant $\mathbf{F}_2$ of the forces acting on this cut. From the equality of action and reaction, $\mathbf{F}_1 = -\mathbf{F}_2$.

It will be much more convenient for later discussions if we get rid of the minus sign in the preceding equation. To do so, we observe that the outer normal on part 1, $\hat{\mathbf{n}}_1$, and that on part 2, $\hat{\mathbf{n}}_2$, are also equal (of unit length) and opposite in sign. By convention, we can thus count as positive the direction of the outer normal for each part of the body. With this new convention, we obtain $\mathbf{F}_1 = \mathbf{F}_2$. Observe that $\mathbf{F}_1$ and $\hat{\mathbf{n}}_1$ are generally *not* in the same direction; what the "outer normal convention" means is that $\mathbf{F}_1$ is positive if its projection on $\hat{\mathbf{n}}_1$ is positive, and that otherwise it is negative (or zero).

In SI units, a force is measured in newtons (see Appendix A). Thus, the force per unit area on the cut is $\mathbf{F}_1/A$ where A is the area of the cut. Such a quantity is measured in pascals (or $\mathrm{N/m^2}$). Thus,

$$\mathbf{f}_{1,n} = \frac{\mathbf{F}_1}{A} = \frac{\mathbf{F}_2}{A} = \mathbf{f}_{2,n}$$

where the subscript n indicates the direction of the normal.

In fact, though, the force per unit area may vary on different parts of the cut. Think, for instance, of a piece of chalk being bent, and of a cut perpendicular to its axis (Figure 2-2): it is intuitively obvious that the top of the cut is "in tension" and the bottom is "in compression." In order to obtain an exact

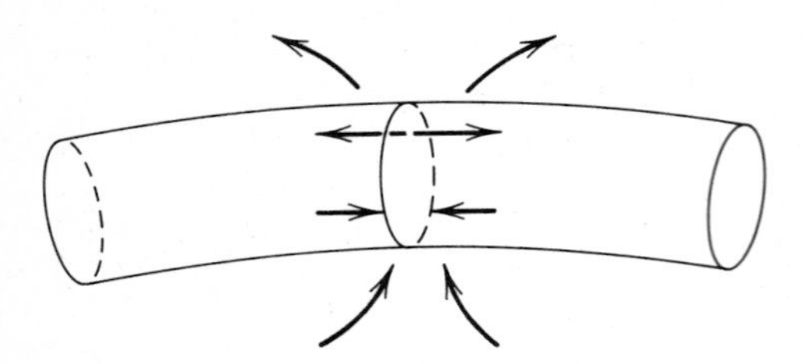

FIG. 2-2 The bending of a piece of chalk (as shown by the curved arrows) results in a tensional stress at the top and a compressional stress at the bottom on a cut normal to the axis.

statement of the situation, we consider only a small part of the cut near P (see Figure 2-1), and then let the area of this part go to zero. In most cases, the forces per unit area, on this part, $\mathbf{f}_{1,n}$ and $\mathbf{f}_{2,n}$, will tend to some limit, say, $\mathbf{p}_{1,n}$ and $\mathbf{p}_{2,n}$. Clearly, $\mathbf{p}_{1,n} = \mathbf{p}_{2,n}$. Thus, the indices 1 and 2 are no longer necessary, and we can write simply

$$\mathbf{p}_n = \mathbf{p}_{1,n} = \mathbf{p}_{2,n} \tag{2-1}$$

$\mathbf{p}_n$ is called the traction at P on $\hat{\mathbf{n}}$—that is, on the cut of outer normal $\hat{\mathbf{n}}$. Note particularly that gravity plays no role in obtaining the relation $\mathbf{p}_{1,n} = \mathbf{p}_{2,n}$, because gravity at a point acts on an infinitesimal mass, $dm = \rho\, dv$. Thus, gravity is a body (or volume) force. Tractions, on the other hand, act on an infinitesimal surface; thus they are surface forces. Accordingly, the nature of these forces and their dimensions are different. To repeat: ρg has the dimensions of a force per unit volume (kg m^{-2} s^{-2}), whereas $\mathbf{p}_n$ has the dimensions of a force per unit area (kg m^{-1} s^{-2}).

The equality of action and reaction is a *very limited* statement: it states only that *at P on cut* $\hat{\mathbf{n}}$, $\mathbf{p}_{1,n} = \mathbf{p}_{2,n}$. From it, we can infer nothing about the traction at any other point in the body, about the traction at P on another cut, about the relation between $\mathbf{p}_n$ and the forces acting on the surface of the body, or about the relation between $\mathbf{p}_n$ and the direction of gravity. In general, these relations are extremely complex.

Since $\mathbf{p}_n$ is generally not normal to cut $\hat{\mathbf{n}}$, it may be decomposed into a component along the normal, p_{nn} (Figure 2-3), and two components along the cut —say, p_{ns} and p_{nt} if we orient the s- and t-axes along the cut. (The n-, s-, and t-axes are mutually orthogonal.) The normal stress is p_{nn} (in geological literature, often σ_n), whereas p_{ns} and p_{nt} are tangential stresses (often τ_{ns} and τ_{nt}, respectively; sometimes τ_t and τ_s, respectively). Since $\mathbf{p}_{1,n} = \mathbf{p}_{2,n} = \mathbf{p}_n$, it follows that $p_{1,ns} = p_{2,ns} = p_{ns}$. These equal and opposite components induce a shear across cut $\hat{\mathbf{n}}$. Tangential stresses are therefore also called shear stresses.

The method given above shows whether p_{nn} is positive or negative in any given case. In the first case we say that p_{nn} is a tension; in the second, a compression. The signs of p_{ns} and p_{nt} are less easily decided. The simplest way to assign signs is to state that a shear stress is positive if it induces a positive shear strain (defined below). In two dimensions, the shear stresses shown by arrows on Figure 2-4 are counted as positive. It will be seen that such stresses tend to decrease the angle between the x- and y-axes. A more rigorous treatment follows.

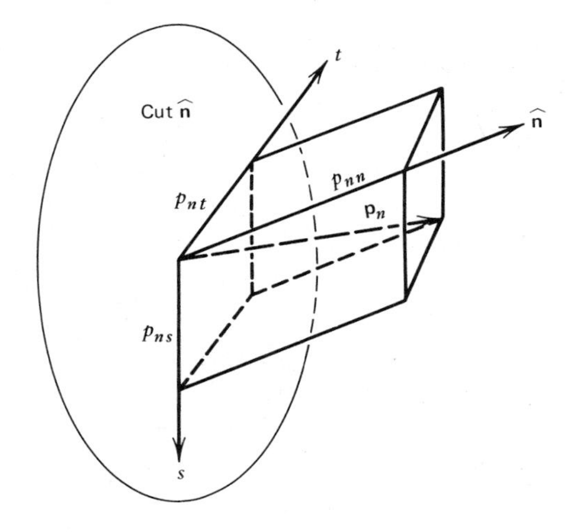

FIG. 2-3 *The traction $\mathbf{p}_n$ and its components p_{nn}, p_{ns}, and p_{nt}.*

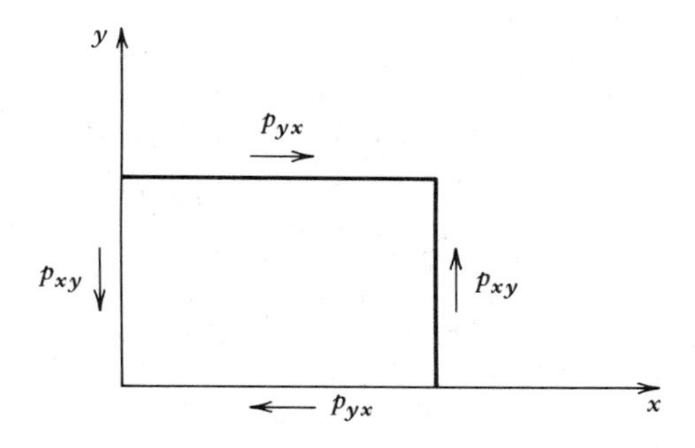

FIG. 2-4 *The shear stresses are positive in the directions shown by the arrows in the two-dimensional case.*

Let x, y, z be the coordinate axes. We define $\mathbf{p}_x$ (see Figure 2-5) as the traction on the face $ABCD$ of the parallelepipedon π—that is, on the cut perpendicular to the x-axis for that part of the body for which the positive x-direction is the outer normal. (Face $EFGH$ does not qualify here, since its outer normal is in the negative x-direction.) On face $ABCD$, the normal stress p_{xx} is positive in the $+x$-direction, the shear stress p_{xy} is positive in the $+y$-direction, and p_{xz} is positive in the $+z$-direction. For $\mathbf{p}_y$ on face $ADEF$ the same conventions apply to p_{yy}, p_{yz}, p_{yx}. The application to $\mathbf{p}_z$ is clear. On face $EFGH$, for which the $+x$-direction is the *inner* normal, p_{xx} is positive in the negative x-direction, p_{xy} is positive in the negative y-direction, and so forth.

As we already know, the traction $\mathbf{p}_d$ on another cut at P of normal $\hat{\mathbf{d}}$ bears no simple relation to $\mathbf{p}_n$. (Of course, $\mathbf{p}_d$ can also be decomposed into, say, p_{dd}, p_{de}, and p_{df}.) The question thus arises whether, at a point, the traction on every cut is independent of the traction on every other cut. Fortunately, this is not the case. We will see that given the traction on any three orthogonal planes at P, we

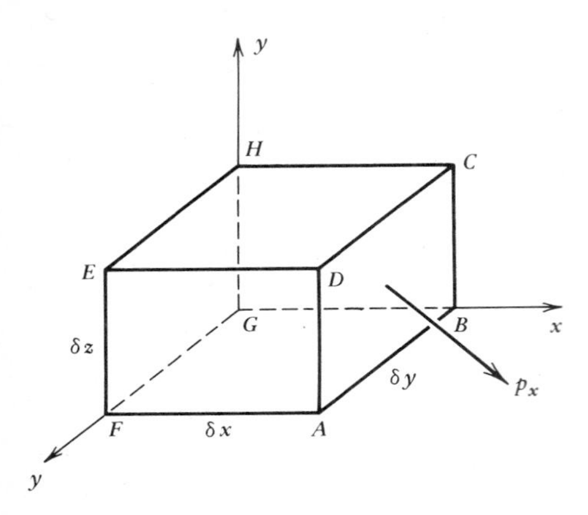

FIG. 2-5 *A rectangular parallelepipedon π of edges $\delta x, \delta y, \delta z$ near the origin, and the traction $\mathbf{p}_x$ on face ABCD.*

can compute the traction on any plane at P—that is, we can find the state of stress at P.

Because the two-dimensional case is simpler, we will examine it first. Thus, we assume that $\mathbf{p}_z = 0$ and that at P in Figure 2-6, we know $\mathbf{p}_x$ and $\mathbf{p}_y$. We want to find $\mathbf{p}_d$; $\hat{\mathbf{d}}$ is the outer normal on face AB of triangle PAB. This triangle is assumed to be infinitesimal. In order for the computation to be correct, $\mathbf{p}_x$, $\mathbf{p}_y$, and $\mathbf{p}_d$ must be drawn as positive. Consider $\mathbf{p}_y$ first: p_{yy} is positive along the outer normal to PA (i.e., in the negative y-direction); as a consequence, p_{yx} is positive along the negative x-direction, as shown. The same reasoning applies to $\mathbf{p}_x$, p_{xx}, and p_{xy}. On the other hand, p_{dd} is positive along the positive d-direction, and hence p_{de} is positive along the positive e-direction, as shown. It is also necessary that the relative orientation of the $x - y$ axes and that of the $d - e$ axes be the same: that is, that a counterclockwise rotation of $\pi/2$ brings x to y and d to e.

The triangle PAB is in equilibrium under the influence of the forces shown. To find p_{dd}, we project all the forces on the d-axis; their sum must be zero. We first note that in two dimensions, the stresses and tractions are forces per unit length rather than forces per unit area. We start with p_{xx}, which acts on line PB. The total force on PB exerted by p_{xx} thus is $p_{xx} \times PB$ and is negative. This force makes an angle (xd) with the d-axis. All the other cases are similar, and we find

$$\begin{aligned}&[-p_{xx}\cos(xd) \times PB] - [p_{xy}\cos(yd) \times PB] \\ &\qquad - [p_{yx}\cos(xd) \times PA] - [p_{yy}\cos(yd) \times PA] + [p_{dd} \times AB] = 0\end{aligned}$$

But $PB/AB = \cos(xd)$ and $PA/AB = \cos(yd)$, hence

$$\begin{aligned}p_{dd} = p_{xx}&\cos(xd)\cos(xd) + p_{xy}\cos(yd)\cos(xd) \\ &+ p_{yx}\cos(xd)\cos(yd) + p_{yy}\cos(yd)\cos(yd)\end{aligned}$$

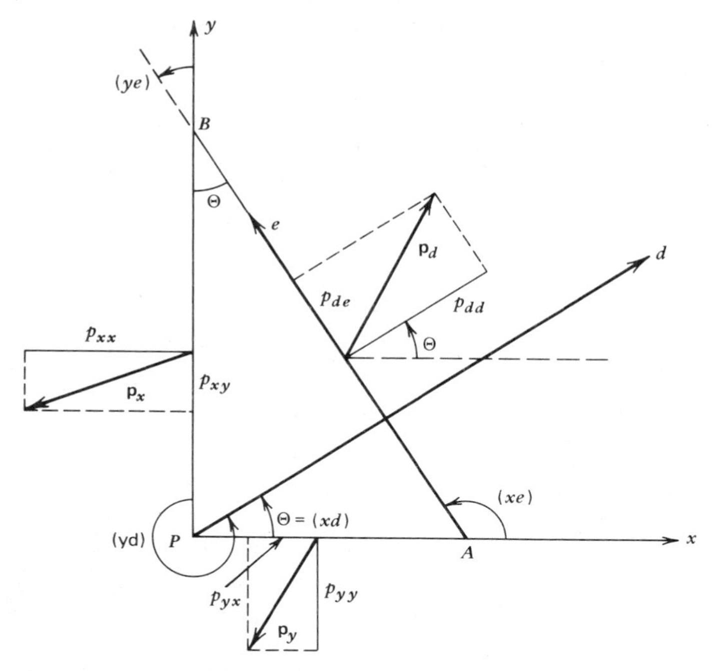

FIG. 2-6 *The triangle PAB. The tractions* $\mathbf{p}_x$, $\mathbf{p}_y$, *and* $\mathbf{p}_d$ *on each side are assumed to be positive, as are their components. The angles (xd), (xe), (yd), (ye), and* Θ *are also shown.*

Similarly, for p_{de},

$$
\begin{aligned}
-[p_{xx}\cos(xe)\times PB]-[p_{xy}\cos(ye)\times PB]-[p_{yx}\cos(xe)\times PA]\\
-[p_{yy}\cos(ye)\times PB]+[p_{de}\times AB]=0
\end{aligned}
$$

Hence,

$$
\begin{aligned}
p_{de}=p_{xx}\cos(xe)\cos(xd)+p_{xy}\cos(ye)\cos(xd)\\
+p_{xy}\cos(xe)\cos(yd)+p_{yy}\cos(ye)\cos(yd)
\end{aligned}
$$

These expressions for p_{dd} and p_{de} are cumbersome, but they have the advantage of clearly exhibiting the underlying symmetry. Furthermore, the method used to derive them can easily be extended to three dimensions (see below). By sacrificing symmetry, we can greatly simplify these expressions. Counting all angles clockwise, we note that

$$(xd)=\Theta, \qquad (yd)=\gamma=3\pi/2+\Theta$$

and

$$(xe)=\Theta+\pi/2, \qquad (ye)=\Theta$$

Hence, $\cos(yd) = \sin\Theta$ and $\cos(xe) = -\sin\Theta$. Finally, as we shall see (Equation 2-5), $p_{xy} = p_{yx}$. These relations enable us to write

$$p_{dd} = p_{xx}\cos^2\Theta + p_{yy}\sin^2\Theta + p_{xy}\sin 2\Theta \tag{2-2}$$

$$p_{de} = (p_{yy} - p_{xx})\sin\Theta\cos\Theta + p_{xy}\cos 2\Theta$$

where Θ is the angle from the x-axis to the d-axis.

At first sight, it may appear surprising that the above equation contains terms in $\cos^2\Theta$, $\sin^2\Theta$, and so forth. Where do these squared terms come from? Essentially, they stem from the fact that, say, p_{xx} is projected on the d-axis. The angle between the x- and d-axes is Θ, which gives rise to $p_{xx}\cos\Theta$. In addition, p_{xx} acts on side PB, whereas p_{dd} acts on side AB. These sides also make an angle Θ with each other, which gives rise to another $\cos\Theta$ factor. The squared term in $\cos^2\Theta$ is thus explained; the explanation for the other such terms is similar.

These equations are much more compact than the preceding ones but are not symmetric. They may be used, for example, to find the principal axes of stress (see below) or the orientation of the line on which the shear stress is maximum. To be strictly correct, we should have included in the balance of forces a force proportional to the area of the triangle (the two-dimensional analog of gravity). However, the area of the triangle is of order $dx \times dy$, whereas the sides are of order dx. The "area force" is thus negligible compared to the other forces.

We now can extend our method to three dimensions. Assuming that we know $\mathbf{p}_x$, $\mathbf{p}_y$, and $\mathbf{p}_z$ at point P, we wish to find $\mathbf{p}_d$ where $\hat{\mathbf{d}}$ is arbitrary. For simplicity's sake we will show only how to compute p_{de}; the other computations will be evident. To find p_{de}, we first write the equations of equilibrium of an infinitesimal tetrahedron (Figure 2-7) whose sloping side has $\hat{\mathbf{d}}$ for outer normal; $\hat{\mathbf{e}}$ is, of course, perpendicular to $\hat{\mathbf{d}}$. We again observe that the outer normals on the planes

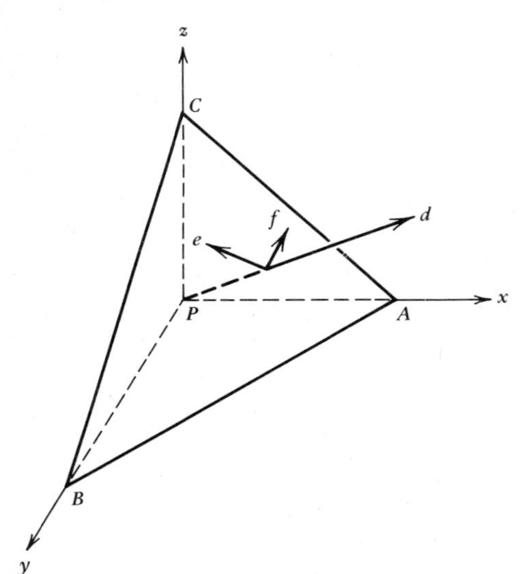

FIG. 2-7 *The tetrahedron PABC, with the x-, y-, and z-axes and the d-, e-, and f-axes.*

of the x-, y- and z-axes are in directions opposite to those axes. This implies that the positive direction of all stresses on these planes is opposite to the direction of the axes. On the other hand, the positive direction of the stresses on the d-plane is in the direction of the d-e-f axes. It is again necessary that both the x-y-z and the d-e-f axes be either right-handed (as shown) or left-handed.

Since we are computing p_{de}, we project all forces on the e-axis. Looking at, for example, p_{xx}, we see that it operates on the area PBC and makes an angle (xe) with the e-direction. All the other cases being similar, we obtain

$$\begin{aligned}
-[p_{xx}\cos(xe)\times PBC]-[p_{xy}\cos(ye)\times PBC]-[p_{xz}\cos(ze)\times PBC]\\
-[p_{yx}\cos(xe)\times PAC]-[p_{yy}\cos(ye)\times PAC]-[p_{yz}\cos(ze)\times PAC]\\
-[p_{zx}\cos(xe)\times PAB]-[p_{zy}\cos(ye)\times PAB]-[p_{zz}\cos(ze)\times PAB]\\
+[p_{de}\times ABC]=0
\end{aligned}$$

But the angle between two planes is equal to the angle between their normals; hence, $PBC/ABC=\cos(xd)$, $PAC/ABC=\cos(yd)$, and $PAB/ABC=\cos(zd)$. Thus,

$$\begin{aligned}
p_{de}=p_{xx}\cos(xe)\cos(xd)+p_{xy}\cos(ye)\cos(xd)+p_{xz}\cos(ze)\cos(xd)\\
+p_{yx}\cos(xe)\cos(yd)+p_{yy}\cos(ye)\cos(yd)+p_{yz}\cos(ze)\cos(yd)\\
+p_{zx}\cos(xe)\cos(zd)+p_{zy}\cos(ye)\cos(zd)+p_{zz}\cos(ze)\cos(zd)
\end{aligned}\qquad(2\text{-}3)$$

The symmetry of this equation is obvious. The other components (e.g., p_{ef}) may be obtained by replacing d by e and e by f in Equation 2-3.

PROBLEMS

2-1 Derive Equation 2-2 from Equation 2-3.

2-2 (**a**) Compute the traction on a horizontal cut through the base of an idealized vertical wall of total mass 450 kg per linear meter. The wall is 18 cm thick. Ignore atmospheric pressure.

(**b**) Using Equation 2-2, compute the normal and the shear stresses on cuts making angles of 20°, 40°, 60°, 80°, and 90° from the horizontal. Plot them against the angle. Sketch the x-, y-, d-, and e-axes, with their directions.

Although p_{xy} may seem to be a cumbersome notation, it has the advantage of drawing attention to the fact that it is the y-component of the $\mathbf{p}_x$ vector, which itself is the traction on the plane of outer normal $\hat{\mathbf{x}}$—and the same holds true for all the other ps. It is also a highly symmetric notation.

We have assumed that only a *force* acts on cut $\hat{\mathbf{n}}$. There also could be a *couple* acting on it, attempting to rotate it around the normal $\hat{\mathbf{n}}$. However, as the area of the cut tends to zero, the couple per unit area normally also goes to zero (Ewing et al. 1957, pp. 2–3), whereas the force per unit area does not. This is why the couple is often not mentioned at all in the problem. (See the discussion of moment tensors in Chapter 7, Section C.)

Definition

The array of numbers

$$\begin{matrix} p_{xx} & p_{xy} & p_{xz} \\ p_{yx} & p_{yy} & p_{yz} \\ p_{zx} & p_{zy} & p_{zz} \end{matrix} \tag{2-4}$$

is called the stress tensor at P.

Symmetry

Only six components of the stress tensor are independent, because

$$p_{xz} = p_{zx}, \quad p_{yx} = p_{xy}, \quad p_{zy} = p_{yz} \tag{2-5}$$

That is, the stress tensor is symmetrical.

Symmetry can be demonstrated by considering the infinitesimal parallelepipedon shown in Figure 2-8, which is assumed to be in equilibrium. (A body is said to be in equilibrium if both the linear acceleration of its center of mass and its angular acceleration about any axis are zero.) We assume that the components of the stress tensor vary with x, y, and z.

Examine the balance of rotational forces around the y-axis. Take first the two faces perpendicular to the x-axis. Three forces act on the left face, $EFGH$, whose area is $\delta y\,\delta z$.

1. $\delta y\,\delta z\,p_{xx}$. This force is very nearly balanced by the force acting on the right face, $ABCD$; the latter force is $\delta y\,\delta z[p_{xx} + (\partial p_{xx}/\partial x)\,\delta x]$, where $(\partial p_{xx}/\partial x)\,\delta x$ expresses the rate of change of p_{xx} as we move along the x-axis, multiplied by the distance traveled. As δx goes to zero, the difference between these two forces also goes to zero. These forces are almost equal and opposite, but each is positive along the outer normal to its respective face.
2. $\delta y\,\delta z\,p_{xy}$. This force is parallel to the y-axis and thus exerts no torque around it. The same is true of the force $\delta y\,\delta z[p_{xy} + (\partial p_{xy}/\partial x)\,\delta x]$ on the opposite face.
3. $\delta y\,\delta z\,p_{xz}$. This force passes through the y-axis, so it exerts no torque. The force on the opposite face, $\delta y\,\delta z[p_{xz} + (\partial p_{xz}/\partial x)\,\delta x]$, is unbalanced, however. Neglecting the term in δx, $T_1 = p_{xz}\,\delta x\,\delta y\,\delta z$ gives us the torque exerted by this force.

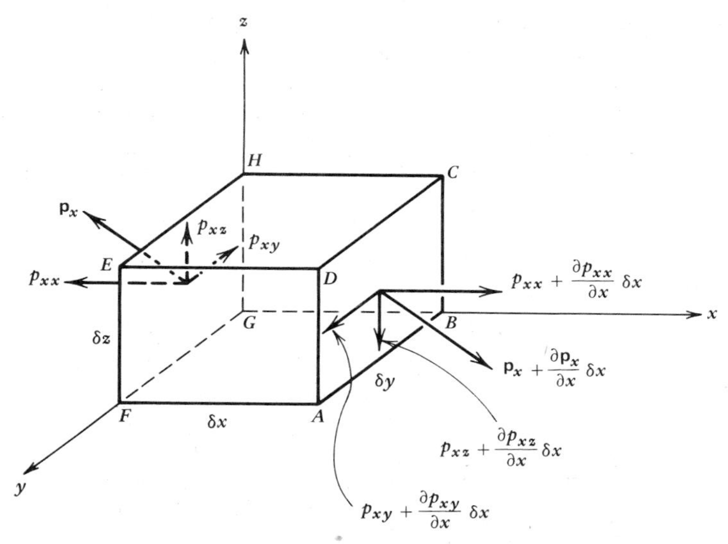

FIG. 2-8 *The tractions* $\mathbf{p}_x$ *and* $\mathbf{p}_x + (\partial \mathbf{p}_x/\partial x)\,\delta x$ *on the two x-faces of the parallelepipedon* π *of edges* $\delta x, \delta y, \delta y$, *and their components.*

Applying the same reasoning to the y-faces, we find that the forces are all balanced to order δy. Finally, for the z-faces:

1. The force $\delta x\,\delta y\,p_{zz}$ is balanced to order δz.
2. The force $\delta x\,\delta y\,p_{zy}$ has no torque.
3. The force $\delta x\,\delta y\,p_{zx}$ on the bottom face has no torque as it passes through the y-axis. However, the force $\delta x\,\delta y[p_{zx} + (\partial p_{zx}/\partial z)\,\delta z]$ on the upper face is unbalanced. Its torque is given by $T_2 = p_{zx}\,\delta x\,\delta y\,\delta z$.

Since the parallelepipedon has no angular acceleration, $T_1 = T_2$, from which it follows that $p_{xz} = p_{zx}$. The same reasoning obviously applies to the other components.

Pressure

At any given point the sum of the normal stresses on *any* three orthogonal planes is constant (i.e., invariant with respect to a rotation of the coordinate axes). Such a quantity, which has only magnitude, is a scalar. (Temperature is an example of a scalar.) The pressure, P, is defined as

$$P = -\tfrac{1}{3}(p_{xx} + p_{yy} + p_{zz}) \tag{2-6}$$

This is the only meaning of the "pressure" in the general case. (In the *special* case of a liquid at rest, $p_{xx} = p_{yy} = p_{zz} = -P$; this is the hydrostatic pressure.) The minus sign is used because the pressure is conventionally taken as positive for a compression, whereas the normal stresses are positive for a tension.

In geology, it is often assumed that the pressure in the earth is lithostatic—that is, $P = \rho g h$, where ρ is the density, g the acceleration of gravity, and h the depth. This equation often provides a good approximation at depths of at least a few kilometers, but it is not always correct near the surface: faults occur because of departure from the hydrostatic state.

Applications

2-1 Two layers of any two materials, solidly glued or welded together, are being bent (Figure 2-9). A point A is at the interface, and the y-axis at A is perpendicular to the interface. Which (if any) component(s) of the stress tensor is (are) continuous across the interface, and why? Is the pressure continuous, and why? (A function is said to be *continuous* at a point if the value of the function is the same regardless of the direction from which the point is approached. This is not an exact mathematical definition, but it is physically correct.)

Solution

Since the y-axis at A is perpendicular to the interface, the infinitesimal cut $\hat{\mathbf{y}}$ is along the interface (at A). Regardless of the nature of the two materials, the traction on cut $\hat{\mathbf{y}}$ on the first material (part 1) is thus equal and opposite to that on the second material (part 2). Because we count them as positive in opposite directions, it follows that $\mathbf{p}_y$ is continuous across the interface. Its components, p_{yx}, p_{yy}, and p_{yz}, thus are also continuous.

Consider now cut $\hat{\mathbf{x}}$, perpendicular to the interface. In each material, $\mathbf{p}_x$ is of course continuous, but there is no reason to assert that $\mathbf{p}_x$ in material 1 is the same as $\mathbf{p}_x$ in material 2. To visualize this concept think of the simple case of a rubber band stretched *in vacuo* in the x-direction (Figure 2-10). In the rubber band, $p_{xx} \neq 0$, but *in vacuo* it must of course be zero. Thus p_{xx} is discontinuous across the interface. On the other hand, $\mathbf{p}_y$ is null at the surface of the rubber band, since no force of any kind can act on it from the

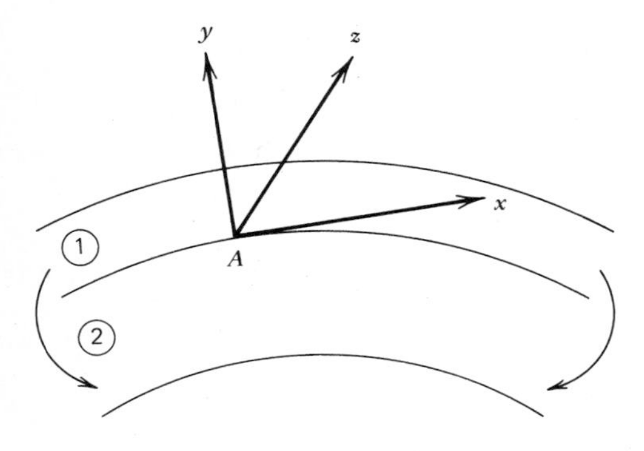

FIG. 2-9 *Two layers, of materials 1 and 2 glued together, are being bent.*

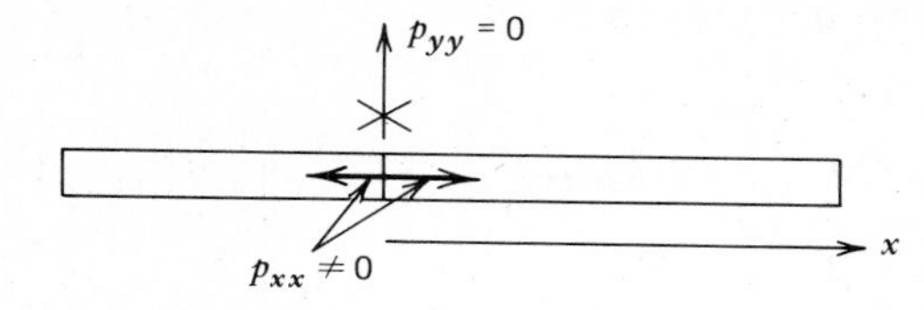

FIG. 2-10 A rubber band stretched in the x-direction.

vacuum. Thus all its components vanish at the surface: $p_{yy} = p_{yz} = p_{yx} = 0$. They are also zero *in vacuo* and are thus continuous across the interface. Similarly, $\mathbf{p}_z = 0$. Since $p_{xx} \neq 0$ whereas $p_{yy} = p_{zz} = 0$, we obtain $P \neq 0$ at the surface in the rubber band but $P = 0$ *in vacuo*. Reverting to the problem at hand, we can now see clearly that p_{yy} is continuous across the interface but that p_{xx} and p_{zz} generally are discontinuous. It follows that P generally is discontinuous also.

The fact that these considerations are independent of the material properties, of the state of rest or motion, and of the presence or absence of gravity, cannot be overemphasized.

2-2 Setting aside the influence of the air, is the pressure always zero at the earth's surface?

Solution
This case is similar to that of the rubber band being stretched, for the following reason. If we take the y-axis as vertical, then the traction $\mathbf{p}_y$ on cut $\hat{y}$ is null, just as in the case of the rubber band. But the tractions $\mathbf{p}_x$ and $\mathbf{p}_z$ on cuts perpendicular to the earth's surface generally are not null: there may be, say, a tension along the E–W direction and a compression along the N–S direction. Thus, the components of $\mathbf{p}_x$ and $\mathbf{p}_z$ generally are not null. In particular, p_{xx} and p_{zz} usually are not null. Thus, in general,

$$P = -\tfrac{1}{3}(p_{xx} + p_{yy} + p_{zz}) \neq 0$$

2-3 What is the exact meaning of the statement: "The lithosphere in a subduction zone is in tension"?

Solution
Strictly speaking, this statement is meaningless. "Tension" and "compression" have meaning only if the normal to the cut on which the tension or the compression acts is defined. What actually is meant (see Figure 3-28) is that the normal stress in the direction of subduction is a tension.

PROBLEMS

2-3 Consider the surface of an outcrop that is planar and horizontal. Take a point P on the surface, and draw the z-axis vertical through P. Which, if any, component(s) of the stress tensor is (are) unconditionally zero at P, and why? (Disregard the effects of the air.)

2-4 Does the definition of the stress tensor depend on the "nature" of the body (elastic, viscous, etc.)? Why or why not?

2-5 Consider an elastic solid that is compressed and twisted by forces acting on its sides. The upper surface consists of two parts: a heavy mass rests on the left part, whereas the right part (r) is free and horizontal. Take a point P on r with the z-axis vertical. Do any components of the stress tensor at P vanish, and why?

2-6 Consider a slowly flowing liquid. Choose a point P at the surface, and take the x-axis perpendicular to the surface at P. Which components of the stress tensor are zero at P, and why? (Disregard the influence of the air.)

Principal Stresses

It may be shown that at every point in a body, regardless of the state of stress, there exist three orthogonal planes on which the traction is purely normal—that is, on which there are no shear stresses. The intersections of these planes define the principal axes of stress, and the stresses on them are called the principal stresses. In engineering literature these stresses are designated σ_1, σ_2, σ_3 and the convention $\sigma_1 > \sigma_2 > \sigma_3$ is used (Jaeger 1969, p. 13). If two of the principal stresses are equal, only the plane on which the third principal stress acts is determined. If all three principal stresses are equal, there is no shear stress in any direction; this is the hydrostatic case. In the two-dimensional case there are, of course, only two principal stresses, at right angles to each other, and the values of the normal stresses are greatest and least along the principal axes of stress.

In geophysics and engineering a traction is generally taken as positive along the *outer* normal, and hence a tension (normal stress) is taken as positive. In geology, compressive normal stresses are generally taken as positive: in other words, traction is counted positive in the direction of the *inner* normal. Still, $\sigma_1 > \sigma_2 > \sigma_3$; thus, σ_1 is the largest principal compressive stress, σ_2 the second largest, and so forth. (We will not follow this geologic convention.) Finally, it should be noted that the word "pressure" is sometimes used by engineers to mean "compressive normal stress."

Application

2-4 In preparation for the more realistic case in which strains are measured, we consider the following two-dimensional problem: If the normal stresses in three arbitrary but coplanar directions at a point are known, what are the magnitude and direction of the principal stresses and of the maximum shear stress?

Solution

We begin by orienting the x-axis along the direction of one of the principal stresses. We do not know this direction, nor do we know whether the principal stress along the x-axis is larger or smaller than that along the y-axis. Thus, we designate these two principal stresses $p_{xx}{}^{\mathrm{M}}$ and $p_{yy}{}^{\mathrm{M}}$ (where the superscript M stands for "maximum"). Finally, we use Θ_1, Θ_2, and Θ_3 to designate the angles from the unknown x-axis to directions, 1, 2, and 3, respectively (Figure 2-11). Note that whereas these angles are unknown, the angles between the directions are known.

Since $p_{xx}{}^{\mathrm{M}}$ and $p_{yy}{}^{\mathrm{M}}$ are the principal stresses, $p_{xy} = 0$. Moreover, $\sin^2\Theta = 1 - \cos^2\Theta$. Equation 2-2 thus yields, for each direction,

$$p_{11} = p_{xx}{}^{\mathrm{M}}\cos^2\Theta_1 + p_{yy}{}^{\mathrm{M}}(1 - \cos^2\Theta_1) \tag{2-7}$$

$$p_{22} = p_{xx}{}^{\mathrm{M}}\cos^2\Theta_2 + p_{yy}{}^{\mathrm{M}}(1 - \cos^2\Theta_2) \tag{2-8}$$

$$p_{33} = p_{xx}{}^{\mathrm{M}}\cos^2\Theta_3 + p_{yy}{}^{\mathrm{M}}(1 - \cos^2\Theta_3) \tag{2-9}$$

Hence,

$$\begin{aligned} p_{11} - p_{22} &= p_{xx}{}^{\mathrm{M}}(\cos^2\Theta_1 - \cos^2\Theta_2) - p_{yy}{}^{\mathrm{M}}(\cos^2\Theta_1 - \cos^2\Theta_2) \\ &= (p_{xx}{}^{\mathrm{M}} - p_{yy}{}^{\mathrm{M}})(\cos^2\Theta_1 - \cos^2\Theta_2) \end{aligned}$$

But (Peirce 1929, p. 76)

$$\cos^2 a - \cos^2 b = -\sin(a+b) \times \sin(a-b)$$

Hence,

$$p_{11} - p_{22} = (p_{yy}{}^{\mathrm{M}} - p_{xx}{}^{\mathrm{M}})\sin(\Theta_1 + \Theta_2)\sin(\Theta_1 - \Theta_2)$$

Figure 2-11 shows that $\Theta_2 = \Theta_1 + \alpha_2$. Hence,

$$\Theta_1 - \Theta_2 = -\alpha_2, \qquad \Theta_1 + \Theta_2 = 2\Theta_1 + \alpha_2$$

where, as previously noted, α_2 is known but Θ_1 is not. We thus obtain

$$p_{11} - p_{22} = (p_{yy}{}^{\mathrm{M}} - p_{xx}{}^{\mathrm{M}})\sin(2\Theta_1 + \alpha_2) \times \sin(-\alpha_2)$$

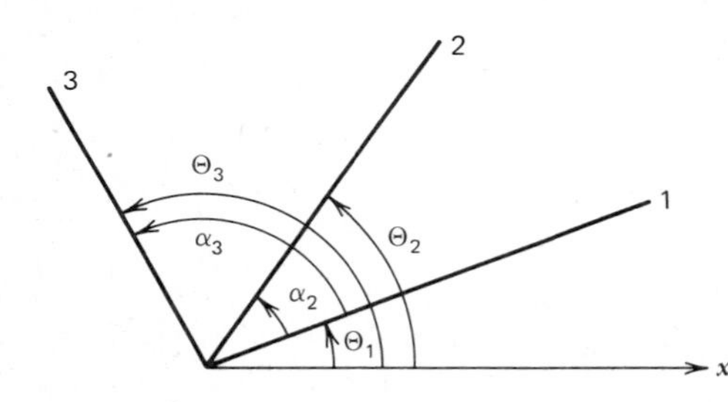

FIG. 2-11 *The directions 1, 2, and 3, the angles Θ_1, Θ_2, and Θ_3, and the angles α_2 and α_3.*

or

$$p_{11} - p_{22} = \left(p_{xx}{}^{\text{M}} - p_{yy}{}^{\text{M}}\right) \sin\alpha_2 \times \sin(2\Theta_1 + \alpha_2)$$

Similarly,

$$p_{11} - p_{33} = \left(p_{xx}{}^{\text{M}} - p_{yy}{}^{\text{M}}\right) \sin\alpha_3 \times \sin(2\Theta_1 + \alpha_3)$$

Dividing the last two equations by each other yields

$$\frac{\sin(2\Theta_1 + \alpha_2)}{\sin(2\Theta_1 + \alpha_3)} = \frac{(p_{11} - p_{22})\sin\alpha_3}{(p_{11} - p_{33})\sin\alpha_2} \tag{2-10}$$

Because all the terms on the right are known, the equation may be solved for Θ_1. (It may be necessary to use graphs or a computer, but the solution can be obtained.) Thus, we have determined the direction of the x-axis, and along with it, that of the y-axis. We obtain the values of Θ_2 and Θ_3 from those of Θ_1, α_2 and α_3.

We now insert the values of Θ_1, Θ_2, and Θ_3 into any two of the three equations 2-7, 2-8, and 2-9, and solve for $p_{xx}{}^{\text{M}}$ and $p_{yy}{}^{\text{M}}$.

The maximum shear stress, $p_{de}{}^{\text{M}}$, can now be obtained from Equation 2-2, where $p_{xy} = 0$: that is,

$$p_{de} = \left(p_{yy}{}^{\text{M}} - p_{xx}{}^{\text{M}}\right) \sin\Theta \times \cos\Theta$$

where the angle Θ from x to d remains to be determined. Clearly, p_{de} is maximum for

$$\frac{dp_{de}}{d\Theta} = 0 = \left(p_{yy}{}^{\text{M}} - p_{xx}{}^{\text{M}}\right)\left(\cos^2\Theta_{\text{M}} - \sin^2\Theta_{\text{M}}\right)$$

that is, for $\Theta_{\text{M}} = \pi/4$ or $3\pi/4$. Thus, $\cos\Theta_{\text{M}} \times \sin\Theta_{\text{M}} = 0.5$, and

$$p_{de}{}^{\text{M}} = 0.5\left(p_{yy}{}^{\text{M}} - p_{xx}{}^{\text{M}}\right) \tag{2-11}$$

which solves the problem.

PROBLEM

2-7 The three normal stresses given below have been measured along the directions indicated.

N 50°E: 1.464 MPa

N 5°E: −0.649 MPa

N 55°W: 0.2395 MPa

Find the direction and magnitude of the principal stresses and of the maximum shear stress.

B. The Strain Tensor

Definition

One might intuitively define *strain* as a measure of the relative change in the length or the shape of an object. More precise definitions follow.

Relative Change in Length Consider, first, a rod that is not subject to any force and is 50 cm long in the direction of the y-axis. Forces—any forces, not necessarily "pulls"—are applied to it, and its length increases to 50.2 cm, still in the y-direction. The relative lengthening is $(50.2 - 50)/50$. [One might ask whether the relative lengthening is not $(50.2 - 50)/50.2$; in fact, this question does not arise, because we are considering only infinitesimal strains.) Now consider a small element of length, initially δy, that has been lengthened in the y-direction by δv (here, as a first approximation, $\delta y = 50$ and $\delta v = 0.2$). If $\delta y \to 0$, then, normally, $\delta v / \delta y$ goes to a certain limit. The strain e_{yy} is defined as

$$e_{yy} = \frac{\partial v}{\partial y} \approx \frac{\delta v}{\delta y} \tag{2-12}$$

where the ∂ indicates that we only consider the change of v with y. If the rod is uniform and a tensional force acts only at the ends, $e_{yy} = 0.2/50 = 4 \times 10^{-3}$ for any point along the rod.

PROBLEM

2-8 Apart from a traction on its ends, what simple physical situation could lead to a lengthening of a cylindrical rod by applied forces?

In general, however, the situation is more complex. Consider the case shown in Figure 2-12. In a planar body at rest are the points P, A, and B which are infinitely close together. After some forces have been applied P has not moved but A has moved to A' and B has moved to B'. Assume that the lengths of AA' and BB' are much smaller than those of PA and PB, respectively.

Designating u and v as the components of the displacement along the x- and y-axes, respectively, we find, just as in the case of the rod, that

$$e_{xx} = \frac{\text{(final-initial) length in } x\text{-direction}}{\text{initial length in } x\text{-direction}}$$

or

$$e_{xx} = \frac{PA'_x - PA}{PA} = \frac{AA'_x}{PA}$$

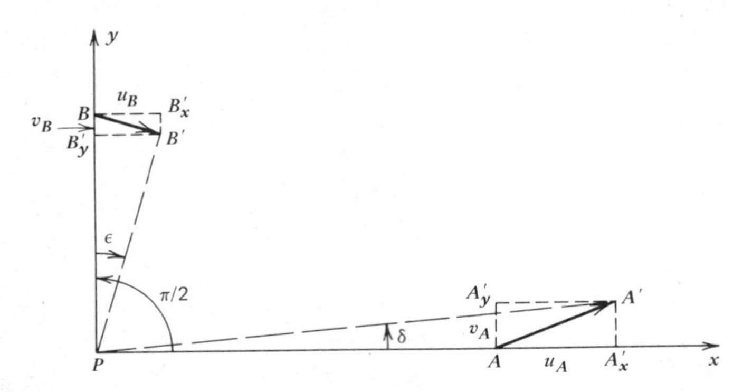

FIG. 2-12 *Displacements of points A and B near the origin, their components, and the resultant angles.*

But $AA'_x = u_A$. P has not moved, so $u_P = 0$. Thus, $AA'_x = u_A - u_P = \delta u_A$, while $PA = \delta x$. Thus,

$$e_{xx} = \frac{u_A - u_P}{\delta x} = \frac{\delta u_A}{\delta x}$$

(Remember that we suppose that points P, A, and B are infinitely close together.) We then obtain

$$e_{xx} = \frac{\partial u}{\partial x}$$

It is important to note that the length PA' is not relevant here. Only its projection on the x-axis, PA'_x, is significant, because the variation of u with x alone defines e_{xx}. It is to distinguish this variation from a general variation of u that the symbol ∂ (partial derivative) is used.

In exactly the same way, we obtain

$$e_{yy} = \frac{\text{(final-initial) length in } y\text{-direction}}{\text{initial length in } y\text{-direction}}$$

or

$$e_{yy} = \frac{PB'_y - PB}{PB} = \frac{BB'_y}{PB} = \frac{\delta v_B}{\delta y}$$

which is exact only for $\delta y \to 0$, in which case, it yields

$$e_{yy} = \frac{\partial v}{\partial y}$$

If the third dimension comes into play, we find, by induction,

$$e_{zz} = \frac{\partial w}{\partial z}$$

To summarize,

$$e_{xx} = \frac{\partial u}{\partial x} \qquad \text{(2-13)}$$

$$e_{yy} = \frac{\partial v}{\partial y}$$

$$e_{zz} = \frac{\partial w}{\partial z}$$

These three quantities are the normal strains; they are nondimensional.

Relative Change in Shape (or Distortion) Consider, first, the simple case shown in Figure 2-13, in which the solid lines indicate the cross section of a parallelepipedon resting on the $x - y$ plane. Suppose that as a result of external forces, the cross section is deformed to the rhombus shown by dashed lines, and that in the process all the points move parallel to the x-axis. The area of the cross section has not changed, but the shape has. The angle ε shown on the figure is a measure of this change of shape, or distortion, which is called the shear strain. (It will be seen below that the shear strain is $\frac{1}{2}\varepsilon$.) This type of shear, called simple shear, is not of much practical interest, but it does illustrate the idea of change of shape.

Now consider the more general case shown in Figure 2-12. BPA is a right angle whereas $B'PA'$ is not. The shear strain measures this angular change. Since BB' is small compared to PB, and AA' is small compared to PA,

$$\varepsilon \approx \tan \varepsilon = \frac{B'_yB'}{PB} \approx \frac{BB'_x}{PB}$$

But $BB'_x = u_B$. P has not moved, so $u_P = 0$. Thus, $BB'_x = u_B - u_P = \delta u_B$, while $PB = \delta y$. Thus,

$$\varepsilon = \frac{u_B - u_P}{\delta y} = \frac{\delta u_B}{\delta y}$$

At the limit when $\delta y \to 0$,

$$\varepsilon = \frac{\partial u}{\partial y}$$

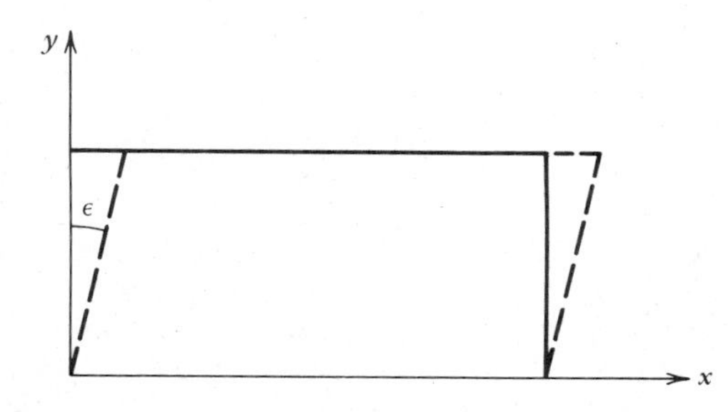

FIG. 2-13 *Simple shear; the angle ε is a measure of the shear strain. Solid lines indicate the undeformed state; dashed lines indicate the deformed state.*

Again, the variation of u with y is the only variation being considered. Exactly the same reasoning leads us to

$$\delta \approx \tan \delta \approx \frac{AA'_y}{PA} = \frac{v_A - v_P}{\delta x} = \frac{\delta v_A}{\delta x}$$

and at the limit,

$$\delta = \frac{\partial v}{\partial x}$$

The shear strain in the (x, y) plane is defined as

$$e_{xy} = \frac{1}{2}(\delta + \varepsilon) = \frac{1}{2}\left(\frac{\partial u}{\partial y} + \frac{\partial v}{\partial x}\right) = \psi$$

Note that $\delta > 0$ if $\delta v > 0$ (if $\delta x > 0$), and that the same is true for ε. Thus, $e_{xy} > 0$ if the angle from the x-axis to the y-axis *decreases* from $\pi/2$ to $(\pi/2) - \psi$, when $\psi > 0$. If, instead, the situation is as shown in Figure 2-14, $\partial u/\partial y < 0$ and $\partial v/\partial x > 0$. (Note that here, $e_{xx} < 0$). In such a case, we might have $e_{xy} = 0$—in which case the body has simply rotated around P.

The three shear strains are thus

$$e_{xy} = \frac{1}{2}\left(\frac{\partial u}{\partial y} + \frac{\partial v}{\partial x}\right) \tag{2-14}$$

$$e_{yz} = \frac{1}{2}\left(\frac{\partial y}{\partial z} + \frac{\partial w}{\partial y}\right)$$

$$e_{xz} = \frac{1}{2}\left(\frac{\partial u}{\partial z} + \frac{\partial w}{\partial x}\right)$$

Both the normal and the shear strains may be written in the same form:

$$e_{ij} = \frac{1}{2}\left(\frac{\partial u_i}{\partial x_j} + \frac{\partial u_j}{\partial x_i}\right) \tag{2-15}$$

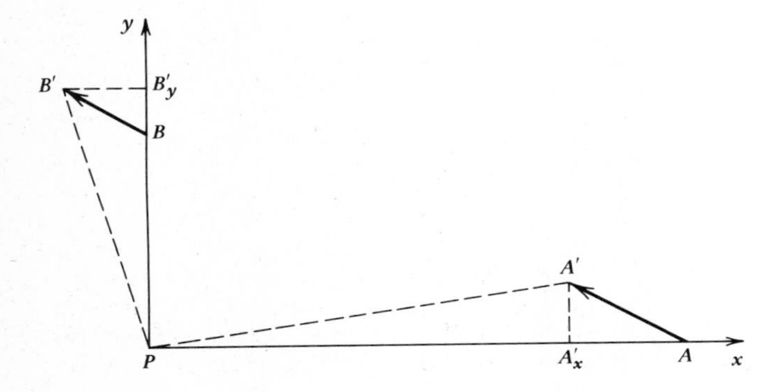

FIG. 2-14 *Displacements of points A and B near the origin in a case different from that shown on Figure 2-12.*

for

$$i=1:\ u_i=u, \qquad x_i=x$$
$$i=2:\ u_i=v, \qquad x_i=y$$
etc.

and similarly for j.

Geologists often use the convention

$$\varepsilon_{ij}=-\frac{1}{2}\left(\frac{\partial u_i}{\partial x_j}+\frac{\partial u_j}{\partial x_i}\right)$$

according to which a *decrease* in length becomes positive, as does an *increase* in angle. Engineers use the notation $\gamma_{xy}=2e_{xy}$, which makes it impossible to write Equation 2-15.

Finally, note that e_{xy} involves no change in volume. Figure 2-15 shows that if the rectangle $PABC$ gets distorted into $PA'B'C'$, the two areas remain equal, since AA', BB', etc., are infinitesimal. In three dimensions the volume is preserved.

In the general case (Figure 2-16), if P, A, and B have all moved but $AA' \ll PA$, $BB' \ll PB$, and so forth, we have, approximately,

$$e_{xx}\approx\frac{(x\text{-displacement at end})-(x\text{-displacement at origin})}{\text{original length along } x}$$

Thus,

$$e_{xx}\approx\frac{AA'_x-PP'_x}{PA}$$
$$e_{xx}\approx\frac{u_A-u_P}{\delta x}=\frac{\delta u_A}{\delta x}$$

and, at the limit, $e_{xx}=\partial u/\partial x$. Similarly, the shear strains may be computed as follows.

$$\tan\delta\approx\frac{(y\text{-displacement at end})-(y\text{-displacement at origin})}{\text{original length along } x}$$

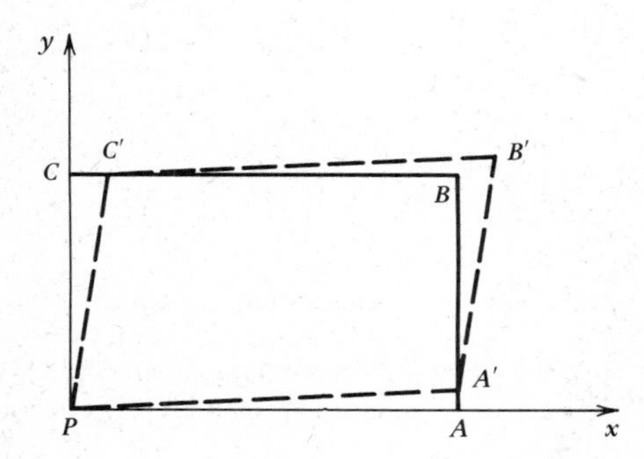

FIG. 2-15 *The area of a rectangle PABC deformed in shear to PA′B′C′ does not change.*

or

$$\delta \approx \tan\delta \approx = \frac{AA'_y - P'_xP'}{PA}$$

since $P'_xP = AA''$. Hence,

$$\delta = \frac{v_A - v_P}{\delta x} = \frac{\delta v_A}{\delta x}$$

and, for $\delta x \to 0$,

$$\delta = \frac{\partial v}{\partial x}$$

with the same holding true for ε. Thus,

$$e_{xy} = \frac{1}{2}\left(\frac{\partial v}{\partial x} + \frac{\partial u}{\partial y}\right) \approx \frac{1}{2}\left(\frac{v_A - v_P}{\delta x} + \frac{u_B - u_P}{\delta y}\right)$$

These expressions are very similar to the ones used to compute the shear strain in California and other earthquake-prone areas by careful measurement of the change with time of the distance between points. The main difference is that in the field, it is generally impossible to choose PA and PB at right angles to each other. At the end of this section, we will see how this difficulty can be overcome.

The preceding theory supposes that the deformations are small and that the strains are infinitesimal. Is such a theory valid in the earth's deep interior, where the change in volume caused by compression is important? Yes and no! In the case of an elastic wave passing through the earth's interior, for example, the attending displacement is small and the resultant strains are much smaller than 10^{-3}, so the infinitesimal strain theory will work very well. If, in contrast, we wish to compare the density at a depth of 2000 km with that near the surface, then we will find the strains to be larger than 10^{-1} and will need to use finite strain theory

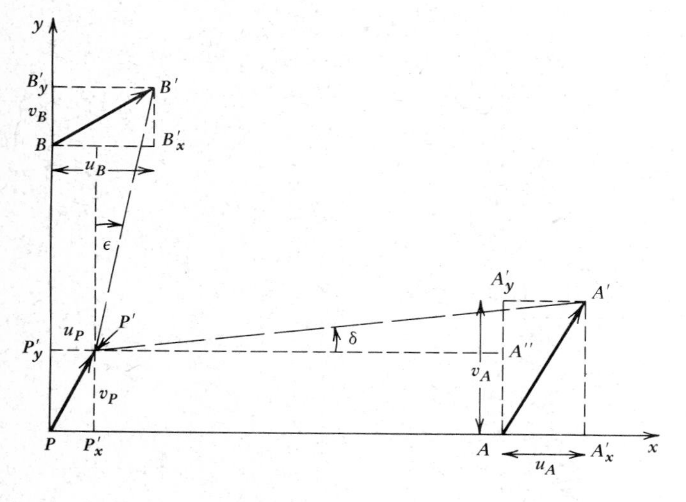

FIG. 2-16 *Displacements of the origin P and of points A and B to P′, A′, and B′.*

(Murnaghan 1951; Birch 1952), which is not discussed in this text. One might at first think that finite strain theory would be needed in one other case—that of viscous flow (Chapter 12), such as folding. Here, however, the displacements are replaced by the velocities, and, similarly, the strains are replaced by the (time) rates of strain. The latter remain infinitesimal, and the strains at any point are the integrals of the rates of strain over the time interval.

Symmetry

Because both terms measure the angular change in the (x, y) angle, $e_{xy} = e_{yx}$.

Dilatation

Just as in the case of stresses, the sum of the normal strains across any three orthogonal planes at a point is constant: it is the dilatation (or dilation) θ. Thus,

$$\theta = e_{xx} + e_{yy} + e_{zz} = \frac{\partial u}{\partial x} + \frac{\partial v}{\partial y} + \frac{\partial w}{\partial z} \tag{2-16}$$

Note the difference between this equation and Equation 2-6, which relates the pressure to the stresses. To see why θ is physically a dilatation, examine Figure 2-17: the length δx (before strain) has become $\delta x + \delta x\, e_{xx}$, and the same holds true for δy. Thus the difference between the strained area and the unstrained one is

$$\delta S = \left[\delta x(1 + e_{xx})\,\delta y(1 + e_{yy})\right] - \delta x\,\delta y$$

and the relative change of area (or change of area per unit area) is

$$\frac{\delta S}{\delta x\,\delta y} = e_{xx} + e_{yy} + e_{xx}e_{yy}$$

But because e_{xx} and e_{yy} are supposed very small, $e_{xx}e_{yy}$ is negligible, and so the relative change in area is given by $e_{xx} + e_{yy}$. In three dimensions the relative change in area becomes the relative change in *volume* or dilatation.

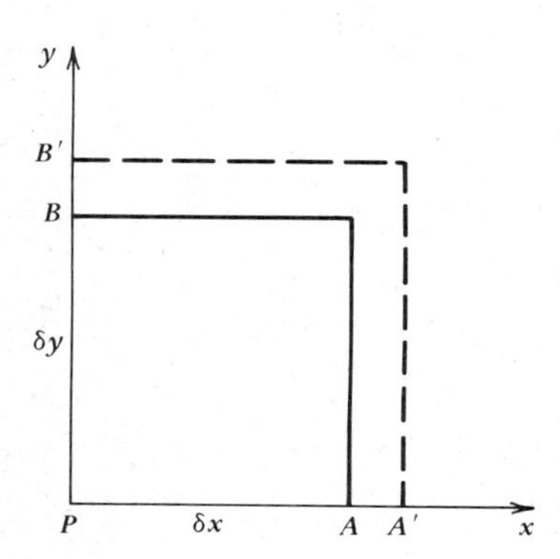

FIG. 2-17 *Why θ is a dilatation.*

Principal strains

Like a stress, a strain is a tensor, and at any point there exist three orthogonal planes on which the strain is purely normal. These planes define the principal axes of strain and the principal strains, respectively. It is thus meaningless to speak of the shear strain a *point* since at this point there are three planes on which shear strains vanish.

Applications

2-5 Points A, B, and C form a triangle. The lengths (a_0, b_0, c_0) and directions of the sides have been carefully measured at time t_0. Several years later, at time t_1 these lengths are remeasured and found to be a_1, b_1, and c_1. Suppose that there had been no strain of any kind at time t_0. At time t_1, what is the maximum shear strain, and along what directions does it occur?

Solution
This application is practically the same as Application 2-4. To find the solution, we need only compute the normal strains along the three sides and then replace p_{ij} with e_{ij} throughout the previous application.

2-6 We will now solve a fairly realistic case that shows how the maximum shear strain can be determined, say, near the San Andreas Fault. Because geological strains are small $(\sim 10^{-6})$, it will be necessary to carry out our computations with a very high degree $(\sim 10^{-6})$ of precision. Accordingly, the data also must be highly precise. The solution is long and tedious, but it involves nothing more difficult than trigonometry and simple algebra.

Data
In 1908, side c of a triangle was determined to measure exactly 36.5 km and to be oriented exactly N 120°E (from A to B). The angle A is exactly 65°, the angle B exactly 35°.

In 1920, the measurements were repeated with the following results.

$$c' = 36.50004 \text{ km}$$

$$B' = 35.0004°$$

$$A' = 65.001°$$

The direction of AB is essentially unchanged.

Find the direction and magnitude of the maximum shear strain, assuming that there was no strain in 1908.

Solution
First, we determine the configuration of the triangle in 1908. From the law of sines $(a/\sin A = b/\sin B = c/\sin C)$ and $C = 180° - 65° - 35°$, we ob-

tain (distances in kilometers)

$$a = 33.590550, \qquad b = 21.258504$$

The same computations for 1920 yield

$$a' = 33.59101, \qquad b' = 21.25884$$

The normal strains are thus

$$e_{aa} = \frac{(33.591015 - 33.59055)}{33.59055}$$
$$\cong 1.36 \times 10^{-5}$$

Similarly,

$$e_{bb} \cong 1.54 \times 10^{-5}$$
$$e_{cc} \cong 1.10 \times 10^{-6}$$

In line with the precision of the data used, only the first two digits of the strains are likely to be correct.

Figure 2-18*a* shows the configuration of triangle *ABC*. We now assume that this triangle is infinitesimal, so that all measurements can be assumed to have been made at the same point. Of course, this is only an approximation.

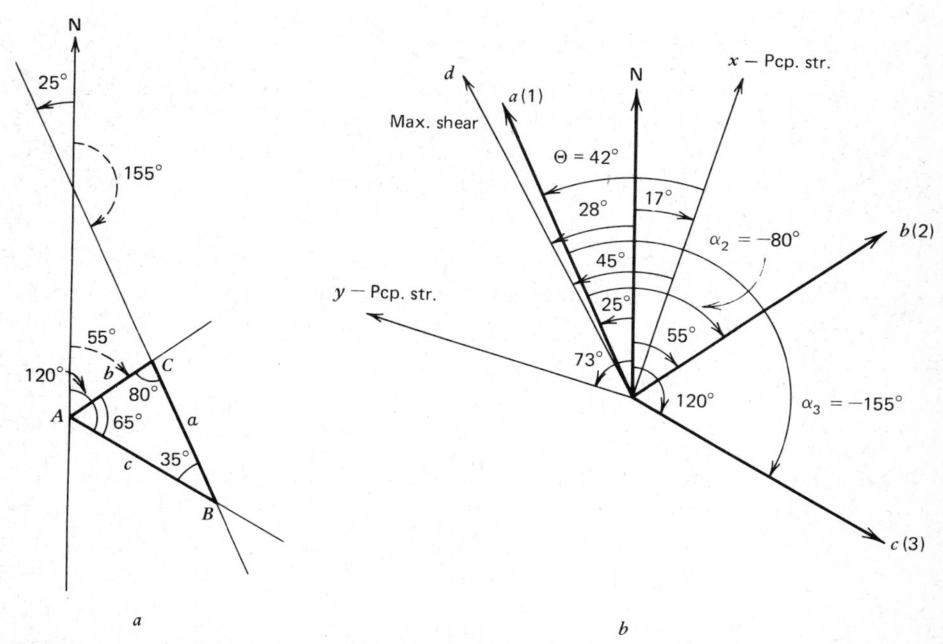

FIG. 2-18 *(a) Sketch of triangle ABC, showing the angles involved. (b) Sides a, b, and c have been moved to a common point. Heavy lines indicate the known directions— a or 1, b or 2, c or 3. The results of the computations are shown by thin lines.*

Figure 2-18*b* shows the sides translated to the same point, along with the angles α_2 and α_3 from side a (or 1) to the other sides. The values of α_2 and α_3 can be found immediately; note that they are negative (indicating clockwise rotation).

We now use Equation 2-10, replacing all the ps with es. Remember that this equation gives the angle (Θ_1) from the x-axis taken as one of the principal axes of stress to side 1. By replacing the stresses with strains, we can determine the angle (also Θ_1) from one of the principal axes of strain, referred to as x, to side 1 or a. Specifically, we replace p_{11} with e_{aa}, p_{22} with e_{bb}, and p_{33} with e_{cc}. Thus,

$$\frac{\sin(2\Theta_1 - 80°)}{\sin(2\Theta_1 - 145°)} = \frac{(1.36 - 1.54)\sin(-145°)}{(1.36 - 0.11)\sin(-80°)}$$

or

$$\frac{\sin(2\Theta_1 - 80°)}{\sin(2\Theta_1 - 145°)} = -0.083$$

We now can easily find the solution:

$$\Theta_1 \cong 42°$$

Hence,

$$\Theta_2 = \Theta_1 + \alpha_2 = -38°, \qquad \Theta_3 = \Theta_1 + \alpha_3 = -103°$$

From the values of Θ_1 and the direction of side a, we determine that the x principal axis of strain is directed at N 17°E. The y-axis thus lies in direction N 73°W. The direction of the maximum shear strain, $e_{de}{}^{\mathrm{M}}$, is at 45° to x and y: that is, at N 28°W and also N 62°E. The direction of $e_{de}{}^{\mathrm{M}}$ is very close to the direction of side a.

In order to compute the value of $e_{de}{}^{\mathrm{M}}$, we must first compute the values of $e_{xx}{}^{\mathrm{M}}$ and $e_{yy}{}^{\mathrm{M}}$. By replacing the stresses with strains in Equations 2-7 and 2-8, we obtain

$$e_{11} = e_{xx}{}^{\mathrm{M}} \cos^2\Theta_1 + e_{yy}{}^{\mathrm{M}}(1 - \cos^2\Theta_1) \tag{2-17}$$

$$e_{22} = e_{xx}{}^{\mathrm{M}} \cos^2\Theta_2 + e_{yy}{}^{\mathrm{M}}(1 - \cos^2\Theta_2) \tag{2-18}$$

Having determined all the quantities except $e_{xx}{}^{\mathrm{M}}$ and $e_{yy}{}^{\mathrm{M}}$, we can compute the latter to be

$$e_{xx}{}^{\mathrm{M}} = 2.4 \times 10^{-5}$$

$$e_{yy}{}^{\mathrm{M}} = 8 \times 10^{-7}$$

Finally, we obtain $e_{de}{}^{\mathrm{M}}$ from an equation similar to Equation 2-11:

$$e_{de}{}^{\mathrm{M}} = 0.5(e_{yy}{}^{\mathrm{M}} - e_{xx}{}^{\mathrm{M}}) \tag{2-19}$$

Thus,

$$e_{de}{}^{\mathrm{M}} = -1.16 \times 10^{-5}$$

The fact that $e_{de}{}^{\mathrm{M}}$ is negative means that the angle between the principal strain directions as defined here has increased between 1908 and 1920. (There are other possible choices for the principal strain directions—e.g., x at N(180° − 73°)E = N 107°E and y at N 17°E, but they are not consistent with the previous scheme of notations.)

PROBLEMS

2-9 In 1960 you measured the following distances: $AB = 21.56825$ km exactly in an E–W direction, and $AC = 10.25386$ km exactly in a N–S direction. B is east of A and C is north of A.

Upon returning in 1975, you find that A has moved to A' and B has moved to B'. AA' is 6 cm long, in the direction N 123°W; BB' is 17 cm long, in the direction N 155°E. Similarly CC' is 15 cm long and its direction is N 77°E.

(a) Compute, as best you can, e_{xx}, e_{yy}, and e_{xy} in the triangle ABC. *Hint*: A drawing may be helpful.

(b) Compute the strain rates both per year and per second. (They are written $\dot{e}_{xx}, \dot{e}_{yy}, \dot{e}_{xy}$.)

2-10 Refer back to Application 2-1, which examines a body made up of two layers of different materials. Is any component of the strain tensor continuous across the interface of the two layers. Why or why not? If not, what quantity (or quantities) is (are) continuous, and why? (*Hint*: Think of steel and rubber as the two materials.)

C. Constitutive Equation

In Sections 2A and 2B, we ignored the material properties of the body considered. Here, we will assume that the body is homogeneous and isotropic: that is, that its properties are independent of both spatial coordinates and direction. In addition, we will assume that the stresses in the body are linear combinations of the strains (Hooke's law): that is, that

$$p_{xy} = ae_{xx} + be_{yy} + ce_{zz} + de_{xy} + fe_{yz} + ge_{zx}$$

and that a similar relation holds true for all other stresses. Under these conditions it may be shown that the principal axes of stress are identical with the principal axes of strain, and that the stress-strain relations become

$$p_{xy} = 2\mu e_{xy}, \quad p_{yz} = 2\mu e_{yz}, \quad \text{etc.} \tag{2-20}$$

where μ is the rigidity. Thus, the shear stresses are proportional to the shear strains. On the other hand, the relations for the normal stresses are

$$p_{xx} = \lambda\theta + 2\mu e_{xx} \tag{2.21}$$
$$p_{yy} = \lambda\theta + 2\mu e_{yy} \qquad \text{etc.}$$

All of these equations may be written together as

$$p_{ij} = \lambda\theta\,\delta_{ij} + 2\mu e_{ij} \tag{2-22}$$

where the indexing scheme used for Equation 2-15 is employed and δ_{ij} (the Kronecker delta) is 1 if $i=j$, and 0 otherwise.

It should be emphasized that the definitions of *stress* and *strain* (Sections 2-A and 2-B) are independent of any observational evidence. In contrast, the linear relation between stresses and strains is based purely on observation. It is true, of course, that perfect elasticity is an idealization. In a perfectly elastic body the strains would vanish instantaneously and completely if the stresses vanish, in which case the mechanical energy used in deforming an elastic body would be completely recoverable—none would be transformed into heat and lost. Clearly, then, perfect elasticity cannot exist, but in many cases the concept yields extremely useful approximations. (See also "Attenuation, Q," in Chapter 3, Section B.)

Equation 2-22 is called the constitutive equation for isotropic elasticity. It implies that each shear stress is proportional to the corresponding shear strain, but that each normal stress depends both on the dilatation and on the corresponding normal strain.

Before discussing the consequences of this equation, we should note that the condition of homogeneity is fairly well satisfied in rocks. In a given rock there may be significant differences from one mineral to another, but on a scale of meters these differences generally tend to vanish—although there are, of course, many different rock-types. Isotropy is generally a reasonable assumption, except in rocks such as shale whose elastic properties may vary by up to 25%, depending on the direction.

The physical meaning of μ is obvious: if it is large, the body deforms little under any given shear stress. It is thus called the modulus of rigidity, or the shear modulus (G in many engineering texts). Much less clear is the meaning of λ, which is sometimes called the first Lamé elastic constant (after the nineteenth-century French mathematician Lamé). Both λ and μ are very difficult to measure directly. For this reason, other elastic constants are measured first, and λ and μ are then computed from them.

Other Elastic Constants

Young's modulus, *E* Young's modulus measures the resistance to extension if no other forces act on the body. Figure 2-19 shows a test piece of constant section A except for the two ends that is subjected to a force $\mathbf{F}$ along the x-axis. (Note that there are no forces acting on the side of the body.) Clearly, $p_{xx} = F/A$

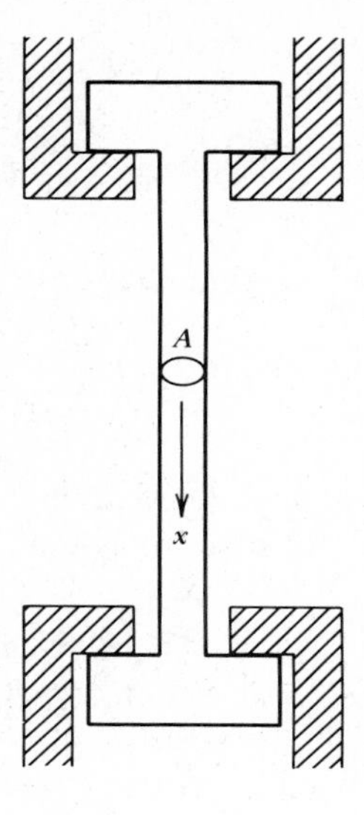

FIG. 2-19 *Schematic representation of a test piece under tension along the x-axis.*

while all other p_{ij}s are zero. For this configuration we define Young's modulus as

$$E = \frac{p_{xx}}{e_{xx}} \tag{2-23}$$

To express it in terms of λ and μ, we write

$$p_{xx} = \lambda\theta + 2\mu e_{xx} \tag{2-24}$$

$$0 = p_{yy} = \lambda\theta + 2\mu e_{yy} \tag{2-25}$$

$$0 = p_{zz} = \lambda\theta + 2\mu e_{zz} \tag{2-26}$$

Adding,

$$\begin{aligned} p_{xx} &= 3\lambda\theta + 2\mu(e_{xx} + e_{yy} + e_{zz}) \\ &= (3\lambda + 2\mu)\theta \end{aligned} \tag{2-27}$$

In order to obtain Equation 2-23 we must express θ in terms of e_{xx}. By symmetry, $e_{yy} = e_{zz}$. Thus,

$$\theta = e_{xx} + 2e_{yy} \tag{2-28}$$

and, from Equations 2-25 and 2-28,

$$\lambda\theta = \lambda\left[e_{xx} + 2e_{yy}\right] = -2\mu e_{yy}$$

Hence,

$$-2e_{yy} = \frac{\lambda e_{xx}}{\lambda + \mu} \tag{2-29}$$

Thus, from Equation 2-28,

$$\theta = e_{xx} - \frac{\lambda}{\lambda + \mu} e_{xx} = e_{xx}\left(\frac{\mu}{\lambda + \mu}\right)$$

Introducing this value into Equation 2-27 yields

$$E = \frac{p_{xx}}{e_{xx}} = \frac{(3\lambda + 2\mu)\mu}{\lambda + \mu} \tag{2-30}$$

Incompressibility, κ This constant measures the resistance to a change in volume under pressure.

$$\kappa = -\frac{dP}{d\theta} \tag{2-31}$$

(Because $P > 0$ when $\theta < 0$, the minus sign is necessary for κ to be > 0.) To relate κ to λ and μ, we again write Equations 2-24, 2-25, and 2-26, but here we assume that neither p_{yy} nor $p_{zz} = 0$. Thus,

$$p_{xx} + p_{yy} + p_{zz} = -3P = (3\lambda + 2\mu)\theta$$

Thus,

$$\kappa = \lambda + \tfrac{2}{3}\mu \tag{2-32}$$

κ is also called the bulk modulus. For simplicity, it is convenient to think of κ as a measure of the hydrostatic pressure necessary to achieve a given decrease in volume, but this particular physical situation need not obtain.

The fact that a body is perfectly incompressible ($\kappa = \infty$) does *not* mean that it is perfectly rigid—hence $\mu \neq \infty$; from Equation 2-32, it then follows that $\lambda = \infty$.

Poisson's ratio, σ (ν in many geological and engineering texts.) Poisson's ratio (after the nineteen-century French mathematician, Siméon Denis Poisson) is the ratio of the lateral contraction to the longitudinal extension, supposing that only one normal stress—here, p_{xx}—acts on the body. Thus,

$$\sigma = -\frac{e_{yy}}{e_{xx}}$$

for Figure 2-19. (The minus sign is justified by the fact that $e_{xx} > 0$ and $e_{yy} < 0$). To relate σ to λ and μ, assume the situation shown in the figure, then use Equation 2-26 to obtain

$$0 = \lambda\theta + 2\mu e_{yy} = \lambda(e_{xx} + e_{yy} + e_{zz}) + 2\mu e_{yy}$$

But, by symmetry, $e_{yy} = e_{zz}$. Thus,

$$\lambda(e_{xx} + 2e_{yy}) = -2\mu e_{yy}$$

Hence,

$$\sigma = \frac{\lambda}{2(\lambda + \mu)} \tag{2-33}$$

In the earth, Poisson's ratio ranges from as low as 0.1 to as high as 0.38 near the surface (Birch 1966, Table 7-15). At hydrostatic pressures equivalent to a depth of ~ 13 km the ranges are from 0.23 to 0.31—except for quartzite for which $\sigma = 0.15$ (Birch 1966, Table 7-16). In the absence of any other information, it is often assumed that

$$\sigma = 0.25 \tag{2-34}$$

which is known as Poisson's relation. It is easy to see that Equation 2-33 implies that, in this case, $\lambda = \mu$. (This is Poisson's second appearance! His name will appear again in the chapters dealing with gravity and magnetism.)

If the body is incompressible, $\theta = e_{xx} + e_{yy} + e_{zz} = 0$; but, by symmetry in Figure 2-19, $e_{yy} = e_{zz}$. Since $\sigma = -e_{yy}/e_{xx}$, we find that

$$\sigma = 0.5 \tag{2-35}$$

Other relations Elementary algebra yields the following useful equations.

$$\lambda = \frac{E\sigma}{(1+\sigma)(1-2\sigma)} \qquad \mu = \frac{E}{2(1+\sigma)} \tag{2-36}$$

$$\kappa = \frac{E}{3(1-2\sigma)} \qquad \frac{\lambda}{\mu} = \frac{2\sigma}{1-2\sigma}$$

Application

2.7 We can now compute the principal stresses and the maximum shear stress from the strain measurements given in Application 2.6.

Solution

In Application 2.6, we found the principal strains $e_{xx}{}^{\mathrm{M}}$ and $e_{yy}{}^{\mathrm{M}}$ and the maximum shear strain $e_{de}{}^{\mathrm{M}}$ in both magnitude and direction ($e_{xx}{}^{\mathrm{M}} = 2.4 \times 10^{-5}$, $e_{yy}{}^{\mathrm{M}} = 8 \times 10^{-7}$, $e_{de}{}^{\mathrm{M}} = 1.16 \times 10^{-5}$). If the assumptions of homogeneity and isotropy are justified, the directions of the principal and shear stresses coincide, respectively, with those of the principal and shear strains.

The magnitude of the maximum shear stress is

$$p_{de}{}^{\mathrm{M}} = 2\mu e_{de}{}^{\mathrm{M}}$$

for $\mu = 2.1 \times 10^{10}$ Pa, we find $p_{de}{}^{\mathrm{M}} = 4.9 \times 10^{5}$ Pa $\cong$ 5 bars, which is a very small value.

The values of p_{xx} and p_{yy} are a little more difficult to obtain. We have

$$p_{xx} = \lambda(e_{xx} + e_{yy} + e_{zz}) + 2\mu e_{xx}$$

$$p_{yy} = \lambda(e_{xx} + e_{yy} + e_{zz}) + 2\mu e_{yy}$$

Unfortunately, e_{zz} is unknown. On the other hand, $\mathbf{p}_z = 0$, so $p_{zz} = 0$, since the traction is null on the earth's surface taken as horizontal. Hence,

$$p_{zz} = \lambda(e_{xx} + e_{yy} + e_{zz}) + 2\mu e_{zz} = 0$$

Thus,

$$e_{zz} = -\frac{\lambda(e_{xx} + e_{yy})}{\lambda + 2\mu}$$

Insertion of this value into the equations for p_{xx} and p_{yy} yields the desired result.

Assuming that $\lambda = \mu = 2.1 \times 10^{10}$ Pa,

$$e_{zz} = -\tfrac{1}{3}(e_{xx} + e_{yy})$$

Using the values of $e_{xx}{}^{M}$ and $e_{yy}{}^{M}$ given above, we obtain

$$p_{xx} \cong 13 \times 10^5 \text{ Pa} \cong 13 \text{ bar}$$

$$p_{yy} \cong 3.2 \times 10^5 \text{ Pa} \cong 3.2 \text{ bar}$$

These values are very small. If the data are real, we can safely assert that the stress in the area is not building up, and thus that no earthquake is impending.

PROBLEMS

2-11 A block of stone is 50 cm long, 2 cm thick, and 3 cm wide. You stand it upright and place atop it a mass of 1200 kg. Upon measuring it again, you find that it is shorter by 0.125 mm. Assume $\sigma = 0.25$. Find E, λ, and μ—all in SI units. Give the units.

2-12 **(a)** After performing the experiment in Problem 2-11, you remove the mass and put the block under a 15 kbar hydrostatic pressure. What is its total change in volume (i.e., from the uncompressed to the compressed state)? Be sure to state the units. (See the conversion table in Appendix A.)

(b) What would be the height of a water column needed to yield a 15 kbar hydrostatic pressure at the earth's surface?

D. Scale Models

Although the subject of scale models in geology lies outside the scope of this text, the preceding analysis casts an instructive light on scale models. We have seen that a traction is a force per unit area. In contrast, the combination of gravity and density results in a force per unit volume. Thus, if the size of an object (i.e., its linear dimensions) is multiplied by a given factor, say, S, its volume is multiplied by S^3, and the force exerted by gravity is also multiplied by S^3. However, the area of any section of the object is multiplied only by S^2. Hence the traction on any section caused solely by the force of gravity is multiplied by $S^3/S^2 = S$. Unless the material of the large object is S times stronger than that of the small one, it will not deform in the same way—indeed, the larger object might break under its own weight. Thus, if the subscripts L and s designate the large and small objects, respectively, we must have $\lambda_L = S \times \lambda_s$, $\mu_L = S \times \mu_s$ to obtain a correct scaling. Galileo appears to have been the first to recognize the need for dimensionally correct scaling, and to point out the need for a corresponding change in material properties: "For this increase in height can be accomplished only by employing a material which is harder or stronger than usual" [Galileo (1638) 1914, p. 130].

Thus a laboratory model of a phenomenon, such as mountain building, in which the decrease in size is of the order of 10^5 (1 cm may represent 1 km) must use a material 10^5 weaker than natural rocks. It is remarkable that more than 250 years after Galileo, this rule was ignored by geologists who devised scale models. A thoroughgoing discussion of scale models in geology, including the correct scaling of time-dependent phenomena, can be found in Hubbert (1937).

3

Body Waves

A. Introduction

Definitions

Sinusoidal waves traveling in the positive x-direction at a velocity c are described by the equations

$$\begin{aligned} y &= A \sin\left[\frac{2\pi}{\Lambda}(x - ct)\right] = A \sin\left[2\pi\left(\frac{x}{\Lambda} - \frac{t}{T}\right)\right] \\ &= A \sin\left[2\pi(kx - \nu t)\right] \end{aligned} \tag{3-1}$$

The motion of the particles is along the y-axis. These waves can be considered in two ways: at one point (i.e. for a given value of x), they are periodic in time; at one instant, they are periodic in space. The distance at any one instant between two crests or two troughs, or between any two adjacent points having the same phase, is called the wavelength, Λ. We will sometimes use the wave number defined as $k = 1/\Lambda$ (which seems preferable to definition $k = 2\pi/\Lambda$). Similarly, the time interval at any one point between two crests, or between two adjacent points having the same phase, is called the period T; its inverse, ν, is the frequency (measured in hertz, Hz). A is the amplitude, the angular frequency is $\omega = 2\pi\nu$. Note that k is to space as ν is to time. Finally, note that following relations follow immediately from Equation 3-1.

$$\Lambda = cT, \qquad \text{hence} \qquad \nu = kc$$

Equation 3-1 assumes that the wave propagates in the x-direction, that the particle motion is in the y-direction, and that the situation is two-dimensional. In a two-dimensional situation nothing changes along a given direction, here the z-axis: all particles along the z-axis move together or, more generally, the motion is independent of z. We can generalize Equation 3-1 slightly by writing

$$D = f(x, t) = A \sin [2\pi k(x - ct)]$$

Nothing is changed except that we may now say that the quantity D propagates at a velocity c along the x-axis and is independent of y and z. This means that all points of the same x move together: they are "in phase." These points are in the (y, z) plane and form wave fronts. For this reason, we say that the equations given above represent the propagation of plane waves.

Body Waves in Infinite Media

In an infinite, homogeneous, solid, elastic, and isotropic medium, only two kinds of waves can propagate: P-waves and S-waves. They travel at velocities given by, respectively,

$$\alpha = \sqrt{\frac{\lambda + 2\mu}{\rho}} \qquad \beta = \sqrt{\frac{\mu}{\rho}} \tag{3-2}$$

The term *P-wave* means "primary wave," since it arrives first; S stands for "secondary wave." It is obvious that if Poisson's relation obtains—that is, if $\sigma = 0.25$ and thus $\lambda = \mu$—then $\alpha = \beta\sqrt{3}$.

The validity of the assumptions of isotropy and homogeneity was briefly discussed in Chapter 2, Section C. Note that in most cases the size of mineral grains is much smaller than the smallest wavelength involved. This may not be the case in laboratory experiments, where frequencies on the order of the kilohertz or higher may be used; in such cases the assumption of homogeneity may be invalid.

In the earth, α and β generally increase downward from the surface in any given rock type, partly because of the reduction in porosity and partly because of the closing of cracks. Values of α, β, and ρ for some common rock types are given in Appendix C2. These values are only approximate. In general, though, α and β are lowest in sedimentary rocks, higher in igneous and metamorphic rocks, and highest in basic igneous rocks.

PROBLEM

3-1 What are α and β in Problem 2-11, if $\rho = 2800$ kg/m^3?

Particle Motion

P-waves may be considered the propagation of a dilatation alternating with a compression; for this reason they are also called compressional waves or dilatational waves. Consider the equation

$$\theta = A \sin [2\pi k(x - ct)]$$

It is identical with the equation $D = A \sin[2\pi k(x - ct)]$ seen above, except that θ replaces D. It thus represents the propagation of a dilatation, θ, along the x-axis independent of y and z. But if a particle dilates, and its neighbors in the (y, z) plane dilate at the same time, the only possible remaining motion is along the x-axis. We thus must conclude that in a P-wave the particles move back and forth in the direction in which the wave propagates. S-waves, on the other hand, represent the propagation of a shear strain. Consider, for example, a shear strain, e_{xz}, propagating in the x-direction. The particle motion necessary to create such a situation is in the planes shown in Figure 3-1—more specifically, along the z-axis, and thus perpendicular to the x-axis. Since x and z are arbitrarily oriented (except for being orthogonal), it is clear that the particles in a shear wave move perpendicularly to the direction of propagation. Shear waves are also called distortional waves and equivoluminal waves.

The motion of a particle in an S-wave may be decomposed into two components: a purely horizontal one, and one in a vertical plane. (This is true regardless of the direction of propagation of the wave.) A wave that contains only the first component is called an SH-wave, one that contains only the second, an SV-wave. An S-wave thus consists of a combination of SH- and SV-waves. This decomposition is advantageous in the earth, where many interfaces, as well as the surface, are horizontal, or nearly so. (In approximately the upper 200 km of the

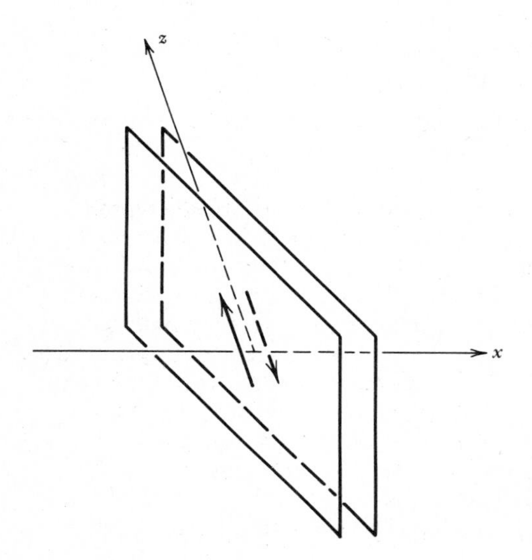

FIG. 3-1 *A shear wave as a propagating shear strain, e_{xz}.*

mantle, the velocities of *SH*- and *SV*-waves may differ slightly due to a slight anisotropy. See Figure 3-15.)

In a perfect liquid or gas, where $\mu = 0$ by definition, only *P*-waves can exist. They are then called acoustic waves, since they can transmit sound. In contrast, only *S*-type waves exist in electromagnetism.

Equation 3-2 shows that, all things being equal, if ρ increases, then α and β should *decrease*. One should thus avoid saying, "*Since* the material has a higher density the seismic velocities are greater." In general, though, heavier materials have higher α and β than lighter ones do, because in most cases λ and μ increase faster than ρ.

Symmetry shows that *P*- and *S*-waves propagate along a straight line in a homogeneous medium. These waves are thus called body waves. From a point source they spread in three dimensions. Points of equal phase, or wave fronts, are thus spheres. Since the energy is conserved, and since the surface of the sphere increases according to the square of its radius, it follows that the energy at any given point is proportional to the inverse of the square of the distance from the source. Furthermore, in any given medium the energy is proportional to the square of the amplitude. Hence, the amplitude of body waves is inversely proportional to the distance from the source.

Finally, Equation 3-2 shows that α and β are independent of the frequency or wave number. This is not the case for other types of waves, and it implies that *P*- and *S*-waves do not change shape as they travel. Both these points are discussed in Chapters 4 and 5).

The High Frequency Approximation

The concept of a seismic ray is entirely analogous to that of an optical ray. We will now see that some care must be exercised when we apply this idea to the earth or to other heterogeneous media.

Consider, first, a very simple situation: a source *S* and a receiver *R* are both located in a homogeneous medium. Clearly, the energy travels in a straight line from *S* to *R*. For simplicity's sake, take *SR* to be horizontal. Now imagine that the velocity "near" *SR* is changed—say, decreased above *SR* and increased below it—but that in a "thin" strip along *SR* it is unaltered. Is it reasonable to assume that the energy will now continue to travel along *SR* only, as if no change had taken place? Common sense suggests that such an assumption would be absurd, and that the word "near" must be clarified. Since the only physical quantity involving distance is the wavelength, we can say that, for example, "near means less than one wavelength away." We thus come to the following statement: "Ray representation is valid as long as the velocity changes are 'small' in approximately one wavelength on either side of the ray. If this condition is not met, the time of arrival given by the ray representation will remain generally valid, but the shape of the wave will be distorted as it travels from *S* to *R*." It remains only to define a "small" velocity change. This concept can be expressed in terms of the distortion: a 1% velocity change per wavelength gives rise to approximately a 1% distortion, which is imperceptible; a 10% velocity change to a 10% distortion, which is quite noticeable, and so forth.

Finally, a long wavelength corresponds to a long period, and a short one to a short period—that is, to a high frequency. Hence, what is "near" for a long wavelength (or period) is "far" for a short wavelength (or a high frequency). Consider a body in which the velocity varies from 9.0 km/s to 9.5 km/s between the depths of 100 km and 150 km, again assuming a ray *SR* traveling in a horizontal direction. Is this change in velocity "large"—that is, is the ray approximation valid? The answer depends on the frequency: for a frequency of 0.1 Hz ($T = 10$ s, $\Lambda \approx 90$ km), the velocity change per wavelength is on the order of 10%; for a frequency of 10 Hz, the velocity change per wavelength is negligible. The ray approximation is thus much better for high frequencies than for low frequencies. This situation, which is quite general, is the reason why the drawing of rays (also called "ray optics" or the "ray approximation") is also known as the "high frequency approximation."

PROBLEM

3-2 One of the simplest ways to determine the elastic constants of a rock is to measure its density and the travel times of *P*- and *S*-waves across a small sample. Suppose that you cut a core 2 cm in diameter and 6 cm long out of a homogeneous hand specimen of dunite. In order to determine its density you determine its weight to be 61.45 g. A compressional impulse given at one end arrives at the other end after 8.6 μs; for a shear impulse, you find a travel time of 14.5 μs. Compute E and μ in SI units. Give the units.

Reflection and Refraction

Figure 3-2 shows a horizontal interface between two bodies of velocities α_1, β_1 and α_2, β_2, on which a *P*-wave is incident from above. (A *P*-wave is indicated by an arrow along the ray, an *S*-wave by a bar across it.) Fermat's "principle of stationary time" (generally but incorrectly called the principle of minimum time) states that a *P*-wave going from *A* to *C* does so in a minimum time (or a maximum time, or a stationary time)—that is, that *O* is "chosen" by the ray so as to minimize the total time *AOC*. (There is no way to maximize it.) This is also true for all other paths shown. From this principle, it is easy to derive Snell's law, which states that the sines of the angles of incidence are proportional to the wave velocities:

$$\frac{\alpha_1}{\sin i_{P1}} = \frac{\beta_1}{\sin i_{S1}} = \frac{\alpha_2}{\sin i_{P2}} = \frac{\beta_2}{\sin i_{S2}} \tag{3-3}$$

The angle of incidence is defined as the angle between the normal and the ray.

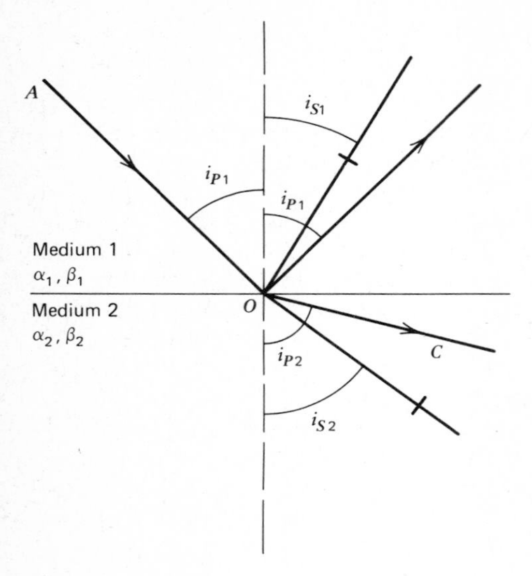

FIG. 3-2 *A P-wave incident from medium 1 and giving rise to reflected and refracted P- and S-waves, and the corresponding angles of incidence. P-waves are shown by arrows, S-waves by bars.*

The waves penetrating medium 2 are said to be refracted or transmitted; the others are reflected. In optics, refraction is the phenomenon that gives rise to, among other things, the illusion that a stick dipped in water is bent and the possibility of using lenses. Since $\beta < \alpha$ in any medium, $i_S < i_P$ also.

PROBLEM

3-3 Derive Snell's law from Fermat's principle.

Horizontal Interfaces

Because interfaces in the earth often are horizontal, or nearly so, this case deserves separate treatment. Consider a *P*-wave traveling in a vertical plane and incident upon a horizontal interface. Its particle motion also lies in a vertical plane, as do those of the *P*- and *SV*-waves that it excites (see Figure 3-2). Those of an *SH*-wave, on the other hand, are by definition horizontal (i.e., perpendicular to the vertical plane). The particle motion of the *P*-wave is thus at right angles to that of an *SH*-wave. To put it another way: the particle motion of the *P*-wave has no component in the direction of the particle motion of the *SH*-wave, and vice versa. Thus, these two waves cannot excite each other. The net result is that an incident *P*-wave generates both *P*- and *SV*-waves, whereas an incident *SH*-wave only generates a reflected *SH*-wave and a refracted *SH*-wave. Although the simplicity of this situation makes it obviously appealing in exploration, *SH*-waves, unfortunately, are not easy to generate.

PROBLEM

3-4 Consider two half-spaces* in contact along a plane. The material constants are as follows.

$$\alpha_1 = 5.6 \text{ km/s}, \quad \sigma_1 = 0.20, \quad \rho_1 = 2.7 \text{ g/cm}^3$$

$$\alpha_2 = 8.1 \text{ km/s}, \quad \sigma_2 = 0.3, \quad \rho_2 = 3.2 \text{ g/cm}^3$$

A P-wave is incident from medium 2 at an angle of incidence of 25°.

Compute all the angles of incidence and draw all the reflected and refracted rays.

Amplitudes and Energy

Snell's law tells us nothing about the amplitudes of the various waves. From the particle motion, we can determine that at normal incidence ($i = 0$) a wave is reflected or transmitted without a change in type. For a wave coming from medium 1 at normal incidence, the amplitude reflection coefficient, R_A, and the amplitude transmission coefficient, T_A, may be computed from the continuity of the traction and the displacements (*not* from the strains—see Problem 2-10) across the interface. They are

$$R_A = \frac{\rho_2 c_2 - \rho_1 c_1}{\rho_2 c_2 + \rho_1 c_1} = \frac{A_r}{A_i}, \qquad T_A = \frac{2\rho_1 c_1}{\rho_2 c_2 + \rho_1 c_1} = \frac{A_t}{A_i} \tag{3-4}$$

where A_i, A_r, A_t are the amplitudes of, respectively, the incident, reflected, and transmitted waves, and c_1 and c_2 may be either α_1 and α_2 or β_1 and β_2, but not a mixture of α and β. If $R_A < 0$, a compression is reflected as a dilatation, and vice versa. The equations given above are approximately correct near normal incidence. The acoustic impedance of the medium is $Z = \rho c$ [Brekhovskikh, 1960, p. 17].

The coefficients for reflection and transmission of energy are R_E and T_E, respectively. In a given medium the energy depends on the square of the amplitude. Since a reflected wave stays in the same medium, $R_E = A_r^2/A_i^2 = R_A^2$. The situation is different for T_E: it is physically evident that the energy flux on the boundary must be preserved, hence $T_E + R_E = 1$. In different media, however, the energy depends both on the square of the amplitude and on the material properties of each medium. The latter differ in medium 1 and 2, hence $T_E \neq T_A^2$. Thus, at normal incidence we have:

$$R_E = \left[\frac{Z_2 - Z_1}{Z_2 + Z_1}\right]^2 = R_A^2, \qquad T_E = 4\frac{Z_1 Z_2}{(Z_1 + Z_2)^2} \neq T_A^2. \tag{3-5}$$

If $Z_1 \ll Z_2$, then $T_E \approx 4Z_1/Z_2$, which is small; similarly, if $Z_2 \ll Z_1$, then

*A half-space is a region of space bounded by a plane.

$T_E \approx 4Z_2/Z_1$. Thus, if two media have very different acoustic impedances, it is difficult to transmit energy from one to the other; they are said to be weakly coupled. Reflection prospecting is based upon the reflection of seismic waves at interfaces between rocks of various types. Since the changes in velocities and in densities generally are small, R_A is also small.

PROBLEMS

3-5 Compute R_A, T_A, R_E and T_E for a *P*-wave incident, from the shale side, upon an interface between an average shale and an average limestone. (See Appendix C2 for the necessary data.)

3-6 Why is it more difficult for a seismograph to detect an explosion in the air than one underground?

Reflection Prospecting

The primary application of reflected waves is the reflection prospecting method. From an economic standpoint, the most important use of this method is oil and gas exploration. This subject is abundantly covered in many textbooks (e.g., Dobrin 1976, Telford et al. 1976). From a more academic viewpoint the most important results of this method have come from the determination of the deep structure of the continental crust and of some very large-scale structures, such as the Appalachians (Project COCORP). Before the advent of this method, many geophysicists viewed the crust as a "layer cake." One simple model consisted of (from the surface downward) a thin veneer of sediments, a contorted "basement," a "granitic layer" extending to roughly 18 km and bounded below by the Conrad discontinuity, and a "basaltic layer" limited in turn by the Mohorovičić discontinuity (or "Moho") at approximately 35 km. Sometimes as many as half-a-dozen layers were put in, sometimes only two. This picture was obtained largely by refraction methods (see below) and through studies of near earthquakes. It is now

FIG. 3-3 *A generalized model of the crust, approximately 32 km thick (i.e., ~ 30–35 km). There are three more or less distinct and extremely variable zones and localized reflectors of more mafic character corresponding to the "Conrad" discontinuity— a term now outdated. (From Smithson 1978. Copyright © by the American Geophysical Union.)*

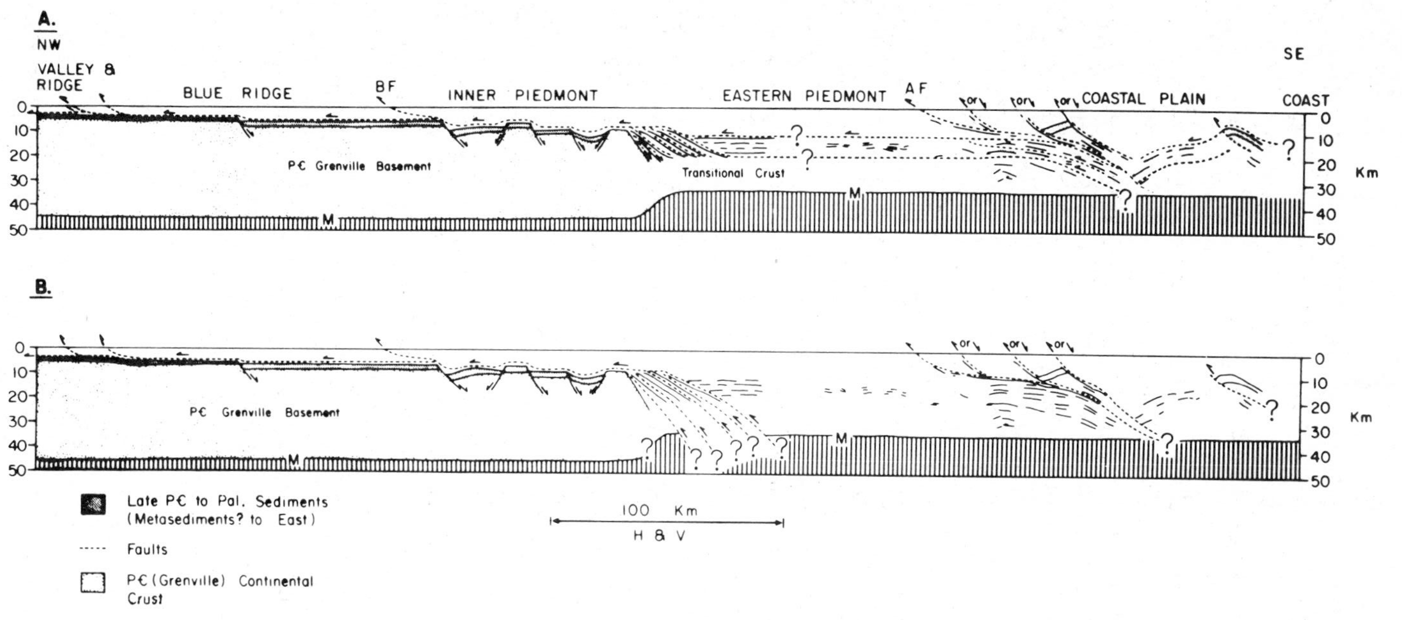

FIG. 3-4 *(a) Generalized cross section of the southeastern United States from the Valley and Ridge to the coast, based on the COCORP data in which a subhorizontal detachment is interpreted to extend beneath the Coastal Plain. (b) An alternate interpretation, in which the thrust surfaces of the Inner Piedmont–Blue Ridge allochthon are interpreted to "root" near the Charlotte belt. The shaded areas are interpreted as late Precambrian / early Paleozoic strata of the continental shelf and associated ocean basin, the lined pattern represents the mantle, and the hachured pattern represents continental crust. Thrust faults are indicated by dashed lines; listric faults east of the Augusta fault are shown as either thrust faults or normal faults. BF-Brevard fault; AF = Augusta fault. (From Cook et al. 1981. Reprinted by permission.)*

clear that the "layer cake" represents at best only an average over tens or hundreds of kilometers. On a smaller scale the picture is less simple, but geologically more realistic (Oliver 1978). Reflection profiles show a very large number of more or less horizontal discontinuities, none of which is clearly more important than others, and none of which is laterally continuous over more than a few kilometers. One picture that has been proposed is shown in Figure 3-3. Here, the upper zone of metamorphic rocks and granitic intrusives is underlain, at a widely variable depth, by a zone of migmatites. In places the latter is abruptly succeeded by a more gabbroic zone; in other places the deep crustal zone consists of intrusives and gneisses, and the transition is very diffuse. Regardless of the details, it is now clear that the idea of the "layer cake" must be abandoned. The term "Conrad discontinuity" is about to suffer the same fate. The Mohorovičić discontinuity, on the other hand, appears to be a widespread and fairly identifiable feature. Its depth over continental areas varies between approximately 30 and 80 km, though the latter is exceptional.

When the reflection method was used to explore the structure of the Appalachians, scientists found (Cook et al. 1981) that the whole mountain chain is a gigantic thrust sheet (Figure 3-4) that had been pushed westward, probably during the collision of the proto–African and proto–North American plates during Carboniferous and Permian times. The profiles also showed that under and to the east of the Appalachians, at a depth of some 6 to 15 km, lay essentially undeformed sediments.

More recently, sophisticated seismic reflection methods have been successfully used to obtain reflections at sea from the Moho (Talwani et al. 1982). Remarkably, such reflections were obtained in almost every area where the methods were used, including the crest of the East Pacific Rise. It was discovered that the distance between the basement and the Moho is approximately 5 km and is almost independent of the age of the ocean floor, and that the sharpness of the reflections did not depend on the age of the floor.

We can estimate the distance over which the change from the crust to the mantle takes place. The frequency used for both continental and oceanic work is close to 8 Hz, and α is on the order of 8 km/s. The longest wavelength (also the only one!) is therefore ~ 1 km. Since the reflection from the Moho is quite sharp, the transition between crust and mantle must occur over a distance $d_{\mathrm{m}} \ll \Lambda_{\mathrm{M}}$: that is, over a fraction of a kilometer. Recent evidence (Hale and Thompson 1982) suggests that a laminated structure consisting of approximately half-a-dozen layers, each ~ 100 m thick and of alternating high and low velocities (~ 7 and 8 km/s), represents a good description of the Moho in continental regions. (The numbers given are only indicative.)

Refraction Prospecting

Reflection prospecting presents nothing remarkable from the point of view of mathematical physics. Much more puzzling from this viewpoint are the waves used in refraction prospecting. Figure 3-5 demonstrates what happens at the

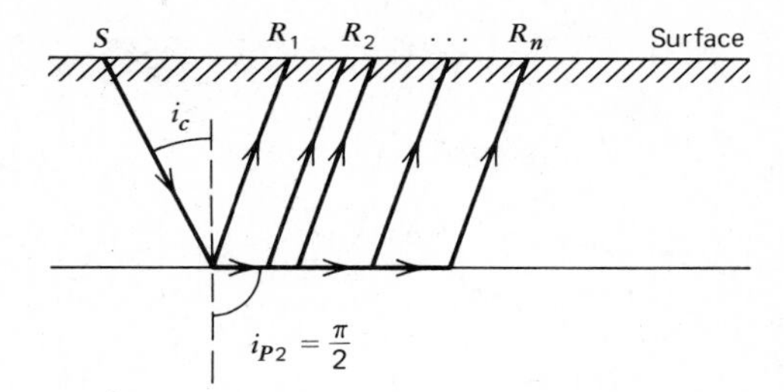

FIG. 3-5 *Incident ray (from S) and some of the "refracted rays," or head waves, arriving at $R_1, R_2 \ldots$ used in refraction prospecting.*

critical angle ($i_c = i_{P1}$)—that is, when $\sin i_{P1} = \alpha_1/\alpha_2$. From Snell's law, $i_{P2} = \pi/2$. From this standpoint the rays are drawn correctly, as minimum-time paths. But it is remarkable to find a single ray propagating along the interface and radiating upward-propagating rays everywhere (only a few are shown here). Common sense suggests that such a phenomenon is impossible—and indeed, that was the view long held by some scientists. Despite this view, refraction prospecting was widely used along the Gulf Coast to locate shallow salt domes, and thus potential oil fields. The principal technique used was called fan shooting (Figure 3-6), which relied on the fact that α in salt is larger than α in sediments (Nettleton 1940, pp. 277–279). A theoretical explanation for the existence of these waves was given only after hundreds of shallow domes had been located (Jeffreys 1926; Muskat 1933). (A thorough, if difficult, treatment can be found in Cagniard 1939 and 1962.) In fact, one of the reasons why such paths are not impossible is that these waves, now called head waves, decay more rapidly than do ordinary body waves.

Unless it is *a priori* certain that the layers are flat, seismic refraction profiles should always be "reversed" (see Figure 3-7). One source (S) and several receivers (R_1, R_2, etc.) should be laid out in one direction, and then the source

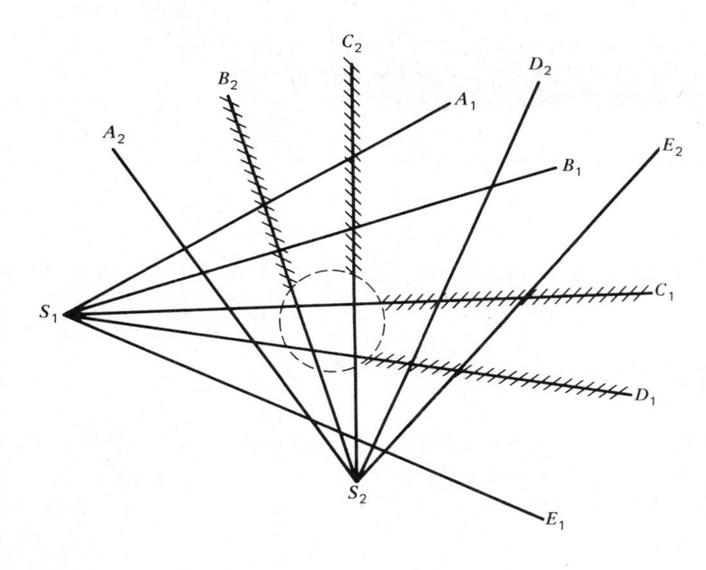

FIG. 3-6 *Two intersecting "refraction fans," from S_1 and S_2, roughly outlining a salt-dome (dashed line). Arrivals along the hachured paths are earlier than normal.*

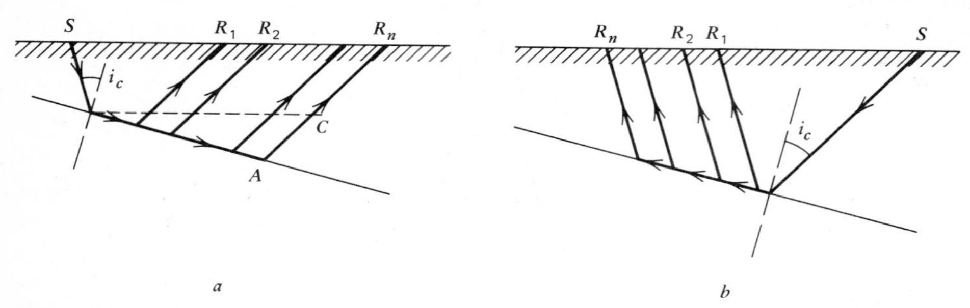

FIG. 3-7 *(a) Down-dip shooting results in an increase in the length of the path through the upper layer as the receiver distance increases. (b) Up-dip shooting results in a decrease in the length of the path through the upper layer.*

should be set at, say, R_n and the direction in which the profile is laid out should be reversed. The reason for this procedure is clear from the figures. In Figure 3-5 the time difference between receivers R_1 and R_n results solely from the increased length of the path along the interface, which permits an immediate determination of the velocity along the interface; this supposes that the interface is horizontal. In Figure 3-7*a*, in contrast, the time difference between R_1 and R_n results partly from the increased path along the interface and partly from the increased path in the slower upper medium (i.e., segment AC). Thus the observed velocity is too small. The opposite reasoning applies to Figure 3-7*b*. The correct velocity can be found through appropriate averaging and simple trigonometry (Dobrin 1976, pp. 303–305).

As has already been mentioned, the refraction method has been and is still extensively used for determining crustal structure both on land and at sea. The results obtained are, of course, averages; but properly interpreted, they yield important information. One of the earliest, and still one of the most important, results was the proof that the oceanic crustal thickness is only on the order of 5 km (Ewing and Press 1955) compared with ~ 35 km for the continental one. This result had been predicted on isostatic grounds (see Chapter 9, Section C) but had remained unconfirmed. The use of the refraction method for exploration purposes is amply covered elsewhere (Dobrin 1976; Telford et al. 1976).

B. Physical Structure of the Earth

Determination of α and β versus Depth

Through observation of travel times, we wish to infer the wave velocities at depth. How can this be done? In more precise terms the situation is the following. Assume that the time and place of a source of seismic waves (explosion or earthquake) are known. Also known are the times of arrival of P-waves (all other

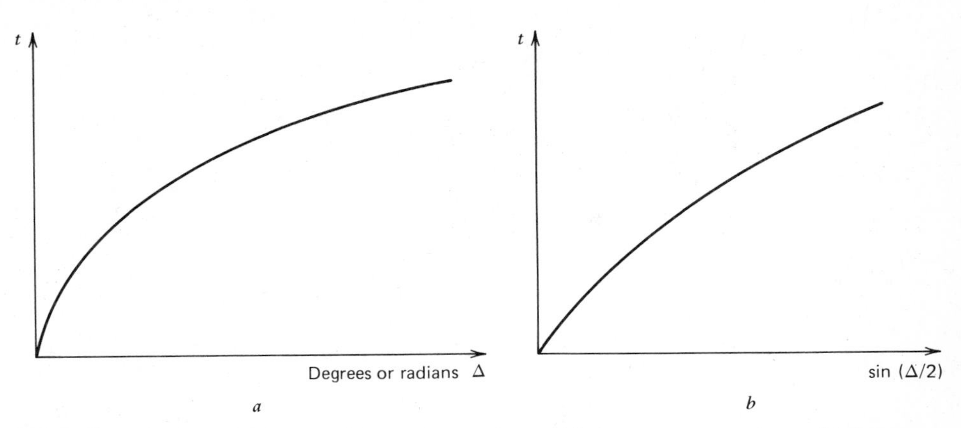

FIG. 3-8 *(a) Sketch of the travel time t versus distance Δ in degrees or radians. (b) Sketch of the travel time t versus* sin(Δ/2).

waves temporarily being disregarded) at a sufficiently large number of seismographic stations so that a travel-time curve, or hodograph (see Figure 3-8*a*) may be drawn with confidence. (In seismology the distance from the point at the earth's surface above the source—the epicenter—to the receiver is traditionally designated Δ. Note that Δ may be measured either along the earth's surface, in kilometers, or by the angle at the center, in degrees or radians.) From these data obtained at the surface, how can we obtain α (or β) at depth?

The first relevant observation that we can make is that between approximately 2000 and 11,000 km (or ~ 100°), the hodograph is convex upward. For simplicity, we will assume that it is convex upward from 0 to 11,000 km. We infer that the waves do not follow the earth's surface—which is, indeed, logical, since they are body waves. To see whether they follow a straight line, we plot t versus $\sin(\Delta/2)$ (see Figure 3-9), since $SR = 2R_e \sin(\Delta/2)$. The resulting diagram (Figure 3-8*b*) is still slightly convex upward. The inference is that the time taken is shorter than it would be if the wave was traveling along SR at a constant velocity—or, more precisely, that as Δ, and thus the maximum depth of the ray,

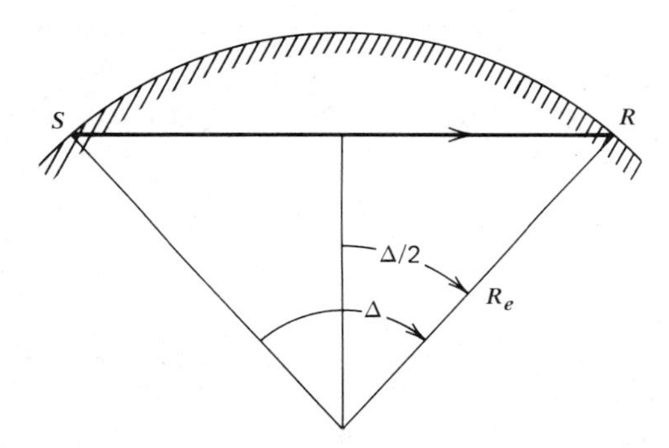

G. 3-9 *Relation between distance traveled and the angle at the center in a homogeneous sphere.*

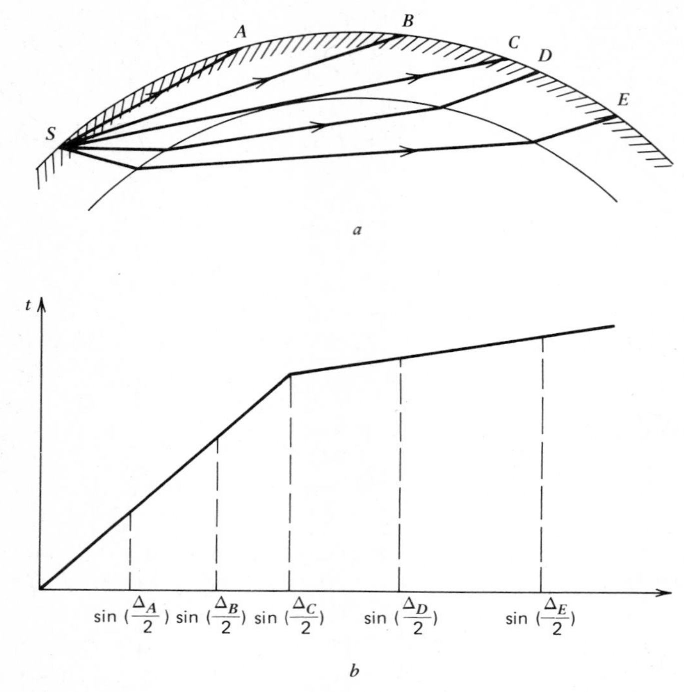

FIG. 3-10 *(a) Ray paths in two concentric homogeneous spheres. (b) Sketch of the travel time t versus* $\sin(\Delta/2)$ *for this case.*

increases, the average velocity also increases. But if the velocity along SR is not constant, Figure 3-9 is incorrect. To see why, assume that the earth consists of two concentric spheres, each with constant velocity (Figure 3-10*a*), and let the velocity in the inner sphere be larger. Then rays such as SA, SB, and SC will correspond to arrivals along one straight line on the travel-time curve, t versus $\sin(\Delta/2)$, and rays such as SC, SD, and SE will correspond to arrivals along another straight line (Figure 3-10*b*). (The second line is not exactly straight, but is close to it.) Clearly, this figure resembles the real situation (Figure 3-8*b*) more closely than does that resulting from a constant velocity throughout the earth. It is easy to see that with three concentric spheres, there would be three segments to the travel-time curve, and so forth. We thus conclude that travel-time curves such as those in Figures 3-8*a* and 3-8*b* result from rays that are convex downward, as

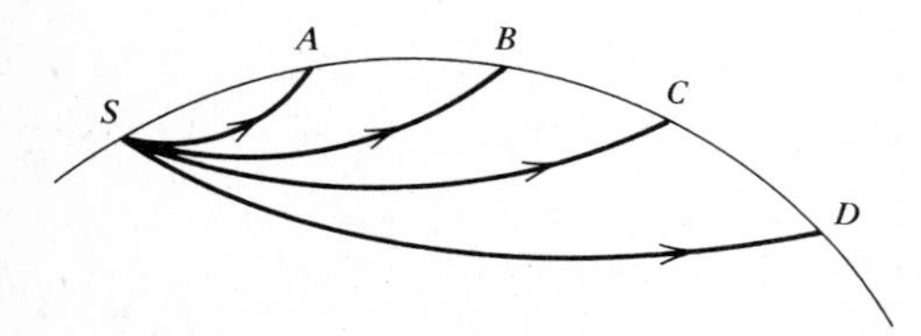

FIG. 3-11 *Ray paths through a sphere in which the velocity increases with depth.*

shown in Figure 3-11: that is, that the velocity must increase continuously with depth for rays emerging at $\Delta < 100°$.

It may be argued that such a continuous change in velocity violates the assumption that the medium is homogeneous. (*Note*: At every depth the velocity is still the same in every direction, so the medium is still isotropic; but it is not the same at every depth, so the medium is not homogeneous.) Strictly speaking, this argument is valid. To see why it is generally unimportant for body waves, we must examine whether the high frequency approximation holds—that is, whether $\Lambda_M \ll d_m$. We may take d_m as being the distance over which α changes by, say, 1%. If we exclude the discontinuities and thus consider the region between a depth of approximately 700 to 2800 km, we see in Figure 3-15 that α changes from ~ 11 to ~ 14 km/s—that is, 3 km/s over 2100 km or ~ 0.15 km/s $\approx$ 1% of 14 km/s in 100 km. Thus, $d_m \approx 100$ km. In order for Λ_M to be smaller than this value, only periods of less than 10 s, and preferably of less than 5 s, can be considered. These periods are those with which classic "body wave seismology" is concerned. For longer periods the tracing of rays becomes increasingly less meaningful.

By gradually increasing the number of layers and decreasing their thickness, we obtain a gradually better approximation to a continuous velocity with depth, and to a smooth hodograph. The real problem, however, is the inverse: Given the travel-time curve, say, $\Delta(t)$, how do we find the variation of velocity with depth, say, $v(r)$? This situation is typical in geophysics: the *direct* problem—in this instance, given the velocities, find the times of arrival—is relatively straightforward, but the *inverse* problem—from the data, infer the model—is much more difficult.

Although the exact method of solution to this problem is mathematically complex, it is instructive to see how it was solved. Consider, first, a seemingly unrelated problem: a ball is released at a certain height z above point B (Figure 3-12) and constrained to roll on a smooth frictionless path that is everywhere downward. We know the time that the ball takes as z increases, and we want to find the path that will produce this time (three such paths are shown). Although this situation is not the same as that found in the earth, there are some similarities: in both cases we know the travel time and want to determine the path. This problem was one of the first solved by the Norwegian mathematician Niels Henrik Abel (1802–1829), in 1823 (Abel 1881). Making use of Abel's method, Herglotz and Wiechert in 1907 first succeeded in determining $v(r)$ from $\Delta(t)$. The method is therefore known as the Herglotz-Wiechert transformation.

Note that one important requirement for the use of this method in its original form is that there must be a vertex at every depth. (The vertex of a ray is its deepest point.) To see that this is not always so, one need only consider the case in which the velocity *decreases* with depth at a certain depth d_1 before increasing again at depth d_2 (Figure 3-13). Here, the ray "next to" ray SB is deflected downward at depth d_1 and is then deflected upward again at depth d_2.

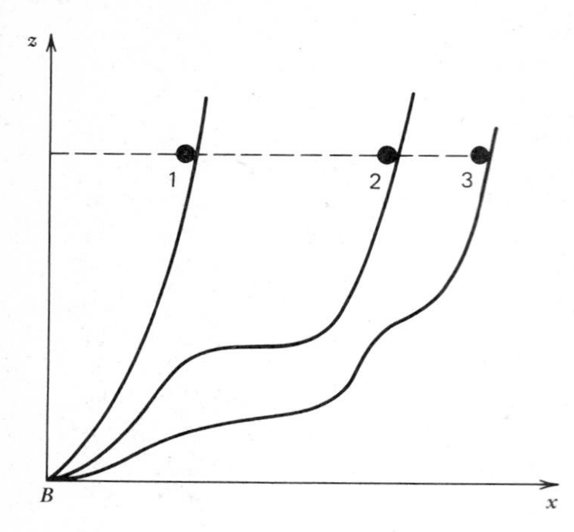

FIG. 3-12 *The rolling-ball problem: find the path to B if the time is known as a function of z.*

Corresponding to the low-velocity zone will be a "shadow zone" (BC), or "zone of no arrival," at the surface. Although many different situations may be imagined, the only relevant point here is that there are *no* rays with vertices between depths d_1 and d_2. In such a case the Herglotz-Wiechert method is not strictly applicable. In practice, what happens is that the existence of the low-velocity layer may not be detected—for instance, because the existence of the shadow zone is missed (see Problem 3-8*b*). For this fundamental reason, the "low-velocity zone" (LVZ) was not identified with certainty for about 50 years after Herglotz's work. (See Chapter 5, Section B.)

PROBLEMS

3-7 Assume that the Earth is homogeneous with $\alpha = 8.1$ km/s. Plot the travel time curve for Δ in km and t in seconds.

3-8 (**a**) Suppose that $\alpha_1 = 8.1$ km/s from the surface to a depth of 100 km, below which $\alpha_2 = 7.5$ km/s. Plot the travel-time curves t versus $\sin(\Delta/2)$ and t versus Δ for $\Delta < 75°$. This problem should be programmed as follows: for the rays entering layer 2, let i_{P1} (using the notation of Figure 3-2) vary by steps of 1°

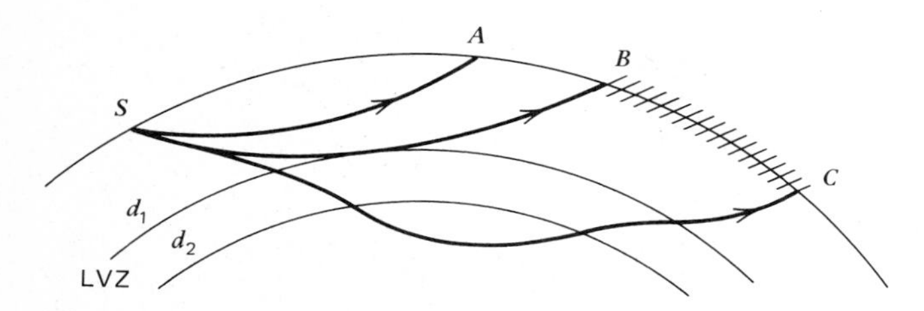

FIG. 3-13 *A layer of decreasing velocity with depth, between d_1 and d_2, causes a shadow zone BC. The velocity increases again below d_2.*

starting with $i_{P1} = 90°$. A table of trigonometric formulae applicable to a triangle may be useful. The earth's radius is ~ 6371 km. (*Hint*: There are very few right angles in this problem!)

(b) Plot the same travel-time curves if α increases back to 8.1 km/s at a depth of 225 km.

The Earth's Core

Once the velocities of *P*- and *S*-waves versus depth were determined, the existence of a core was discovered relatively quickly. It was first recognized that the travel-time curve of *P* does not extend smoothly beyond 103°. This yielded the depth of the core (2885 km). It was then discovered that these *P*-waves are not accompanied by corresponding *S*-waves, which showed that the core is liquid. Proving that a wave does *not* exist is, however, a difficult task, and as late as 1939, well-known seismologists (Gutenberg 1939; Macelwane 1939) expressed doubt about the fluidity of the earth's core.

From these observations of *P* and *S*, the ray paths and the velocity distributions shown in Figures 3-14 and 3-15, respectively, have been inferred. The complexity of the paths results from the combined effects of sphericity and of change in velocity with depth.

The core-mantle discontinuity gives rise to both refraction and reflection: waves reflected there are denoted *PcP*, *PcS*, and so on (see Figure 3-16), whereas rays that have traveled through the core are denoted by *K* (hence *PKP*, *PKS*, etc.). The existence of a solid inner core is now firmly established; *P*-waves through it are noted by *I*, *S*-waves by *J*. Whether the latter have ever been observed is still unclear. Finally, waves reflected once at the surface are denoted *PP*, *PS*, and so on. The proper identification of such waves on actual seismograms requires considerable experience.

The drop in *P*-wave velocity from the mantle to the core (Figure 3-15) results partly from the loss of rigidity, and partly from the change in density and composition. Seismic velocities, of course, only give the ratios $(\lambda + 2\mu)/\rho$ and μ/ρ.

Two zones of rapid velocity change appear to occur in the mantle: the first one near a depth of 350–400 km, and the second one near 650 km (Verhoogen 1980, p. 35). These zones may often be treated as discontinuities. (Why? How is this related to Λ?) The first such zone is generally ascribed to the transition of olivine to a phase of higher density, probably of spinel structure. This change is exothermic. The next one has been ascribed to either of two main causes: the breakdown of olivine–spinel into oxides or a difference in chemical composition. It has been noted that this depth is, within the uncertainty, also the maximum focal depth of earthquakes (see below), but the relationship between these facts remains speculative.

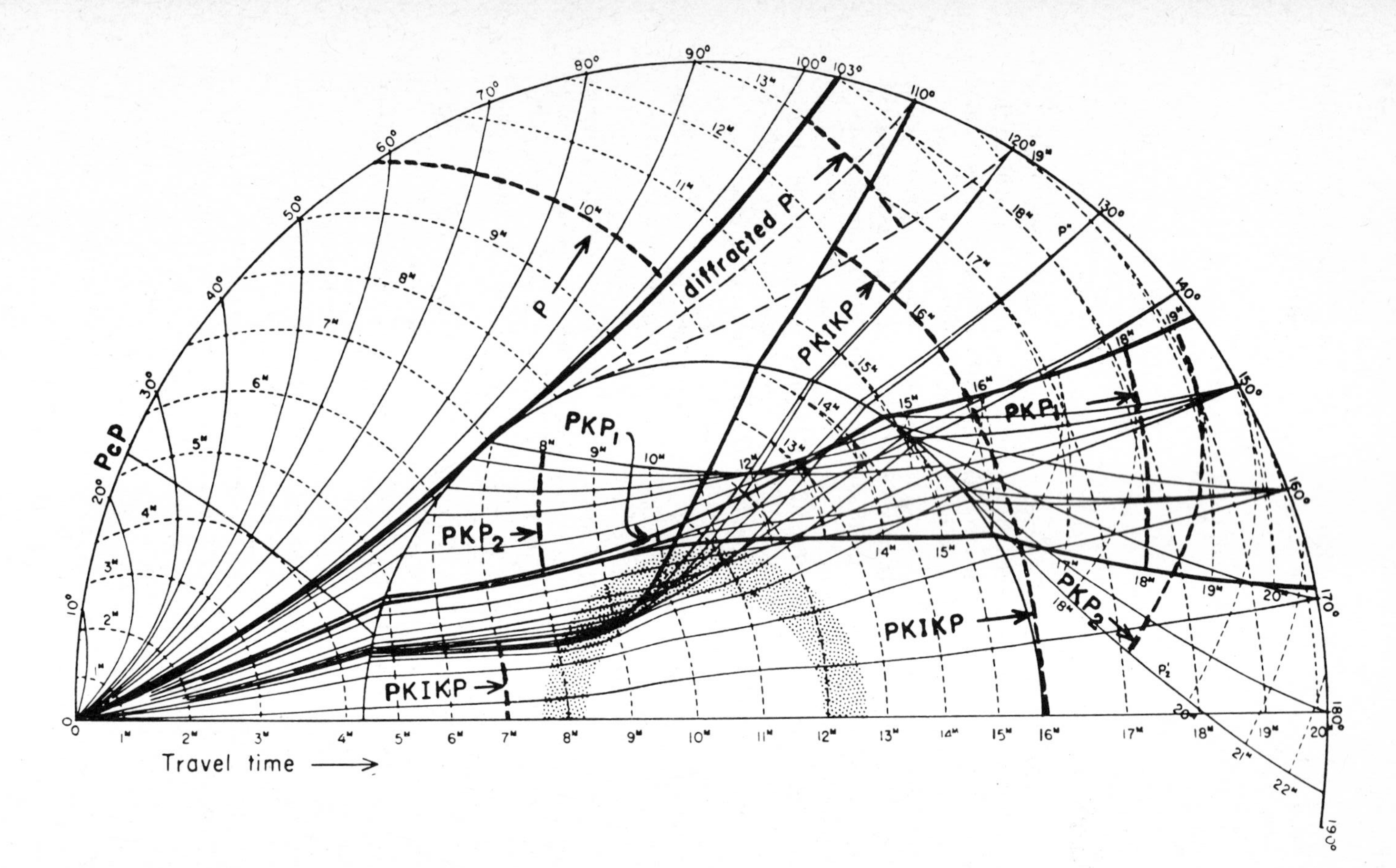

FIG. 3-14 *Selected ray paths and wave fronts for P-waves in the earth. (From Gutenberg 1959. Reprinted by permission. This figure is not exactly compatible with the velocities shown in Figure 3-15.)*

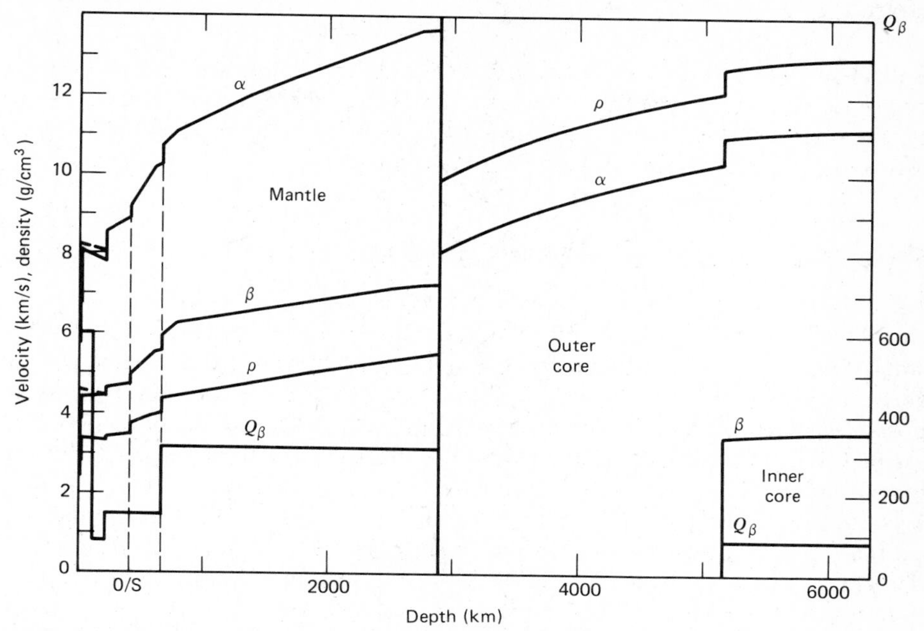

FIG. 3-15 *Velocities of P- and S-waves, densities, and Q_β in the earth. The dashed lines for α and β near the earth's surface are the horizontal velocities. The dashed vertical lines are discontinuities; O/S is the change from olivine to spinel. (From Dziewonski and Anderson 1981. Copyright © by Elsevier Scientific Publ. Co., reprinted by permission of the copyright owners.)*

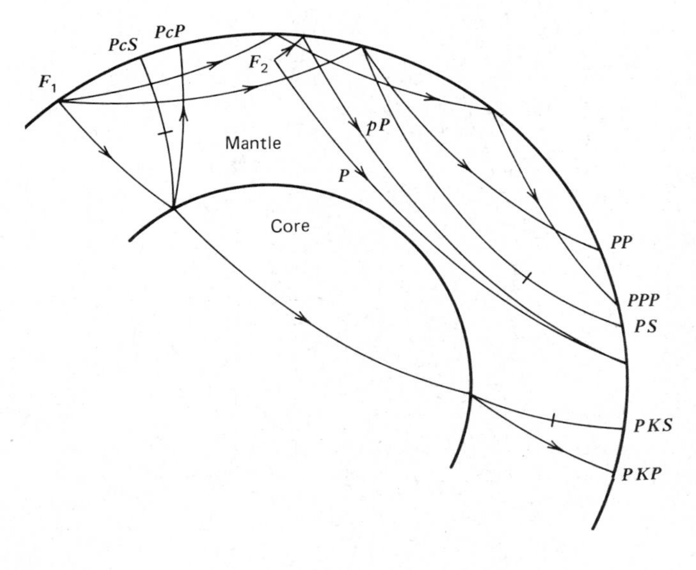

FIG. 3-16 *Some ray paths in the earth from two foci, F_1 and F_2. From F_1: wave reflected at the core-mantle boundary (PcP, PcS) and at the surface once (PP and PS) or twice (PPP) in the usual manner, and passing through the core (PKP, PKS)—but the effect of the inner core is neglected. From F_2: direct P-wave and wave reflected after going upward near the focus (pP). Arrows identify P-waves, bars identify S-waves.*

PROBLEM

3-9 You do not know that the core is liquid. You know that an earthquake occurred at a distance of 131° at time 03-16-25. The *P*-waves arrive at your seismographic station at 03-35-39. Around what time would you look for *S*-waves, using reasonable assumptions? Explain your reasoning.

P- and S-waves of Distant Earthquakes

As a result of the curved ray-paths (see Figure 3-11), *P*- and *S*-waves of earthquakes at a distance greater than about 20° arrive at the surface at a rather small angle of incidence. Consequently, the *P*-motion is much clearer on the vertical seismograph. Under very favorable circumstances it may nevertheless be possible to determine the azimuth of the epicenter from the *P*-motion. We will assume that the horizontal first motions are (except for a constant factor) 0.3 to the south and 0.7 to the east (see Figure 3-17). By vector addition, and taking into account the nature of *P*-wave motions, we find (see Figure 3-18*a*) that the azimuth of the epicenter is either N 67°W or N 113°E. (The *azimuth* is normally defined as the angle from the true north to a direction, counted clockwise. The notation employed here is used by geologists to indicate the strike; it is easier to visualize.) To determine which azimuth is correct, we must imagine a vertical section along the plane from the epicenter *A* to station *S* (Figure 3-18*b*): as the ground moves to the southeast, it moves up. Hence the motion must be upward, and it must be a compression from the azimuth of *A*—that is, from the northwest. Thus the epicentral azimuth is N 67°W.

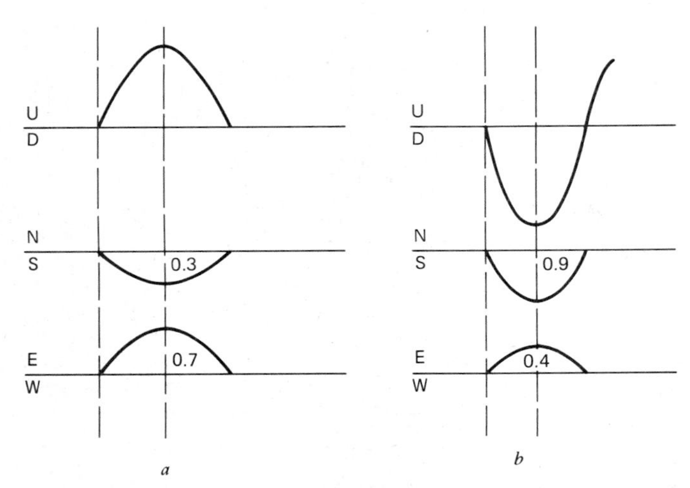

FIG. 3-17 *(a) Idealized records from three identical seismographs recording vertical (up-down), north-south, and east-west ground motions for a compressive first motion. (b) Similar records for a different first motion.*

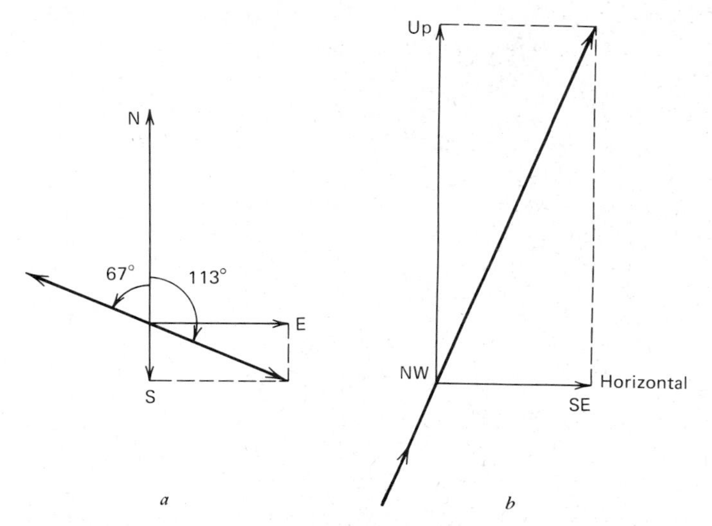

FIG. 3-18 *(a) Vector addition of the horizontal components for Figure 3-17a. (b) The motion upward combined with the horizontal motion gives the direction of the azimuth as N 67° W.*

PROBLEM

3-10 Find the azimuth of the epicenter in the schematic situation shown in Figure 3-17*b*. (All instruments are identical; the first arrival of *P* is shown.)

Deep Focus Earthquakes

The existence of deep focus earthquakes was first recognized by the Japanese geophysicist Wadati, through a beautifully simple method (Wadati 1928) built upon the fact that the interval between the arrival times of *P* and *S* at different stations should vanish at the epicenter for a surface focus. By contouring these intervals, Wadati found that this was not always the case in Japan, where the dense network of seismographic stations made such measurements practical. In some cases there were large residuals. He concluded that some earthquakes had foci (or hypocenters) up to several hundred kilometers deep.

Later, seismologists discovered that a special wave is associated with such foci, one that leaves the focus *upward* (see Figure 3-16) and is reflected at the surface. This wave is denoted *pP* (read "little *p*, big *P*") if both segments of the paths are *P*; otherwise, it may be *pS*, *sP*, or *sS*. Subsequent research has shown that no focus is deeper than 670 km, although the reason for this limitation is still debated (see Section 3E). Since the maximum focal depth is small compared to the earth's radius, most of the paths of *P* and *pP* are very close to each other (see Figure 3-16). It follows that the difference in the arrival times of these two waves depends principally on the focal depth, and not on the distance—which provides a convenient means of estimating the focal depth. It is of some interest to note that the existence of pP results from the spherical shape of the earth. In seismic prospecting and in other cases in which the earth's curvature may be neglected, such a wave cannot exist.

Until recently, only velocities and travel times were generally seen as useful, and practically no use was made of absolute amplitudes. This situation lasted both in earthquake seismology and in seismic prospecting until approximately 1970, with one exception: amplitudes of body waves and of surface waves have been used in magnitude determination since the 1930s (Chapter 7, Section B). Since 1970, great progress in using amplitudes has been made.

Attenuation, Q

Up to now, we have assumed that rocks or other materials are perfectly elastic, while noting that this assumption is not physically realistic. For the same reason, "pure" elastic waves do not exist; consequently, the amplitude of body waves does not decrease *exactly* as the inverse of the distance, some elastic energy must be dissipated—that is, transformed into heat. Thus, the amplitude of each successive wavelength decreases by a small fraction beyond that caused by geometrical spreading. For some engineering and industrial purposes it is desirable to manufacture materials that absorb very little energy per wavelength. Such materials are said to have a high quality factor, or Q.

If we choose to write the amplitude decay in a form somewhat similar to that used in Equation 3-1, we obtain

$$A = A_o \exp\left(-\frac{\pi}{QT}t\right) = A_o \exp\left(-\frac{\pi}{Q\Lambda}x\right) \tag{3-6}$$

Thus after one period (in time) or one wavelength (in space),

$$A_T = A_0 \exp\left(-\frac{\pi}{Q}\right)$$

The amplitude has thus been reduced by the factor $e^{-\pi/Q}$. In most cases of importance, $Q \gg 1$. Thus,

$$e^{-\pi/Q} \approx 1 - \frac{\pi}{Q}$$

Thus, the fractional amplitude loss per cycle is

$$\frac{\delta A}{A} = \frac{A_T - A_o}{A_o} = \frac{\pi}{Q} \tag{3-7}$$

Therefore, $1/Q$ is proportional to the amplitude loss per cycle (or per wavelength). If Q is large, this loss is small, and vice versa. Since the energy, E, is proportional to the square of the amplitude, A, Equation 3-6 gives

$$E = E_o \exp\left(-\frac{2\pi}{QT}t\right) = E_o \exp\left(-\frac{2\pi}{Q\Lambda}x\right) \tag{3-8}$$

Most of the energy dissipated by seismic waves is lost in shear; very little is lost in dilatation. Consequently, the Q of shear waves, Q_β, is smaller than that of P-waves, Q_α ($Q_\alpha \approx 2Q_\beta$). It is an experimental fact that Q is practically independent of wavelength or period within seismology's range of interest, although the reason why this is true is still debated. In the earth's mantle the value of Q_β is on the order of 300 (see Figure 3-15), although it may be as low as 100 and even less in the low-velocity zone (see below) and in near-surface layers (sedimentary rocks). Do not confuse $1/Q$ with the attenuation, γ, which is the relative decrease in amplitude per meter (or kilometer, depending on the unit used for Λ). To find γ, set $x = 1$ in Equation 3-6:

$$\gamma = \frac{A_o - A}{A_o} = 1 - \exp\left(-\frac{\pi}{Q\Lambda}\right) = 1 - \exp\left(-\frac{\pi\nu}{cQ}\right) \tag{3-9}$$

Just as above, we then find (using $e^{-x} \approx 1 - x$ for small x):

$$\gamma \approx 1 - \left(1 - \frac{\pi\nu}{cQ}\right) = \frac{\pi\nu}{cQ} = \frac{\pi}{Q\Lambda} \tag{3-10}$$

where γ, c, Q, and Λ must, of course, all refer to P-waves or all refer to S-waves. Equation 3-10 explains why the dominant period of body waves increases with distance: the amplitudes of the shorter periods (higher frequencies) become too small to be visible, since γ is proportional to ν. This equation also explains why the periods of S-waves tend to be longer than those of P-waves. Both c and Q are smaller for S-waves, and since γ is inversely proportional to their product, the attenuation of S-waves is higher on both counts.

Geological Application

Examination of the records from foci along the Tonga arc, a deep-subduction zone, has revealed (Figure 3-19) that the slab has a high Q. In front of it is a small triangular wedge extending from the trench to the ridge where the value of Q is approximately normal; farther away is a zone of very low Q (Barazangi and Isacks 1971). The region between approximately 80 and 300 km has a low Q and is presumed to coincide with the low-velocity zone (see Chapter 5, Section B). This situation is rather typical near subduction zones.

PROBLEM

3-11 **(a)** Suppose that $2Q_\beta = Q_\alpha = 400$. Compute the attenuation per kilometer for both P- and S-waves of periods 0.3, 1, 3, and 10 s, if $\alpha = 8$ km/s and $\sigma = 0.25$. This case is representative of mantle conditions.

(b) Do the same if $2Q_\beta = Q_\alpha = 25$, $\alpha = 2$ km/s, and $\sigma = 0.35$. This case might occur in sedimentary rocks near the surface.

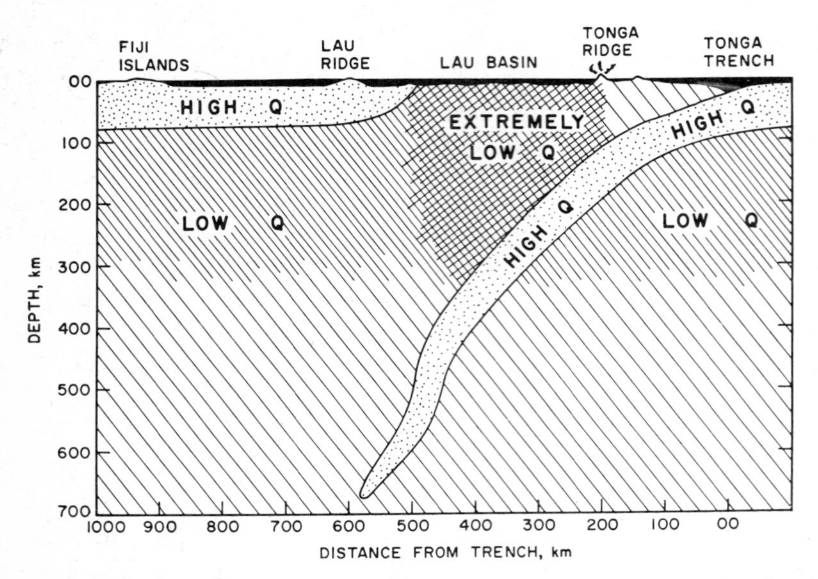

FIG. 3-19 Schematic cross section perpendicular to the Tonga arc, showing the high and low Q zones. (From Barazangi and Isacks 1971. Copyright © by the American Geophysical Union.)

C. Focal Mechanism

All large earthquakes and most small ones are related to faulting. Volcanic earthquakes, caused by an explosive release of gas, are almost always small, even though the explosion may be extremely destructive, because of the very weak coupling between the air and the ground (see Equation 3-5).

Energy Release

Let us suppose that a volume V of rock is stressed to an increasingly higher level, and consider the integral

$$E = \int_V p_{ij} e_{ij} \, dV \tag{3-11}$$

The combination $p_{ij}e_{ij}$ in the integrand is taken to mean

$$p_{xx}e_{xx} + p_{xy}e_{xy} + p_{xz}e_{xz} + p_{yx}e_{yx} + \cdots$$

and represents the elastic energy in the volume dV. If the stresses continue to increase, a situation may be reached in which no more energy can be stored in V, and fracture (or faulting) occurs. A small part of the released energy is transformed into local heat (i.e., frictional energy along the fault); most of it is radiated as seismic waves that eventually are attenuated and transformed into heat.

It is extremely doubtful, however, that (as the foregoing explanation implies) the material would stay elastic until it fractures. Instead, if the deformation on the boundary of the volume V increases gradually, the energy inside the volume may "leak away": that is, the material may deform without fracture. Reiner (1960, pp. 140–144) has compared the process leading to fracture to pouring water into a leaky bucket (see Chapter 12, Section E): if water (energy) is poured in fast enough, the bucket overflows (the material breaks); if not, the water leaks away (the material deforms). What constitutes "slow enough" depends on the nature of the material: wet clay, for example, can be greatly and rapidly deformed without fracture, whereas granite cannot. This difference in material behavior may explain why some segments of the San Andreas fault slip smoothly past each other, whereas others remain locked for long periods and then give rise to large earthquakes when they fail. To pursue the bucket analogy: when the bucket overflows, its sides also fail, at least in part.

Although Equation 3-1 cannot be derived here, it is useful to note that it is dimensionally correct. Let M, L, and T represent the dimensions of mass, length, and time, respectively. Then,

- The left-hand side, E, has the dimensions of energy (i.e., force) $\times$ length, or

$$(MLT^{-2}) \times L = ML^2T^{-2} \text{ (kg m}^2\text{ s}^{-2}\text{ in SI units)}$$

- The right-hand side has the dimensions of (force/area) $\times$ (none) $\times$ (volume), or

$$(MLT^{-2}/L^2) \times (1) \times L^3 = ML^2T^{-2}$$

As required, both sides of the equality thus have the same physical dimensions. This is a simple example of the method of dimensional analysis; the proof that scaling of size requires scaling of material properties (Chapter 2, Section D) is another example of the use of this method.

The Fault-Plane Method

In order to understand how the fault-slip at the focus may be determined, consider first the elementary case shown in Figure 3-20a: a vertical strike-slip fault of infinitesimal extent (point source) moving in an instant (instantaneous source). (These simplifications will gradually be removed.) The fault plane is marked by the solid line FP.

An observer located in either of the quadrants marked C will first see a P-motion toward him—that is, a compression (for him)—whereas an observer in the D quadrants will first see a dilatation. If the fault plane were the dashed line AP, and the motion were in the direction of the dashed arrows, nothing would be

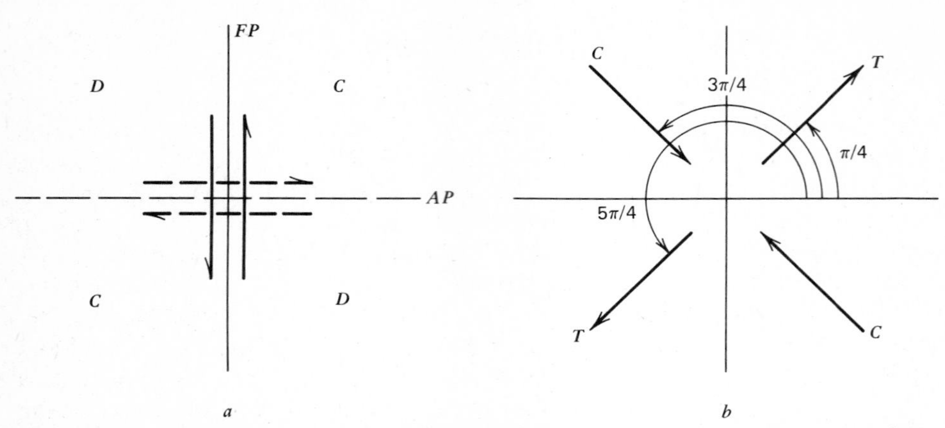

FIG. 3-20 *(a) Map view of a vertical strike-slip fault of infinitesimal extent, the two couples acting on it, the fault plane FP and the auxiliary plane AP, and the compression (C) and dilatation (D) quadrants. (b) The principal tensile stress (or T-axis) and compressive stress (or C-axis) for this case.*

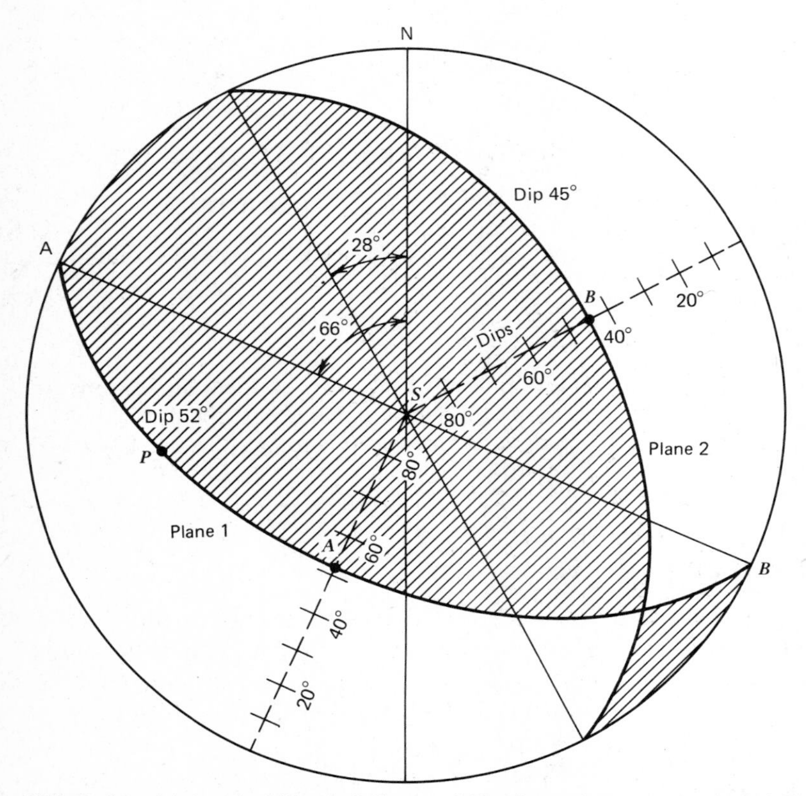

FIG. 3-21 *(a) A typical "football diagram." The D-quadrants are hachured. The strike of each plane is given by the orientation of its diameter. Its dip is determined as explained in the text. (b) A "football diagram" of a shallowly dipping plane (1) and a steeply dipping one (2).*

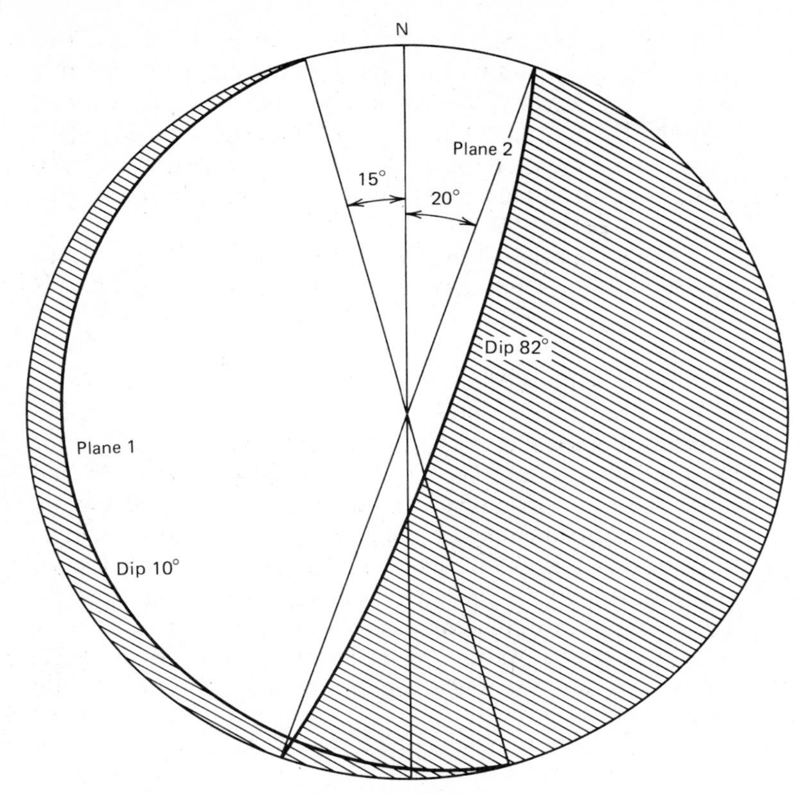

FIG. 3.21 *(Continued)*

changed. There are thus two orthogonal planes—the fault plane (*FP*) along which the motion takes place, and the auxiliary plane (*AP*), perpendicular to the direction of motion.

Usually, of course, *FP* is not vertical, and the direction of motion is neither horizontal (along the strike) nor along the line of greatest slope (along the dip). Nevertheless, the two orthogonal planes should still exist. They intersect the surface of the earth along two intersecting small circles.

For some time after the invention of the fault-plane method by Byerly (1942, pp. 233–239), the results were represented by a stereographic projection of the two circles upon the great circle whose pole is the epicenter. Because the resulting diagrams were difficult to visualize, they have been replaced by "football diagrams" (Figure 3-21)—called "beachball diagrams" by some authors. These diagrams generally represent the equal-area projection of the lower hemisphere of a sphere drawn around the focus (focal sphere). *FP* and *AP* are now two great circles, and the *D*-quadrants are normally hachured. This projection, technically known as Lambert's azimuthal (or zenithal) equal-area projection (after the German physicist and astronomer Johann Heinrich Lambert, 1728–1777), is used in many modern atlases. It is characterized by the fact that the directions from the center of projection are correct, and that the areas on the projection are in a constant ratio to those on the earth (Lambert 1894). Lambert's projection is especially useful for statistical purposes, but it is not conformal: that is, most angles on the projection are not those on the earth's surface. Schmidt first used it in geology, and a comprehensive discussion of its use for this purpose can be found in Turner and Weiss (1963, pp. 46–75).

Here we need only note that each seismographic station—P in Figure 3-21*a*—is plotted in the correct azimuth from the source S, but at a distance from S given by

$$r = R\sqrt{2}\,\sin\frac{i_P}{2}$$

where R is the radius of the "football" and i_P is the angle of incidence at S of the ray emerging at P (see Figure 3-22). The plot of all the compressions and dilatations can then be used to find the two orthogonal planes corresponding to FP and AP. (These planes are great circles on the focal sphere but are not circles on the projection.) Note that, as Figure 3-22 shows, seismographic stations are *not* plotted at those points (A, B, and P) corresponding to their distance from the epicenter S, but at those points corresponding to the straight lines tangent to the rays at the focus (A_e, B_e, and P_e). As this figure makes clear, i_P is the angle between SP_e and the vertical SO. The angle between the straight line SP (not

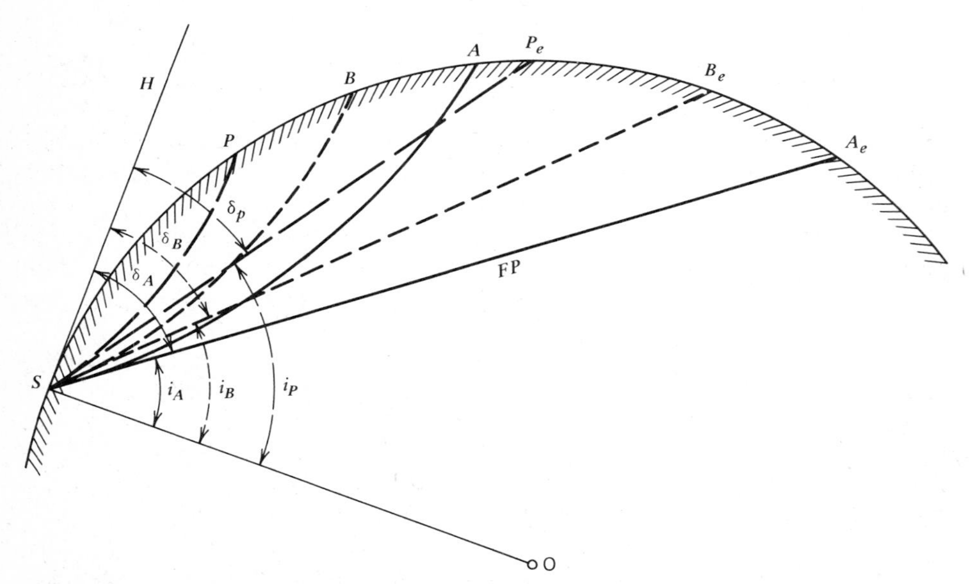

FIG. 3-22 *Great-circle sections through the focus S and the stations A, B, and P. To "straighten out" the rays, plot all stations (A, B, and P) as if they were at* A_e, B_e, *and* P_e*—that is, at distances proportional to* $\cos(i_A/2)$, $\cos(i_B/2)$, *and* $\cos(i_P/2)$.

shown) and the vertical has no physical meaning. The "distance" SP_e is sometimes called the "extended distance" from S to P.

The strike of the fault plane (or of the auxiliary plane) can be immediately found from the azimuth of its diameter. Its dip can best be determined as follows. Plane 1 of Figure 3-21*a* separates stations receiving compressions from those receiving dilatations. Thus, the distance SP of any point P on this circle to the center of projection S obeys the relation $r = R\sqrt{2}\,\sin(i_P/2)$. Obviously, r and i_P are minimum at A where $SA = r_A = R\sqrt{2}\,\sin(i_A/2)$. But the dip, δ, of a plane is measured from the horizontal, whereas i_A is measured from the vertical, and thus $\delta_A + i_A = 90°$. Hence, $\delta_A = 90° - i_A = 90° - 2\sin^{-1}[r_A/(R\sqrt{2})]$. In the figure, $r_A/R \approx 0.46$, hence $\delta_A \approx 52°$. The strike of plane 1 is obviously N 66°W, and the dip is to the SW. For plane 2 we use the same procedure to find that the strike is N 28°W; $r_B/R \approx 0.54$, thus the dip is 45° to the NE. We can, of course, obtain an approximate value by visualizing the figure as if we were looking vertically down at a lower hemisphere. The examples shown in Figures 3-21*a* and 3-21*b* may be interpreted (a) as normal faulting on either plane 1 or plane 2; or (b) as reverse faulting either on plane 1—which would be called thrust-faulting, because of the gentle dip of this plane—or on plane 2, although its steep dip makes this alternative *a priori* less likely. It will be helpful to remember that normal faults correspond to the case where hachured areas are on the "inside" of the circle (i.e., not greatly touching the circumference), and that the opposite is true for reverse faults.

The indeterminacy between FP and AP inherent in the method can be removed in two main ways. First, geological information may indicate which of the two planes is FP. Second, the nature of the two planes can be determined from the precise location of the aftershocks that always follow large shocks (see Section 3E). Aftershocks release some residual elastic energy near the fault, and are thus essentially limited to the fault plane area.

PROBLEM

3-12 How can transcurrent faults be recognized on "football diagrams"?

By using Equation 2-2, we can easily show that Figures 3-20*a* and 3-20*b* are equivalent. We can start, for instance, from Figure 3-20*a* and orient the x- and y-axes along AP and FP. Then,

$$p_{xx} = p_{yy} = 0 \qquad p_{xy} = \tau > 0$$

Orienting the d-axis at $\pi/4 = 45°$ to AP, we find

$$p_{dd} = \tau \sin(2 \times \pi/4) = \tau > 0$$

which corresponds to a tension—T on Figure 3-20*b*. The same result can be obtained by orienting the d-axis at $5\pi/4$ from AP.

Orienting the d-axis at $3\pi/4$ to AP, we find

$$p_{dd} = \tau \sin(2 \times 3\pi/4) = -\tau < 0$$

which corresponds to a compression—C in Figure 3-20*b*. Furthermore, along both directions,

$$p_{de} = \tau \cos(2 \times \pi/4) = \tau \cos(2 \times 3\pi/4) = 0$$

which shows that the T and C directions are the principal axes, and proves that Figures 3-20*a* and 3-20*b* are equivalent. (One may also start from Figure 3-20*b* and obtain Figure 3-20*a*.)

The preceding reasoning implies that the stresses are two-dimensional: that is, that the normal stress perpendicular to the plane of the paper is null. It is intuitively evident that nothing will be changed if a pressure is added at the focus. This can also be demonstrated by means of Equation 2-2.

We will now examine the influence of both the explicit assumptions (point source, instantaneous source) and the implicit one (that the source consists of a fault).

Single Couple versus Double Couple Because of the symmetry of the stress tensor, the shear stress on the fault plane must be balanced by an equal stress on the auxiliary plane (Figure 3-20*a*). When the earthquake occurs, both vanish simultaneously. Motion takes place along the fault plane only, but motion should not be confused with stress (although this was done for many years). Thus, the dynamically proper configuration is the so-called "double-couple" configuration. Data on S-wave motions confirm this theory.

The Extended Source Since an earthquake is produced when a volume of rock releases its strain energy (see Equation 3-1), the volume involved should increase proportionally to the energy, E, released, all things being equal. One might thus expect that the linear dimensions would increase according to $E^{1/3}$. In fact, though, one horizontal dimension normally grows much more rapidly than the other two, because most of the energy is concentrated near the fault. In a very large earthquake (e.g., the 1960 Chile earthquake) the length of the fault break may extend over as much as 1000 km. It is physically unrealistic to expect this length to break instantaneously. One point along the fault fractures first. From there the area of the fracture expands, and the fracture propagates in either one or two directions. It is difficult to obtain data on how fast such a fracture propagates. One would expect the duration of fracture to be approximately equal to the duration of the most intense shaking near the source. But such shaking is much too violent for most conventional seismographs to record in detail, and few people caught in a large earthquake are trained (or inclined) to make such precise and unexpected observations. It appears to have happened once or twice, though. These observations, together with theoretical computations and surface wave observations (see below), have shown that such fractures propagate at a velocity of $v_f \approx 3.5$ km/s—or approximately 0.6β, where β is the S-velocity of the medium involved (Furumoto and Nakanishi 1983). In all likelihood, v_f represents an average velocity; the real phenomenon is likely to involve a succession of slower and faster episodes. The fracture may propagate more slowly when it encounters a segment of the fault that is relatively "hard to break" (i.e., where the coupling between the two sides of the fault is strong); it may speed up when the two sides of the fault are coupled only weakly. Such possible segmentation of large faults and subduction zones into subfaults of varying "asperity" apparently also accounts for the different earthquake patterns observed in different subduction zones (Lay and Kanamori 1981). (See also Chapter 7, Section D.)

Such an extended source moving in space and time has very complex consequences for the radiation pattern of body waves, which is now being used to infer the source mechanism. Here we will only note that, all things being equal, a spatially extended source tends to generate waves whose wavelengths are shorter than, or at most comparable to, its length. The effect of source motion can best be observed with surface waves: it is a Doppler effect—the phenomenon responsible for the red-shift of galaxies ("stretching" their wavelengths as they recede) and for the change in the tone of car's horn perceived by an observer as the vehicle passes by. In the present case, stations in the direction of fracture propagation observe relatively higher frequencies than do those on the opposite side.

Other Difficulties It might appear that short-period instruments would be best suited for fault-plane observations, since they best record the "first motion"; but as shown in Figure 3-23, the opposite is the case. The reason is that, at first, only a small volume of rock is releasing energy, so the amplitudes of the waves are small. These first waves are what short-period instruments record, and they are not easily distinguished from the background "noise." Moreover, there is some question whether this first motion is representative of the major motion during the earthquake.

Vertical instruments are greatly preferred for these studies, as a ground motion upward necessarily represents a compression. Moreover, because of the curvature of the rays, the P-motion tends to be much larger on vertical than on horizontal instruments.

International Implications The fault-plane method has received great attention as a means of distinguishing earthquakes from explosions. When negotiations to ban nuclear explosions began between the Soviet Union and the United States, most U.S. seismologists held that this method would discriminate between earthquakes (with a quadrant distribution of first motions) and explosions (with compressions everywhere). This position was eventually accepted by the Soviets, and the Americans then conducted test explosions. The results did not entirely confirm the theory, and this led to an awkward situation. Seismologists have explained these exceptional observations by the heterogeneity of the earth near an explosion: for example, long fissures may form, leading to very

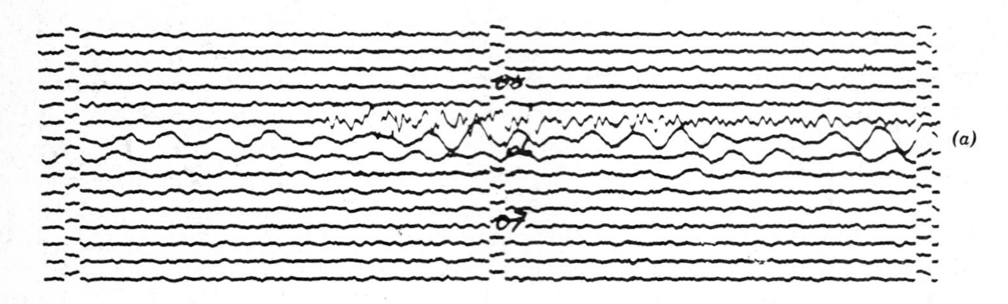

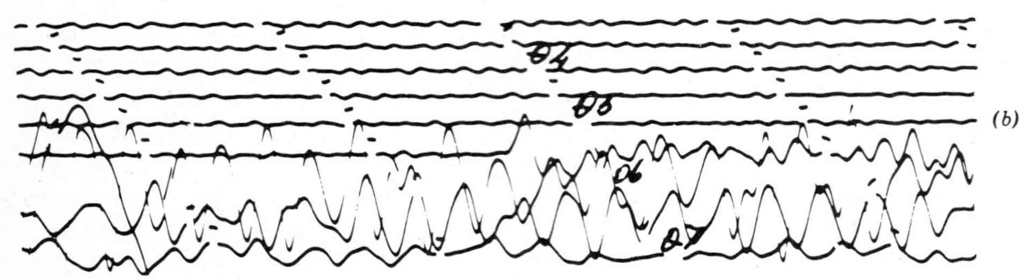

FIG. 3-23 *(a) Short-period vertical seismograph record. (b) Long-period vertical seismograph record of the same earthquake. The first motion is uncertain in a but very clear in b. Traces are deflected once a minute; the numbers indicate the hour. (From Sykes 1967. Copyright © by the American Geophysical Union.)*

anisotropic radiation of seismic energy. Apparently, though, when first motions caused by an explosion are clear, they are always compressive.

The Elastic Rebound Theory Following the great 1906 earthquake in and near San Francisco, horizontal displacements of up to 6.3 m were observed. Precise surveys had been made from 1851 to 1865 and from 1874 to 1892, and they were repeated after the earthquake (Reid 1910, p. 16). These surveys confirmed the theory about energy release given above. The picture that emerges is summarized in Figure 3-24. As points on the left side of the fault move with respect to those on the right while the fault remains locked, the region near the fault suffers increasing distortion. This stage may last anywhere from a few years to, probably, thousands of years. Eventually, the accumulated strain energy may exceed the maximum admissible, and the fault will break. In this process some or most of the elastic energy will be released, and a new cycle may begin. The occurrence of aftershocks testifies to the fact that not all elastic energy is released during the main shock. The following mechanism has been suggested. For a fault to break, the static friction on the fault must be overcome. Once the fault moves, there remains only the dynamic friction, which is lower; and when the energy

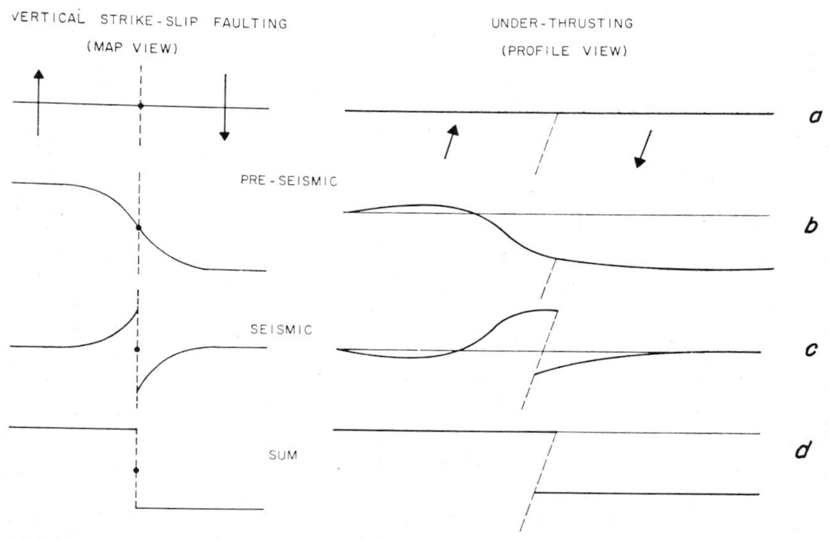

FIG. 3-24 *Maps of vertical strike-slip fault (left) and profile views of thrust faults (right). (a) Situation with no elastic energy stored (assuming that this were possible). (b) Situation just before the earthquake, with maximum stored elastic energy. The amount of stored energy dies off as the distance to the fault increases. (c) During the earthquake the fault breaks. The motion dies off away from the fault. (d) The total deformation, seen immediately after the earthquake, is the sum of b and c. (This supposes that the earthquake has released all the stored elastic energy, exactly.) (From Fitch and Scholz 1971. Copyright © by the American Geophysical Union. Also reproduced in Le Pichon et al. 1976, p. 248.)*

available is no longer sufficient to overcome the dynamic friction, the motion stops. If this explanation is correct, most of the elastic energy remains stored in the rock after the earthquake, and the stress drop during an earthquake represents only a fraction of the stress necessary for fracture. This last point appears to be experimentally correct.

Reid's theory, which was the first clearly to link earthquakes with faulting, is known as the elastic rebound theory. It antedates plate tectonics by more than half a century, and is now generally accepted.

Some Results of the Fault-Plane Method The fault-plane method is a primary source of information about the state of stress in the earth. As we will see, it has helped establish the theory of plate tectonics on a firm basis. Thus, it will be helpful to distinguish between the various plate boundaries on the basis of the plate tectonic model (see Le Pichon et al. 1976, pp. 20–26).

- At accreting (or diverging) plate boundaries (oceanic ridges), earthquakes are very rare, presumably because of the high temperature, and thus low viscosity,

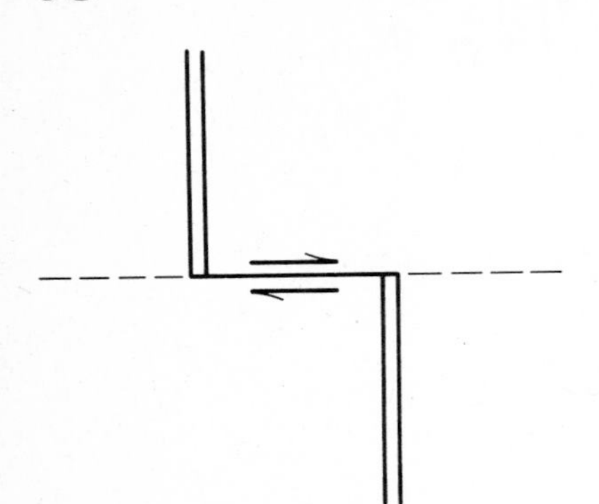

FIG. 3-25 *Transform fault cutting a midoceanic ridge (double line); the active, inter-ridge part is seismic, the rest is not. The ridge offset and the fault motion are in opposite directions.*

of the intrusive material. The East Pacific Rise and the Mid-Atlantic Ridge are typical examples of accreting plate boundaries (see inside front cover).

- Near oceanic ridges, in contrast, earthquakes are numerous along transform faults separating active segments of the ridge. The Romanche fracture zone, which displaces the Mid-Atlantic Ridge by several hundred kilometers, is one of the most conspicuous of these features. Sykes's (1967) observation of the direction of motion in such areas (Figure 3-25) confirmed the prediction of J. Tuzo Wilson about transform faults—that the sense of motion along the fault is the opposite of that expected from a simple consideration of the offset of the ridge, and that only the part of the fault between the two ridge segments is active. Such earthquakes generally are not large ($M \leq 6$) (for a definition of M, the magnitude, see Chapter 7, Section B) nor are they as frequent as those at consuming boundaries.
- At consuming plate boundaries, the process is appreciably more complex. The west coast of South America (see inside front cover) and the arc south of Indonesia (see inside back cover) are examples of such boundaries. It may, however, be noted that not all features described below occur at each such boundary. As the plate approaches the subduction zone, it is first deflected upward (Figure 3-26), as a result of bending. In this region the upper part of

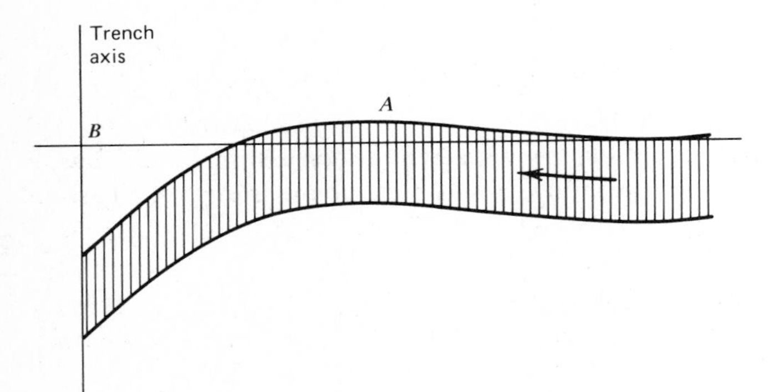

FIG. 3-26 *A lithospheric plate approaching a trench is first deflected upward at A before being subducted.*

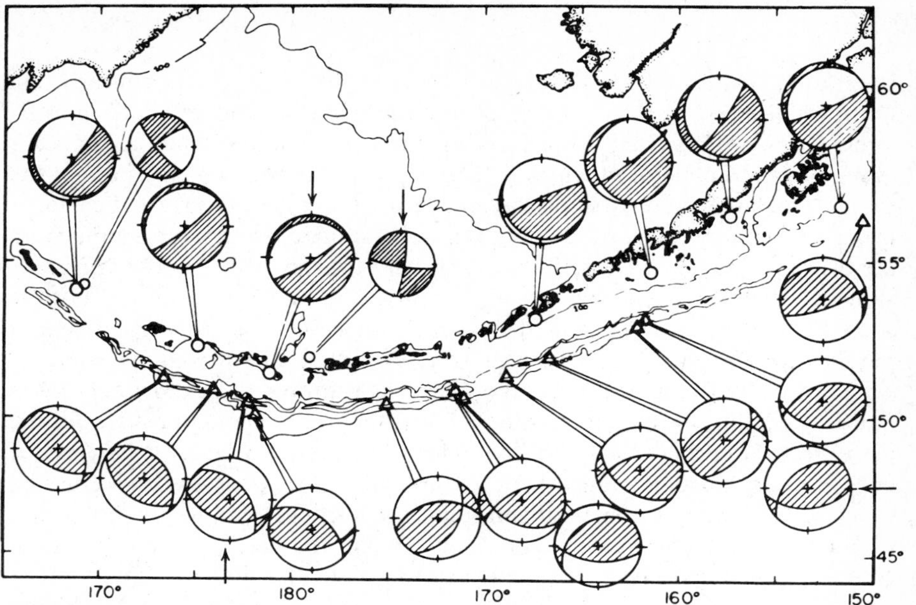

FIG. 3-27 *Focal mechanisms of earthquakes near the Aleutians; the plate moves northwest. Point A in Figure 3-26 corresponds to earthquakes south of the islands. Those under the islands indicate underthrusting to the northwest. (From Le Pichon et al 1976, p. 248; after Stauder 1968. Copyright © by the American Geophysical Union.)*

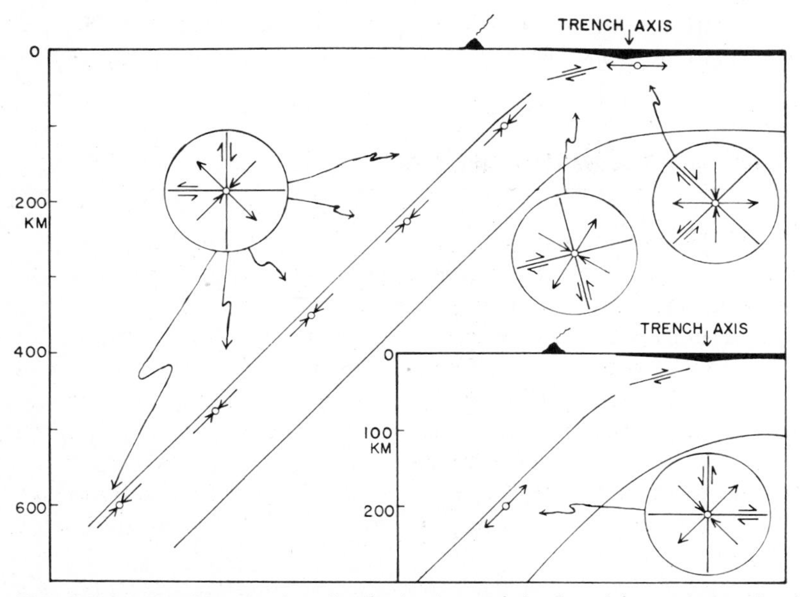

FIG. 3-28 *Focal mechanism distribution in a subducting slab: tension under the trench, thrusting at shallow depth, and generally (but not always), compression at greater depths. (The inset shows a tension at ~ 200 km.) (From Isacks et al. 1968. Copyright © by the American Geophysical Union. Also reproduced in Le Pichon et al. 1976, p. 242.)*

the plate will thus be "in tension"—which, as noted, previously, means that the normal stress in the direction of motion is tensile, and the other normal stresses are not involved. Consequently, the earthquakes in such areas occur along normal faults (see Figures 3-27 and 3-28). At shallow depths under the trench and immediately below the arc, thrust faulting is clearly indicated; it occurs at the top of the subducting lithospheric plate. At greater depths there are often two separate Benioff zones: one near the top of the subducting plate, showing that the plate there is in compression; and one near the bottom showing that it is in tension there. (In both cases one principal stress is parallel to the plate.) The bottom zone has been interpreted by some as evidence of "unbending." Once the plate has ruptured at shallow depth, the faults heal; thus, the curvature imposed near the trench should persist, and the plate should "curl up." Presumably under the influence of gravity, however, it does not do so, but straightens out (unbends) and fractures again in the process. Why most subduction zones are not vertical is, however, unknown. Finally, at depths of greater than about 400 km the plate is in compression. No earthquakes have been recorded at depths of greater than 670 km.

PROBLEM

3-13 Determine the approximate strike and dip of the fault plane and of the auxiliary plane corresponding to the football diagrams indicated by the arrows in Figure 3-27. For each case give the two solutions, indicate the preferred one, and indicate whether the fault is normal or reverse.

D. Epicenter Determination

The difference in velocity between P- and S-waves furnishes an easy method of determining the epicentral distance, Δ, from a seismographic station. Given the travel-time curves of these waves, we can immediately determine Δ from observations. With two stations, only two epicenter locations are possible. The use of three stations would give rise to an overdetermined problem, for the three circles centered on the three stations are highly unlikely to intersect at exactly one point, and some method must be found to select the "best" point. In addition, from the exact times of arrival of, say, P at each station, and from the corresponding epicentral distance, the time of origin, or epicentral time, may be determined for each station. This epicentral time must of course be unique, and, again, a method must be found to determine it.

In practice, only the times of arrival of P at many stations are used, because they can be read more exactly than can those of S. The origin time and epicentral location are determined by a least-squares adjustment (see Chapter 9, Section C). Also, provision must be made for the fact that not all foci are near the surface.

E. Further Results of Body Wave Seismology

Geographic Distribution of Epicenters

A comparison of the maps inside the front and back covers shows that a continent–continent collision (e.g., the Alpine–Himalayan belt) results in a much more diffuse seismicity than that which occurs at other plate boundaries. The area of "high seismicity" (an imprecise expression) is particularly extensive in Tibet and in much of the rest of China. The seismicity of the former area is generally presumed to be the result of the collision of India with Asia, which gave rise to the Himalayas. The origin of the seismicity of the rest of China is still disputed. It is worth noting that some very large and destructive earthquakes in northern China are caused by normal faults.

Similarly, exceptional earthquakes occur far from plate boundaries, possibly because of old faults. It is worth noting that two of the most potentially destructive earthquakes to have struck the conterminous United States fall in this category: one (consisting of two main shocks) in 1811–12 near New Madrid, Mo., the other in 1886 near Charleston, S.C. Figure 3-29 shows the great extent of the areas that suffered strong shaking from these shocks. The property damage caused by the New Madrid earthquake was negligible only because of the sparseness of

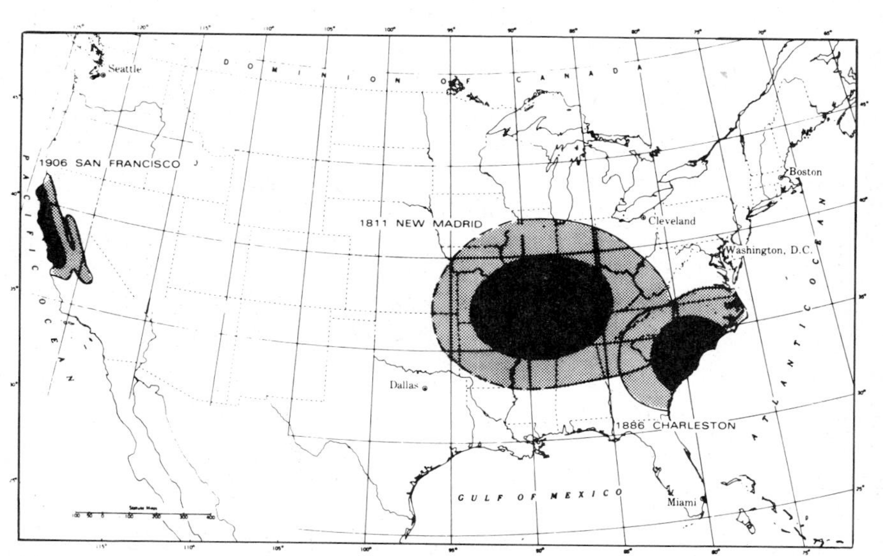

FIG. 3-29 *Isoseismal contours for three major earthquakes in the conterminous United States: New Madrid (1811–1812), Charleston (1886), and San Francisco (1906). The dark areas correspond to intensity VII (considerable damage in badly designed structures); the lighter ones to intensity VI (slight damage only). Note that the first two shocks are not on a present plate boundary, and that the area of intensity VII is smallest for the San Francisco shock, on the San Andreas fault (Earthquake Information Bulletin 13:178).*

the population in the area at the time. Such earthquakes can be extremely destructive, in part because they usually are totally unexpected. In some cases, though, monitoring of the past site of such a large shock will show continuing seismic activity at a level too low to be detected without instruments.

Foreshocks and Aftershocks

A large earthquake is normally followed by a succession of smaller ones, called aftershocks. As noted in Section 3C, aftershocks tend to fall within the area that was fractured by the main shock. They clearly are caused by further local release of elastic energy, perhaps made possible by the weakening of rock where the fault recently moved. Although exceptions occur, as a rule the aftershocks will be at least one magnitude unit smaller than the main shock (see Chapter 7, Section B). Foreshocks are rarely perceptible. Whether recognizable foreshock patterns (detectable only with instruments) exist, and whether they possess more than local validity, are matters of debate (see Chapter 7, Section D).

Deep Focus Earthquakes

Physical Mechanism The physical mechanism of deep focus earthquakes is poorly understood. The difficulty is essentially that at depths of more than, at the most, 50–60 kilometers, friction is large enough to prevent motion on the fault. It has been suggested that a preexisting liquid-filled crack might be the starting point. Near the tip of the crack the stress-energy could be highly concentrated, and the incipient motion would generate enough energy to melt the rock. This shear-molten zone would then both grow and propagate, thus releasing the elastic energy. Sudden phase-changes have also been proposed as an explanation, but this theory has been abandoned for two reasons: in known earth materials under laboratory conditions, phase changes are very sluggish, and it is not clear that they could release shear stresses. Evidence of a volume change accompanying some large deep earthquakes has been obtained by Gilbert and Dziewonski (1975), but its exact meaning is unclear.

Absence below 670 km It has long been recognized that no foci are deeper than 670 km. The reason for this limit has remained obscure. Among the causes suggested have been the following:

- The lack of strength of the material: if the material is incapable of accumulating elastic energy, it will deform without fracturing. However, no one has offered a conclusive explanation of why such a change in property would occur at this depth; it may be caused by a rise in temperature and/or a mineralogical change resulting in a decrease in viscosity. (See Chapter 12, Section E.)

- The lack of motion in the material: if no motion occurred, there could be no deformation, thus no strain, and so forth. Although the exact form of convection pattern in the earth's mantle is a subject of debate, it is generally agreed that the material below 700 km is also in motion.

In summary, it would appear that the significance of this observation will not become clear until the mechanism of deep focus earthquakes is understood. The following parallel is offered by way of speculation. Before the advent of plate tectonic theory, it was known that deep focus earthquakes occurred practically all around the Pacific Ocean except for the coast of North America. But this was an isolated fact that could not be accounted for by any model. With the advent of plate tectonic theory, the recognition came that the west coast of North America is the only part of the Pacific plate bounded by a transform fault, and this explained the lack of deep earthquakes. We may speculate that when a new, more comprehensive model of the earth emerges, the absence of earthquakes below 670 km will seem as natural as the lack of deep foci along the west coast of North America (see Kuhn 1970, pp. 111–135).

4

Surface Waves and Harmonic Analysis

A. Propagation of Surface Waves

Whereas body waves can generally be treated by ray theory (also known as the high frequency approximation) because they propagate *through* the body, the motion of surface waves is confined to the vicinity of the surface. To see how this is possible, we need only think of waves at sea. It appears intuitively obvious that the motions of such waves die off with depth. Furthermore, at equal amplitude at the surface, the motion of a long wave should die off more slowly with depth than should that of a short wave. We would say that a short wave propagates "nearer" the surface than a long wave does. This is *not* a matter of amplitude, but exclusively of wavelength and thus of period (as the wavelength increases, so does the period). Furthermore, in a homogeneous half-space the amplitude decrease per wavelength must be constant, since the medium, so to speak, treats all wavelengths alike: in other words, in such a case there is no length scale, and thus all wavelengths are equivalent. As soon as a layer is introduced, a length scale—the layer thickness—comes into the model, and different wavelengths may behave differently.

Because surface waves are confined "near the surface," their energy spreads out only in two dimensions. (On a half-space these waves would spread out as

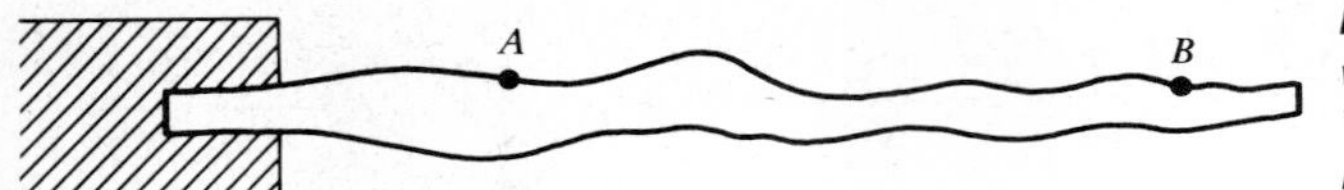

FIG. 4-1 *An arbitrary rod with two points A and B. The rod can be supported in any way.*

circles; body waves, in contrast, would spread out as half-spheres). Consequently, the energy of surface waves is proportional to the inverse of the distance, and their amplitude to the inverse of the square root of the distance, whereas the amplitude of body waves is inversely proportional to the distance (see Chapter 3, Section A). It follows that at a large enough distance, surface waves have a greater amplitude than body waves, and may thus be the only waves whose arrival is detectable.

Another important consequence of near-surface propagation is that surface waves are very little excited by deep earthquakes. This is a consequence of the reciprocal theorem (or reciprocity theorem) formulated by Rayleigh (1945, vol. 1, p. 93 and 150–157). An example of this theorem is provided in Figure 4-1, which depicts two points, A and B, on a rod of any shape (the cross section and/or the material may vary), that may be supported in any way. Now, suppose that a force F acts at A and produces a displacement d at B. Rayleigh showed that this same force acting at B will produce a displacement d at A. In the case of surface waves, this means that because a surface displacement causes little motion at depth, a displacement at depth will cause little motion at the surface. (A common —but not always correct—example of the reciprocity theorem would be: "If you don't hear your neighbor's stereo, he won't hear yours either.") The absence (or small amplitudes) of surface waves is thus an obvious feature of deep earthquakes.

PROBLEM

4-1 When is the trivial example given above correct? When is it clearly incorrect?

B. Love and Rayleigh Waves

In the earth two kinds of surface waves can propagate. The first kind is called the Love wave, after the British mathematical physicist A. E. H. Love (1863–1940), who first gave the theory for their existence in 1911 (Love 1967, pp. 145–178). Love waves are characterized by the fact that their particle motion is horizontal and perpendicular to the direction of propagation. For this reason they are denoted LQ (L for "long"; Q for *quer*, German "transverse"). They are thus similar to SH-waves, except for the fact that the displacement of Love waves decreases with depth. A condition for their existence is that there be at least a general increase of β with depth; there may be low-velocity layers, but β near the surface must be less than β at great depth—that is, the depth at which motion practically ceases. This condition is met in the earth.

Because Love waves are essentially "surface *SH*-waves," they are relatively little excited by explosions. On the other hand, they are greatly excited by properly oriented transcurrent faults. This feature is helpful in the identification of "suspicious events"—a diplomatic euphemism for nuclear explosions. Furthermore, because Love waves represent a type of shear wave, they cannot propagate in fluids: they may propagate *under* the ocean, but their particle motion is null *in* the ocean. As a result, two paths, one purely continental, the other totally or partly suboceanic, would yield the same Love-wave velocities if β were the same for each path. This condition, however is not strictly met.

The existence of Rayleigh waves was predicted by Lord Rayleigh (J. W. Strutt) in 1887. These waves can propagate at the surface of a homogeneous solid or of a solid in which the velocity shows a general increase with depth. The particle motion at the surface is "retrograde elliptical" (Figure 4-2): after the particle has passed its vertical maximum, it moves backward toward the epicenter, then downward, and so on. The corresponding idealized picture on a set of three identical seismographs (one vertical and two horizontal ones) is shown in Figure 4-3. The combination of the horizontals indicates an epicentral azimuth N 34°W or N 146°E; after the crest on the vertical, the motion is to the SE, which thus determines the N 146°E azimuth. It should be noted, however, that the particle motion at the surface of a stack of layers may be retrograde or prograde.

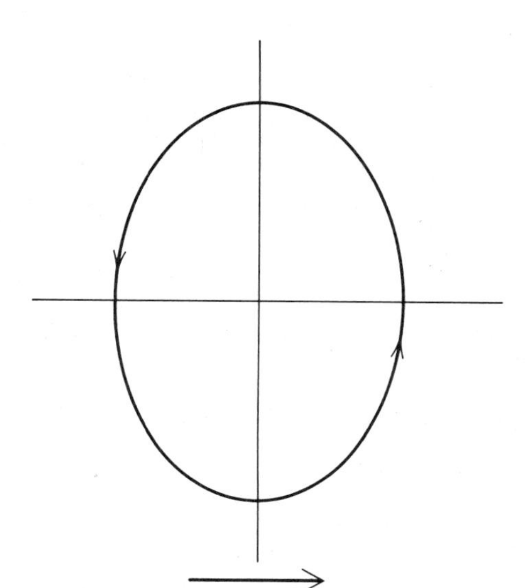

FIG. 4-2 *Particle motion of a Rayleigh wave at the surface, and (large arrow) the direction of propagation of the wave.*

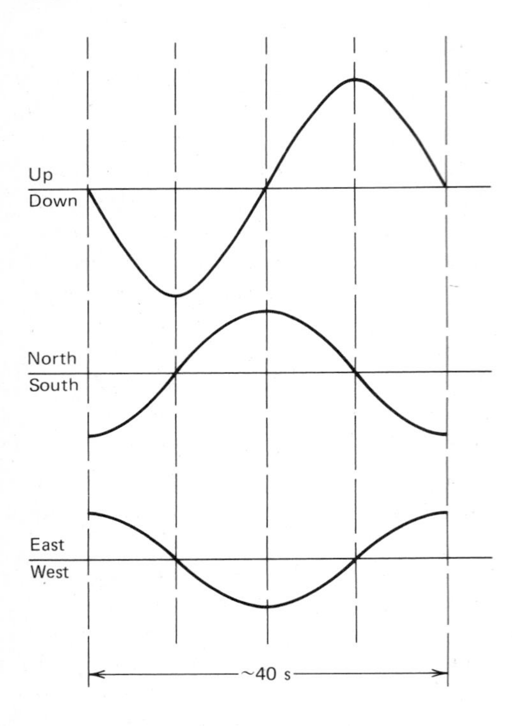

FIG. 4-3 *Idealized records of Rayleigh waves on vertical, N–S, and E–W instruments for a wave coming from the SE.*

PROBLEM

4-2 Draw idealized records (similar to Figure 4-3) for Rayleigh waves if the epicenter is N 75°E of the station.

As we shall see, there are difficulties in using this method in practice, but when applied *near* the beginning of the train of Rayleigh waves it is quite reliable. The fact that the "beginning" itself is often rather hard to define, because of dispersion (see Chapter 5), implies that surface waves change shape as they travel. A more precise definition of this statement follows.

C. Harmonic Analysis

The most important result of surface wave studies was the determination of the variation of β with depth, including the discovery of the low-velocity zone (LVZ). To understand how such determination can be made, one must be familiar with Fourier analysis and synthesis—a concept that is also useful in determining the relationship between the output and input of an instrument and the effects of filtering. We thus must make a somewhat lengthy excursion into harmonic analysis before returning to surface waves. (For excellent treatments of this subject see Guillemin 1949, Ch. 7; Lanczos 1956, Ch. 4; Lee 1960, Ch. 2.)

Linear Systems

In what follows we will consider only systems that possess all of the following properties.

1. If the input is multiplied by a constant k, the output is also multiplied by k (linearity with respect to input).
2. If the input is delayed by δt, the output is delayed by the same amount.
3. The output depends only on the input, not on the time—that is, the system is time-invariant. (This statement will be clarified later.)
4. In most cases the output does not precede the input—that is, the system is causal. This is always the case in systems that operate in real time; it need not be the case in systems operating on prerecorded data (e.g., data recorded on magnetic tape).

A system that possesses these four properties is linear. The first three properties are necessary for conventional addition and multiplication—and thus, of course, integration. The common idea that most systems are linear is erroneous, for several reasons. With respect to multiplication of the input: pushing on an open door will close it, but pushing on a closed door ... pushing too hard on something may break it. With respect to time invariance: today's customs are not those of 10 years ago. Social systems are often exceedingly nonlinear. Far from being common, then, linearity is often difficult to obtain.

PROBLEM

4-3 Give several examples (at least three of each) of linear and nonlinear systems.

Fourier Analysis and Synthesis

Consider a function of time $f(t)$ that is generally nonzero in an interval (t_1, t_2) and zero elsewhere. François Marie Charles Fourier (French administrator and mathematical physicist, 1768–1830) showed that any such function may be represented by an infinite sum of sine waves of different amplitude, frequency, and "time of origin." The question is sometimes raised, "Why are sine waves given such importance?" The reason is that they, alone among all wave types, maintain their shape after passing through any linear system; all other wave types (e.g., square, triangular, sawtooth) may suffer distortion—that is, change shape.

Rather than prove Fourier's statement, we will take a simple case and extend it heuristically. Figure 4-4*a* depicts the function $f(t)$, which is known only in the interval (t_1, t_2). It is evident by inspection that in the case shown in Figure 4-4*a*, there is a rather large component at the frequency represented (with arbitrary amplitude) by the solid line in Figure 4-4*b*, but a very small component at the frequency represented by the dotted line. We thus might turn to the well-known geological technique of cross-correlating $f(t)$ with sine waves of various frequencies, to determine their cross-correlation coefficients. However, we still do not know how to position the various sine waves along the time axis. Several steps are involved in the solution of this problem.

Step 1 Rather than consider a sum of sine waves, we consider a sum of cosine waves, each of a given frequency.

Step 2 Rather than position each of these cosine waves by giving its origin time, we do so by giving its phase angle, ϕ. If each wave is given by the equation

$$y = C\cos(2\pi\nu t + \phi) \tag{4-1}$$

it follows that for $t = 0$ (i.e., at the origin),

$$y = C\cos\phi$$

Figure 4-5 shows a wave of period $T = 5$ s, or $\nu = \frac{1}{5}$ Hz, of phase angle $\phi \approx 60° = \pi/3$, and of amplitude C. We see that y is maximum for $2\pi\nu t + \phi = 0$:

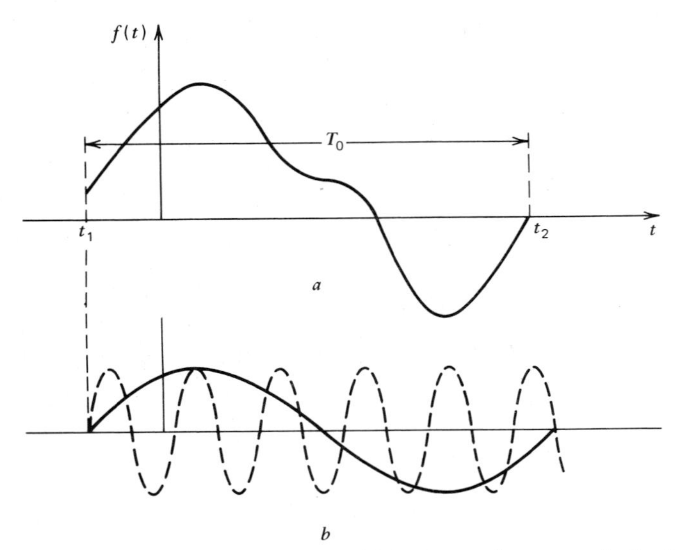

FIG. 4-4 *(a) A known function f(t) that is mostly nonzero in the interval (t_1, t_2). (b) Two sine waves of the same amplitude but different frequencies.*

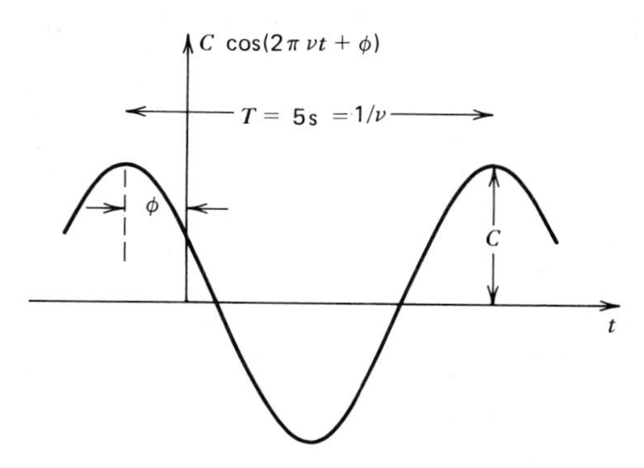

FIG. 4-5 *The wave $C\cos(2\pi\nu t+\phi)$, its period T, and its phase angle ϕ.*

that is, for $t=-\phi/2\pi\nu$. In the case of Figure 4-5, $t=-5\pi/(3\times 2\pi)=-5/6$ s, and thus is before the origin. If the period were 8 s, the same phase angle would imply a maximum 4/3 s before the origin.

Step 3 Rather than attempt to determine C and ϕ directly for a given frequency, we first transform Equation 4-1, $y=C\cos(2\pi\nu t+\phi)$, into

$$y=(C\cos\phi\times\cos 2\pi\nu t)-(C\sin\phi\times\sin 2\pi\nu t) \tag{4-2}$$

Setting

$$A=C\cos\phi \quad \text{and} \quad B=-C\sin\phi \tag{4-3}$$

yields

$$y=A\cos 2\pi\nu t+B\sin 2\pi\nu t$$

If we know A and B, we can find C and ϕ from Equation 4-3, since

$$C=(A^2+B^2)^{1/2} \tag{4-4}$$

and

$$\phi=\tan^{-1}\left(\frac{-B}{A}\right) \tag{4-5}$$

C is positive by hypothesis.

Contrary to what might be thought, ϕ is not indeterminate to within $\pm n\pi$ (n: integer) but to within $\pm 2n\pi$. The reason is obvious from Equation 4-3: if $A>0$, then ϕ is in the first or fourth quadrant; if $B>0$, then ϕ is in the third or fourth quadrant. Hence, by examining the values of A and B, we can determine ϕ within $\pm 2n\pi$.

Step 4 We now can determine A and B for any frequency. (Remember that A, B, C, and ϕ vary with ν). We assume that

$$f(t)=\sum_{n=0}^{\infty}\left[A_n\cos\left(\frac{2\pi}{T_o}nt\right)+B_n\sin\left(\frac{2\pi}{T_o}nt\right)\right] \tag{4-6}$$

where $T_o = t_2 - t_1$: that is T_o is the time interval (Figure 4-4), *not* the period. Then,

$n = 1$ yields the fundamental mode for which the period, T, is equal to T_o

$n = 2$ gives the first harmonic, $T = T_o/2$

$n = 3$ gives the second harmonic, etc.

We will not demonstrate that this equation is correct. Instead we assume that it is and find A_n by multiplying both sides by $\cos[(2\pi/T_o)rt]$ (r: integer), and integrating over the interval.

We thus obtain integrals of two forms, corresponding to the cosine and the sine terms of Equation 4-6. The value of the first integral—say, I—is

$$\mathrm{I} = \int_{t_1}^{t_2} \cos\left(\frac{2\pi}{T_o} nt\right) \cos\left(\frac{2\pi}{T_o} rt\right) dt$$

Setting $2\pi n/T_o = a$ and $2\pi r/T_o = b$, we find (Peirce 1929, p. 49)

$$\mathrm{I} = \int_{t_1}^{t_2} \cos at \times \cos bt \times dt = \left[\frac{\sin(a-b)t}{2(a-b)} + \frac{\sin(a+b)t}{2(a+b)}\right]_{t_1}^{t_2} \qquad \text{for } a \neq b$$

But

$$\sin(a \pm b)t_2 = \sin(a \pm b)t_1$$

because

$$a \pm b = \frac{2\pi}{T_o}(n \pm r)$$

that is, because the interval (t_1, t_2) contains an integral number of periods. Thus the integral vanishes. When $n = r \neq 0$ $(a = b)$, we find

$$\mathrm{I} = \int_{t_1}^{t_2} \cos^2 at\, dt = \left[\frac{1}{2a}(at + \sin at \cos at)\right]_{t_1}^{t_2}$$

The last term vanishes for the same reason as previously. The first term yields

$$I = \frac{t_2 - t_1}{2} = \frac{1}{2} T_o \tag{4-7}$$

If $n = r = 0$, Equation 4-6 becomes simply

$$\int_{t_1}^{t_2} f(t) \cos(0)\, dt = A_o$$

or

$$A_o = \int_{t_1}^{t_2} f(t)\, dt \tag{4-8}$$

which is the average value of $f(t)$ in the interval.

The value of the second integral—say II—is

$$\text{II} = \int_{t_1}^{t_2} \sin\left(\frac{2\pi n}{T_o} t\right) \cos\left(\frac{2\pi r}{T_o} t\right) dt$$

With the same notations as before, this yields

$$\text{II} = \int_{t_1}^{t_2} \sin at \times \cos bt\, dt = \left[-\frac{\cos(a-b)}{2(a-b)} - \frac{\cos(a+b)t}{2(a+b)} \right]_{t_1}^{t_2}$$

If $a \neq b$, these terms vanish, as previously. If $a = b$, we find

$$\text{II} = \int_{t_1}^{t_2} \sin at \times \cos at\, dt = \left[-\frac{1}{4a} \cos(2at) \right]_{t_1}^{t_2}$$

which also vanishes. The net result of the multiplication and integration is thus, from Equation 4-7,

$$\int_{t_1}^{t_2} f(t) \cos\left(\frac{2\pi n}{T_o}\right) dt = \frac{1}{2} A_n T_o$$

or

$$A_n = \frac{2}{T_o} \int_{t_1}^{t_2} f(t) \cos\left(\frac{2\pi n}{T_o} t\right) dt \tag{4-9}$$

except for A_o (see Equation 4-8).

To find B_n we multiply Equation 4-6 by $\sin[(2\pi/T_o)rT]$ and integrate over the interval (t_1, t_2); we obtain

$$B_n = \frac{2}{T_o} \int_{t_1}^{t_2} f(t) \sin\left(\frac{2\pi nt}{T_o}\right) dt \tag{4-10}$$

$$B_o = 0$$

In practice, $f(t)$ generally is not known analytically but is known at a certain number of points in the interval. These points should be equispaced in time. Let the time interval be δt, and let there be p points; then $T_o = t_2 - t_1 = (p - 1)\,\delta t$ (see Figure 4-6, where $p = 6$). Since an integral is the limit of a sum, we find, in place of Equation 4-9,

$$A_n \approx \frac{2\,\delta t}{T_o} \sum_{i=1}^{p} f_i \cos \frac{2\pi n t_i}{T_o} \tag{4-11}$$

$$B_n \approx \frac{2\,\delta t}{T_o} \sum_{i=1}^{p} f_i \sin \frac{2\pi n t_i}{T_o}$$

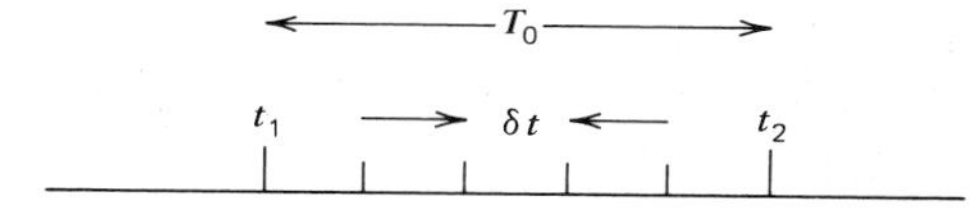

FIG. 4-6 *The time interval (t_1, t_2) is divided into $p - 1 = 5$ smaller intervals δt. The function is known at $p = 6$ points.*

where $t_i = i \times \delta t$ and $f_i = f(t_i)$. We can find C_n and ϕ_n from Equations 4-4 and 4-5. If f is read (or known) at more points in the same interval, p increases but δt decreases correspondingly. The values of A_n and B_n thus remain essentially unchanged. Although Equation 4-11 is a fairly poor approximation of the integrals in Equations 4-9 and 4-10, it serves to demonstrate the principle of the method. Better approximations are classical (see, for example, Kunz 1957, pp. 137–165).

At this point, a simple example may help. If $t_1 = 10$ h, 20 m, 13 s and $t_2 = 10$ h, 21 m, 15 s, then $T_o = (t_1, t_2) = 62$ s. The fundamental thus has a period of 62 s and a frequency of 1/62 Hz. The first higher harmonic has a period of $62/2 = 31$ s and a frequency of 2/62 Hz; the second one, 62/3 s and 3/62 Hz, and so on. Successive harmonics are thus equispaced in frequency at a spacing of $\delta\nu = 1/62$ Hz $= 1/T_o$; they are not equispaced in period. Finally, it makes no sense, *in this context*, to inquire what the value of C is when n is not an integer, or, say, at 1/50 Hz. C_n and ϕ_n are called the amplitude spectrum and the phase spectrum, respectively. They are *not* continuous spectra, but line spectra, since they have *no* value at frequencies between those of the harmonics.

If we double the length of the interval by adding zeroes on each side, the sums in Equation 4-11 remain unchanged, but T_o doubles. Thus, any specific A_n has been halved. It is true that there are now $2p$ points, but only at p of them is f_i not equal to 0, and although δt has not changed, T_o has doubled. In order to avoid this decrease, we temporarily accept the definition

$$A'_n = \frac{T_o}{2}(A_n) \approx \delta t \sum_{i=1}^{p} \left[f_i \times \cos(2\pi n t_i / T_o) \right] \tag{4-12}$$

and a similar definition for B'_n. If we now double the time interval, the value of A'_n does not change. More correctly, the value of A' associated with a specific frequency does not change, but its index doubles. In the example given above, the frequencies were originally spaced $\delta\nu = 1/62$ Hz $= 1/T_o$ apart. As T_o doubles to 124 s, $\delta\nu$ becomes 1/124 Hz. Thus, the fundamental has a frequency of 1/124 Hz, the first harmonic 2/124 Hz, the second one 3/124 Hz, and so forth. If we continue to double T_o, $\delta\nu$ is correspondingly divided by 2. At the limit, with T_o going from $-\infty$ to $+\infty$, $\delta\nu \to d\nu$. A (and B) thus becomes a continuous function of frequency, and n/T_o in the discrete case becomes ν in the continuous case. Thus, after the prime is dropped, Equation 4-12 yields

$$A(\nu) = \int_{-\infty}^{\infty} f(t) \cos(2\pi\nu t)\, dt \tag{4-13}$$

$$B(\nu) = \int_{-\infty}^{\infty} f(t) \sin(2\pi\nu t)\, dt$$

$C(\nu)$ and $\phi(\nu)$ can then be found from Equations 4-4 and 4-5. For symmetry, one usually writes $F(\nu)$ in place of $C(\nu)$. Thus, given $f(t)$, Equation 4-13 yields $F(\nu)$ and $\phi(\nu)$. The following conventions often are adopted.

- Lowercase letters represent a function of time or a function "in the time domain."
- Uppercase letters represent the amplitude spectrum, which is "in the frequency domain."
- Greek letters represent the phase spectrum, which is also "in the frequency domain."

If, as is generally the case in practice, $f(t)$ is not known analytically but only given at p discrete points, the integrals in Equation 4-13 become sums (just as in Equation 4-11), yielding

$$A(\nu) \approx \delta t \sum_{i=1}^{p} f_i \cos(2\pi\nu t_i) \tag{4-14}$$

$$B(\nu) \approx \delta t \sum_{i=1}^{p} f_i \sin(2\pi\nu t_i)$$

Especially when expressed in terms of $F(\nu)$ and $\phi(\nu)$, Equations 4-13 and 4-14 are said to decompose f into its Fourier components, or to perform a Fourier analysis of f. It is not necessary to introduce zeroes in f_i, since the corresponding terms in the sum or in the integral vanish. The counterpart of Equation 4-6 is, similarly,

$$f(t) = 2\int_0^\infty A(\nu)\cos(2\pi\nu t)\,d\nu + 2\int_0^\infty B(\nu)\sin(2\pi\nu t)\,d\nu \tag{4-15}$$

It is also possible to write this equation in terms of $F(\nu)$ and $\phi(\nu)$, in which case it is said to synthetize $f(t)$ from its Fourier components. A much more symmetric form of Equations 4-13 and 4-15 can be obtained by introducing imaginary exponentials and negative frequencies, but we shall not do so here.

$A(\nu)$ and $B(\nu)$ can now be computed at any frequency. $F(\nu) = C(\nu)$ and $\phi(\nu)$ are the (continuous) amplitude and phase spectra, respectively.

These ideas should clarify a number of expressions commonly used and phenomena often observed. For example:

- Why does the same note "sound different" on, say, a violin and an oboe? Because the harmonic content—that is, the relative amplitude of the various harmonics—is different on different instruments.
- Why are different artists able to extract different "sounds" from the same violin? Because the harmonic content is partly determined by the player, and certain harmonic contents are more pleasant to the ear.
- What is the "character" of a wave? Its amplitude and phase spectrum. A "new arrival" at a certain time means that the amplitude and/or phase spectrum change(s) rapidly near that time.

PROBLEMS

4-4 If $A = -3.7$ and $B = -4.3$, find C and ϕ (the latter in radians and in degrees).

4-5 Given $y_o \ldots y_n$ = 1.0, 2.2, 2.5, 3.2, 2.6, 1.0, 0.0, -0.9, -2.0, -1.0, 0.0, 0.9, 1.9, 0.1, -1.0, -2.0, -0.5, 0.0 (all the other $y_i = 0$), and $\delta t = 0.2$ s, write a program to find the amplitude and the phase (in radians) for the Fourier components of frequencies 0.25, 0.50, 0.75, ..., 2.0 Hz. Plot y versus t and C versus ν. Do you observe a relation between the largest value of C and the dominant period? Why? (This is not a realistic case, but it demonstrates the application of the principle.)

5

Surface Waves Revisited

A. Dispersion

Introduction

As previously noted (Chapter 3, Section A), the velocity of body waves is independent of their frequency, and for this reason their shape does not change as they propagate. The meaning of this statement can now be clarified. Since all frequencies that make up these waves—often referred to as Fourier components—travel at the same speed, their phase relations do not change with distance. According to Fourier synthesis, the shape of the resulting waves does not change with distance, either.

The case is different with surface waves. The motion of long-period waves dies off more slowly with depth than does that of short-period ones (Chapter 4, Section A). Hence, long-period waves are more influenced by the velocities of deep layers, and these velocities are higher. Long waves thus travel faster and arrive first at a given point; as time passes, the periods at this point gradually decrease (Figure 5-1). (Strictly speaking, the term *period* cannot be used for a phenomenon that is not exactly periodic, but the meaning of the approximation is clear.)

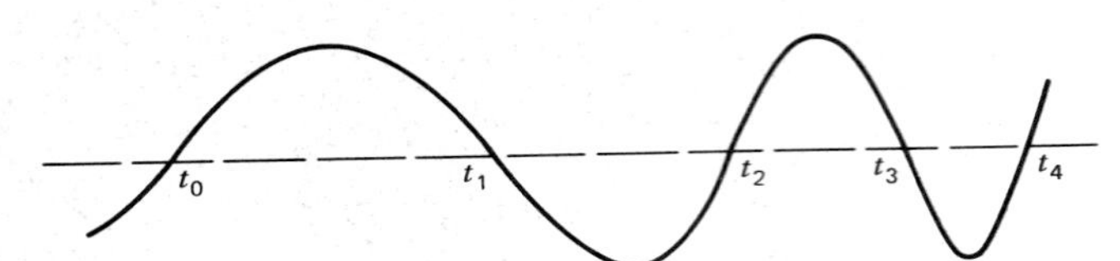

FIG. 5-1 *Idealized wave with continuously varying "period."*

These waves propagate "near" the surface, but what constitutes "near" must be evaluated in terms of the wavelength involved. Their period may exceed 100 s, and since their velocity is on the order of 3.5 km/s, their wavelength may exceed 350 km. This explains why features much deeper than the Moho and as deep as the bottom of the low-velocity zone (see below), or even deeper, can be detected by these waves.

Two differences between surface waves and light waves should be noted. Light waves are more-or-less dispersed as they travel *through* a homogeneous transparent body, as a consequence of the properties of the body; surface waves, in contrast, are dispersed because of the variation of β with depth (α plays only a minor role). Furthermore, a light source cannot operate in a time interval much shorter than the period of vibration of the wave ($\sim 10^{-15}$ s). A source of light waves is thus essentially a steady-state phenomenon compared to the period of the wave; in seismology, the source is a transient one compared to the period of surface waves. (A steady-state phenomenon is one that is time-independent: the state of the system does not vary with time. A transient phenomenon is one of limited duration. A periodic phenomenon is one that is repeated after a certain time interval—the period.)

Phase and Group Velocities

Since there are an infinity of Fourier components, each one possesses only an infinitesimal amplitude and transports only infinitesimal energy, and thus its motion cannot be observed. What is observed must, *ipso facto*, possess a finite energy. Thus, at any one time a seismogram—or earthquake record—consists of a *band* of Fourier components near a certain frequency. This band does not travel at the speed of any given Fourier component, but at the group velocity, U, which is thus the velocity at which the energy of that frequency band travels. It follows that U may be obtained by dividing the epicentral distance by the time taken by the waves. U is lower than c, the phase velocity, or velocity of the individual Fourier component. Both, of course, vary with frequency. It may be shown that

$$U = c - \Lambda \frac{dc}{d\Lambda} = c + k\frac{dc}{dk} = \frac{d\nu}{dk} \qquad (5\text{-}1)$$

Plots of either c or U versus ν or T are referred to as dispersion curves. The equation $U = d\omega/dk$ is frequently used. It differs from Equation 5-1 by a factor of 2π, because it uses the definition $k = 2\pi/\Lambda$ rather than $k = 1/\Lambda$, which we use.

To determine the group velocity from a seismogram, in principle we need only read the times of zero crossings (which can be determined more exactly than the times of the crests or troughs) t_1, t_2, t_3, t_4, counted from the time at the epicenter (see Figure 5-1). Then $T_2 \approx t_3 - t_1$, $T_3 \approx t_4 - t_2$, etc. The "time of arrival" of the wave with period T_2 is approximately $(t_3 + t_1)/2$; that of period T_3, similarly, $(t_4 + t_2)/2$. In fact, though, group velocity curves obtained by this method often show considerable scatter because of other arrivals and inhomogeneity of the path.

It may be shown that, in theory there are an infinity of dispersion curves, each corresponding to a particular "normal mode." The curve that we have been discussing corresponds to the fundamental mode (see the discussion involving Equation 4-6), which is generally the most important one. The situation is somewhat similar to that of a vibrating string, only more complex. The first few higher modes are sometimes observed. They are not in a simple harmonic relation with the fundamental mode.

PROBLEM

5-1 Determine the group velocity curves for some of the seismograms shown in Figure 5-2—in particular, *LR* for *a* and *b*. Compare them and explain the reason for the difference.

B. Results

Data

Typical dispersion curves for Rayleigh and Love waves are shown in Figures 5-3 and 5-4. The reason for the great decrease in the group velocity of Rayleigh waves for periods shorter than ~ 18 s along oceanic paths (Figure 5-3*b*) is that a part of their energy travels as *P*-waves. In oceanic areas, their energy is thus partly in the solid earth and partly in the ocean, where $\beta = 0$ and α is small. The phase velocity curves (Figure 5-3*a*) do not show this influence in any obvious way. Figures 5-3*a* and 5-3*b* also show a systematic variation of the phase and group velocities in oceanic areas correlated with the age of the lithosphere (see below).

Figures 5-3*c* and 5-3*d* show phase and group velocities of Rayleigh waves for recently active tectonic areas and for shields, respectively. The difference in U between them is obvious—for example, the minimum value of U occurs at different periods—and the oceanic case again differs markedly from either of them. Figure 5-4 is a comparable graph for Love waves.

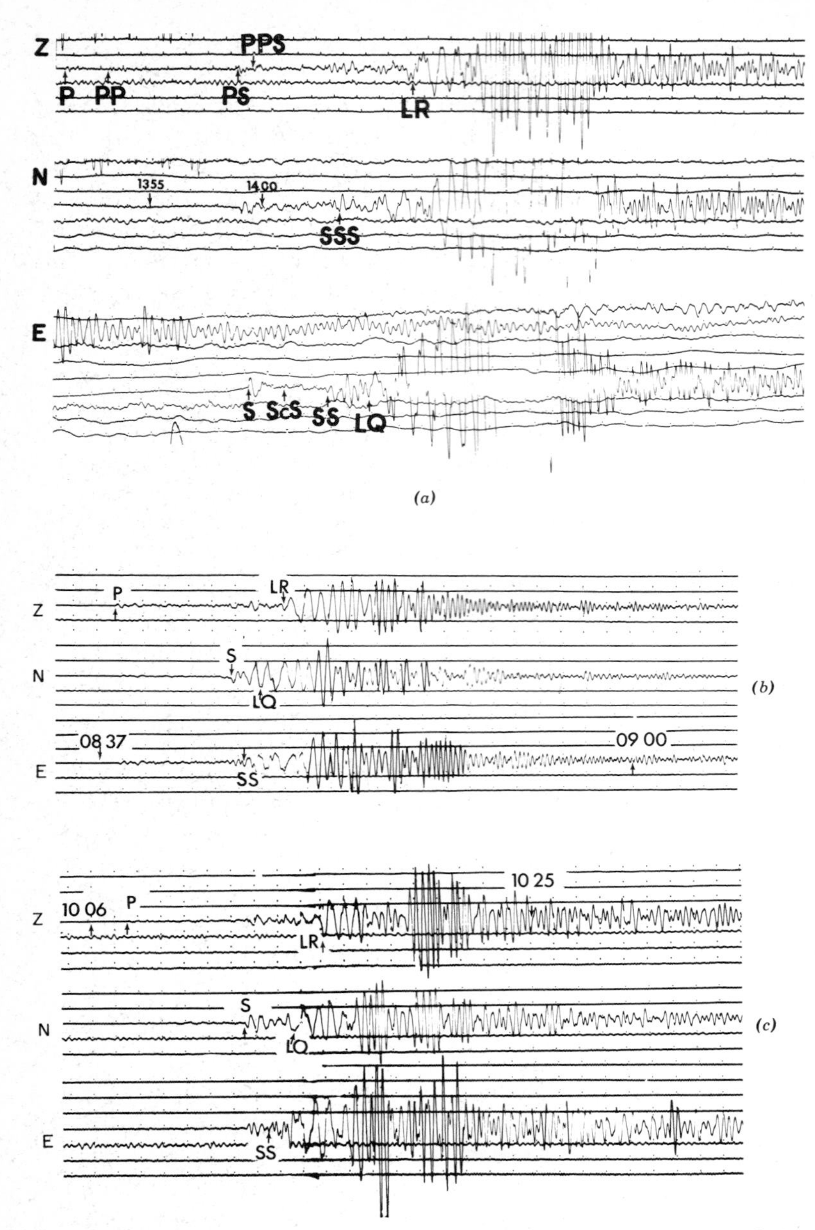

FIG. 5-2 *(a) Vertical, N–S, and E–W long-period records at Golden, Colorado, of an earthquake in the Arctic Ocean, $\Delta = 54°$. Path through shields (Greenland, Canada). 0 = 13:42:02. (b) Same records at Palisades, N.Y., of an earthquake in the North Atlantic Ocean, $\Delta = 29.4°$. Oceanic path. 0 = 08:31:59. (c) Same station, records of earthquake in Central America, $\Delta = 30.2$. Mixed path. 0 = 10:01:35. The gradual decrease in the period of surface waves is obvious in most cases. Note that after a variable time interval, waves scattered or reflected by inhomogeneities start arriving; one should thus cease reading the periods when they become irregular. (From Simon 1981. Reproduced with permission from Earthquake Interpretations, A Manual for Reading Seismograms, by Ruth B. Simon. Copyright © 1981 by William Kaufmann, Inc., Los Altos, California 94022. All rights reserved.)*

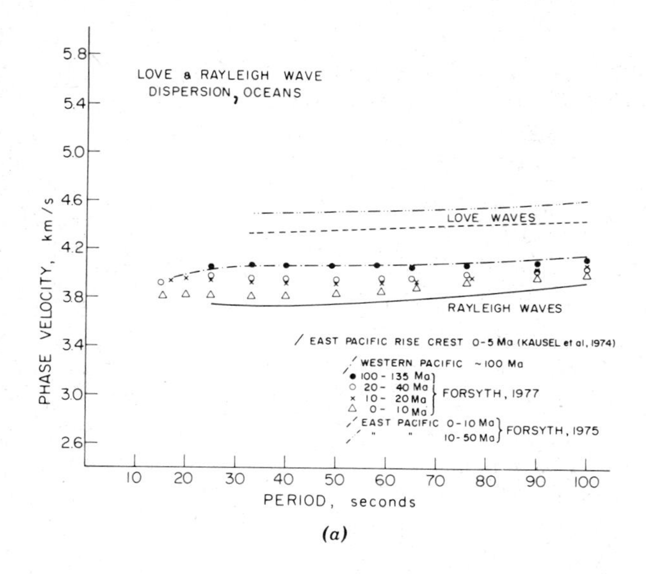

(a)

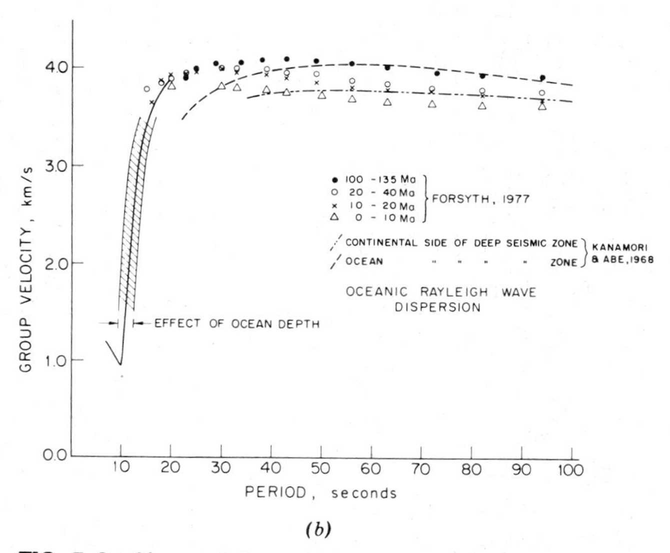

(b)

FIG. 5-3 *Observed dispersion curves for Rayleigh waves. (a) Phase velocity of Rayleigh waves for oceanic paths. The influence of the age is clearly visible. (b) Group velocity of Rayleigh waves for oceanic paths. The influence of the age is clearly visible here, also. (c) Phase and group velocities of Rayleigh waves for the Alps and the Andes— that is, areas of recent tectonic activity. (d) Phase and group velocities of Rayleigh waves for shield areas. (From Kovach 1978. Copyright © by the American Geophysical Union.)*

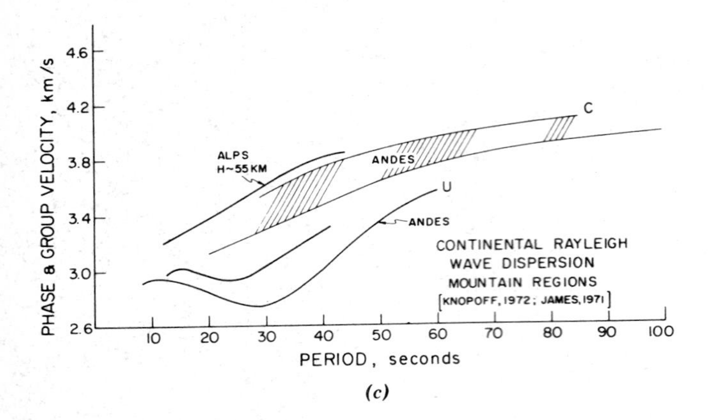

(c)

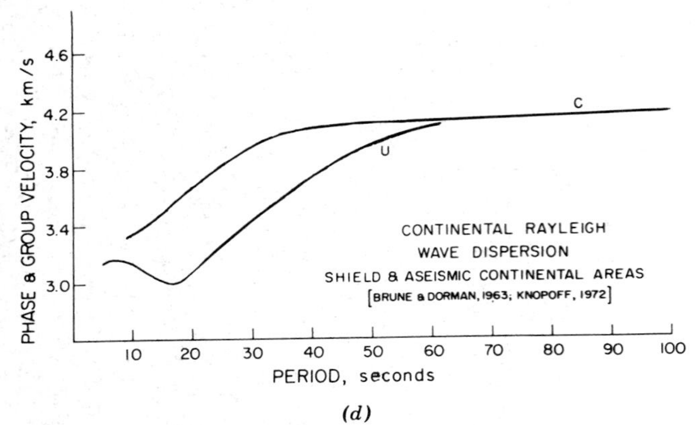

(d)

FIG. 5.3 *(Continued)*

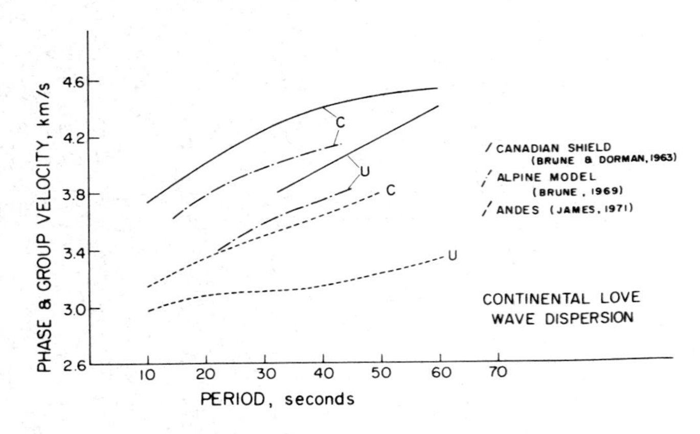

FIG. 5-4 *Phase and group velocities of Love waves for the Canadian shield, the Alps, and the Andes. (From Kovach 1978. Copyright © by the American Geophysical Union.)*

PROBLEM

5-2 From the phase velocity curve for the Western Pacific for 100 Ma given in Figure 5-3*a*, compute the group velocity curve using either a graphical or a numerical approximation of Equation 5-1. Compare your results with Figure 5-3*b*.

Interpretation

For computational reasons it is necessary to assume that the earth consists either of homogeneous concentric spherical shells or, if the wavelengths considered are much smaller than the radius of the earth, of plane layers. The shells or layers can have any desired thickness. We will refer to such an assemblage of layers or shells as a "structure."

The direct problem (given the structure, find the dispersion curve) is relatively easy to solve on a computer. The inverse problem (from the dispersion curves, find the structure) was long solved by trial-and-error. A likely structure was selected as a first guess, and a first dispersion curve computed. Generally, this curve did not agree with the observed dispersion curve. The structure was then modified, a second dispersion curve was computed, and so forth. Eventually, given enough skill on the part of the researcher, a good match would be obtained. The procedure was gradually systematized, and reliable inversion schemes have now been devised (see Tarantola and Vallette 1982*a* and 1982*b*; see also Chapter 9, Section C).

From the analysis of a large number of dispersion curves, it has been concluded that β starts to decrease at a depth of approximately 80 km and starts increasing again at around 200–250 km (Figure 5-5). The detection of this low-velocity zone (LVZ) has probably been the single most important result of surface wave studies. As the figure shows, oceanic areas have, *on the average*, thin high-velocity "lids" and thick and pronounced LVZs, whereas shield areas have thick "lids" and much less pronounced LVZs. Areas of recent tectonic activity have very pronounced but relatively thin LVZs and thick lids.

Most, and perhaps all, of the discontinuities in β shown in the figure are presumed not to be real. They may correspond to regions where β changes rapidly —for example, at the top of the LVZ—or they may simply represent a convenient way to model a continuous change of β with depth. The characteristics of the LVZ point to important geodynamic characteristics of the region (see below).

Emerging from the foregoing discussion are several significant facts. One of these, the differences among shields, oceans, and tectonic areas, has already been examined. Two others, the increase in c and U with the age of the oceanic lithosphere and the existence of the low-velocity zone, demand further explanation.

From our knowledge of plate tectonics, it appears reasonable to ascribe the increase in c and U to the gradual thickening of the lithosphere as it travels away from the ridge. Quantitative computations have confirmed this surmise. This thickening results from a gradual cooling of the lithosphere as it loses its heat through the ocean bottom (see Chapter 13, Section D).

Whereas differences in composition can play only a minor role in explaining the increase in c and U with age, they cannot *a priori* be excluded from an explanation of the LVZ, for differentiation does result in changes in velocities. However, the major change in composition and in velocity occurs at the Moho; hence, it is not obvious what complex change in composition could be widespread enough to give rise to the LVZ. It appears more reasonable to ascribe the existence of the LVZ primarily to the rise in temperature. Under young tectonic areas and oceans the material may reach partial melting at a relatively low temperature ($\sim 1200°$ C), partly because of the hydration of the material and partly because of its composition. The opposite situation may prevail under shields, where the relatively low increase of temperature with depth further contributes to the absence of partial melting.

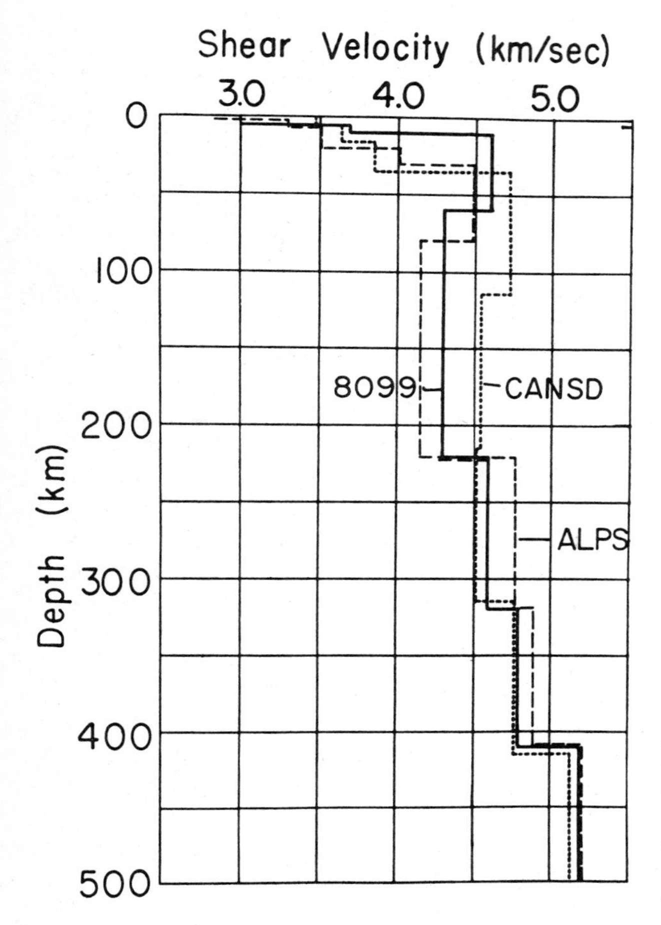

FIG. 5-5 *Shear velocity versus depth for the average Pacific Ocean (curve 8099), the Canadian shield (curve CANSD), and the Alps. (From Dorman 1969. Copyright © by the American Geophysical Union.)*

It has been shown (O'Connell and Budiansky 1977) that partial melting causes a lowering of the velocity and an increase in the attenuation (or a decrease in Q). However, some dislocation mechanisms have the same results (Minster and Anderson 1981). These same mechanisms may account for a concomitant lowering of the viscosity. The LVZ may thus be the asthenosphere, but there is no conclusive evidence that the zones of low velocity, low Q, and low viscosity coincide. Acceptance of this coincidence, however, leads to the implication that the LVZ enables the plates to move somewhat independently of the motions occurring in the mesosphere (below the asthenosphere). This idea has been debated on two principal grounds:

- That there are essentially no data on the motions in the mesosphere, and the prospects of acquiring adequate data are slim.
- That although the viscosity of the asthenosphere is relatively low, this "relatively" low value is still many orders of magnitude greater than that of any liquid that we know of—it is probably comparable to that of glass. Whether a liquid of such viscosity could decouple the motions of the lithosphere from those in the mesosphere remains uncertain.

It should be emphasized that all of the above considerations refer to β. Little is known concerning the variation of α with depth near the LVZ, for the following reasons.

- Insofar as surface waves are concerned, Love waves are essentially "surface *SH*-waves," (Chapter 4, Section B) so they are affected by β only, and not by α. It can also be shown that Rayleigh waves are very little influenced by α.
- Insofar as body waves are concerned, we cannot use the Herglotz-Wiechert transformation (Chapter 3, Section B) to find the velocity distribution in a LVZ.

Finally, it should be pointed out that surface waves have long been used to estimate the energy released by earthquakes (see Chapter 7, Section B). More recently, they have also been used to deduce focal mechanisms (Kanamori and Given 1981) and to help determine density variation with depth.

6

Further Examination of Linear Systems

Although further examination of linear systems is not essential to geophysics, such systems are so commonly encountered in such a large number of fields of science that a more detailed study of them appears warranted simply for its unifying value. As we shall see (Chapter 9, Section C), the usefulness of linear systems is not restricted to seismology or instrumentation—they can also be used, for example, to relate relief with gravity.

The student wishing to save time may skip this chapter without missing any purely geophysical information.

A. The Input-Output Relationship

We observe phenomena by means of instruments. The input of an instrument may be a force, a displacement, an electric current, or any of a number of other things, and the output may be equally varied. Here we will temporarily restrict our focus to a case in which the input and the output are both displacements. Our example will be a simple damped pendulum provided with a pen, or some other means of recording the relative displacement between the mass and the frame of

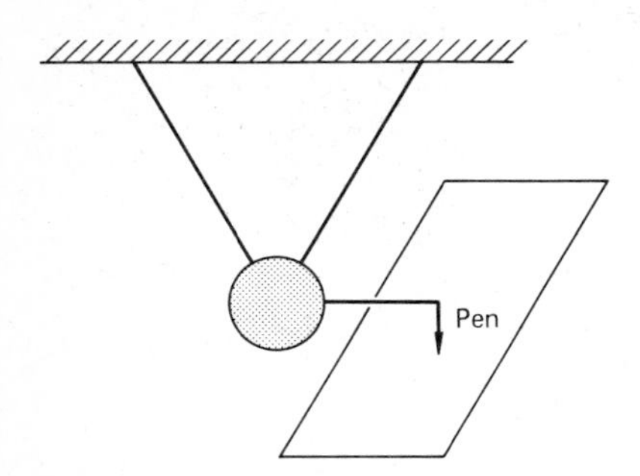

FIG. 6-1 *Highly schematic diagram of a pendulum seismograph (see also Figs. 7-4, 7-5, and 7-6).*

the instrument (Figure 6-1). As the ground moves during an earthquake, the frame, attached to the ground, moves in the same manner; this is the input. The output is the motion of the pen relative to the paper—that is, the seismogram. We wish to determine the relation between the seismogram and the ground motion, in order, for example, to be able to determine the latter. As before, we assume that the system is linear.

Let the actual ground motion be $f(t)$. It is a continuous function, but we will assume initially that it consists of impulses of various "heights" of infinitesimal duration separated by an interval, δt. As $\delta t \to 0$, a continuous function will result. (This is not a rigorous exposition of the subject; for a clear and rigorous treatment, see Lee 1960, pp. 323–331). Suppose that an impulse of unit height and duration δt produces the response $\delta t \times h(t)$, as shown in Figure 6-2. Here $h(t)$ is the unit impulse response, or simply the impulse response. It is zero for $t < 0$, and we assume that it is also zero for $t > t_e$: that is, that the memory of the system does not extend beyond time t_e.

For simplicity, we will assume that $f(t)$ consists of three impulses, f_1, f_2, and f_3, at t_1, t_2, and t_3, respectively (Figure 6-3a).

The outputs, g_1, g_2, and g_3, resulting from each impulse are shown in Figure 6-3b. Thus at any one instant—say, t_r—the total output, $g(t_r)$, is the sum of g_1, g_2, and g_3 at that instant. (Of course, some of the outputs may be null.) Looking first at g_1, we see that $g_1(t_r)$ may be obtained as shown in either Figure 6-3b or Figure 6-3a. For the latter, $h(t)$ has been "flipped over" and its origin set at t_r; if we multiply the impulse f_1 at t_1 by the value of this "flipped" $h(t)$ at t_1, we obtain g_1 at t_r. Similarly, g_2 may be obtained as shown in Figure 6-3b or as

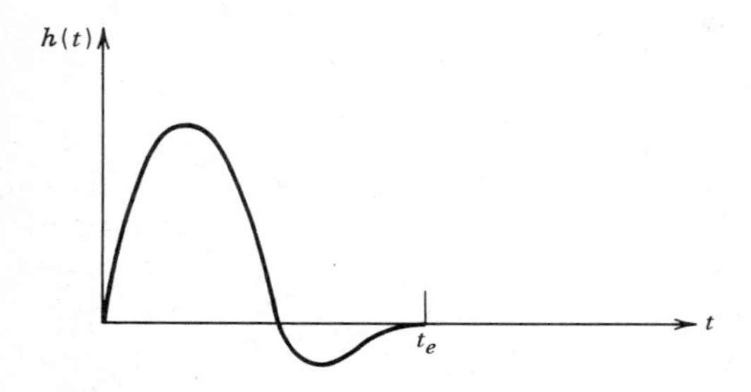

FIG. 6-2 *A unit impulse response.*

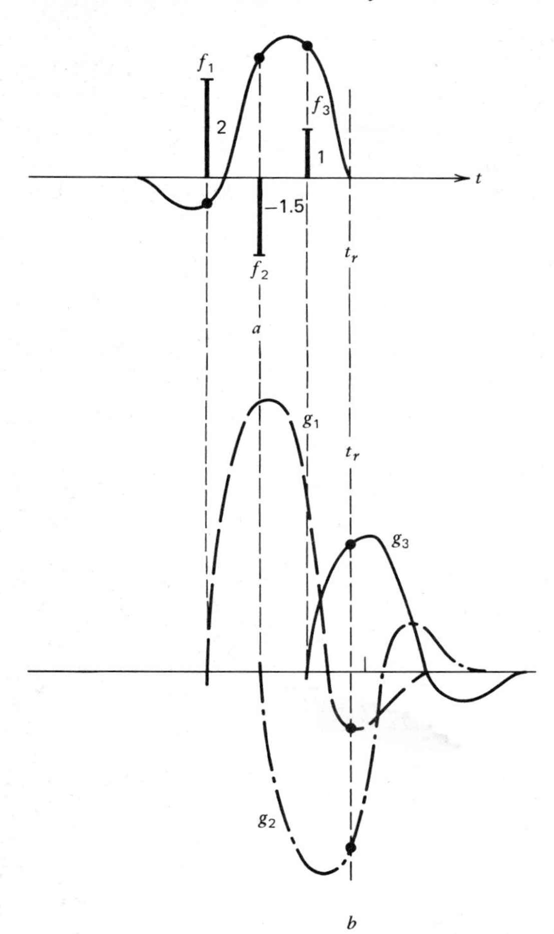

FIG. 6-3 *(a) An input f(t) consisting of impulses f_1, f_2, and f_3; the time t_r at which the output is desired; and the "flipped over" impulse response. (b) The outputs g_1, g_2 and g_3 from the first three impulses.*

shown in Figure 6-3*a*—if the latter, we must multiply the impulse f_2 at t_2 by the value of the "flipped" $h(t)$. The output g_3 is now evident.

But "flipping" $h(t)$ means replacing $h(t)$ with $h(-t)$, and changing the origin to t_r means replacing $h(t)$ with $h(t_r - t)$.

Thus, at t_r,

$$g(t_r) = g_1(t_r) + g_2(t_r) + g_3(t_r)$$
$$= \{[f(t_1) \times h(t_r - t_1)] + [f(t_2) \times h(t_r - t_2)] + [f(t_3) \times h(t_r - t_3)]\}\ \delta t$$

If there are p impulses in $f(t)$, it follows that at any time t (rather than t_r),

$$g(t) = \delta t \sum_{i=1}^{p} f(t_i) \times h(t - t_i) \tag{6-1}$$

At the limit, for a continuous function $f(t)$,

$$g(t)=\int_{-\infty}^{t} f(\tau)h(t-\tau)\,d\tau \tag{6-2}$$

In fact, the lower limit of the integral could just as well be t_e, but it is simpler and more convenient to write $-\infty$. For the same reason, since $h(t)=0$ for $t<0$ (because of causality), we generally write the upper limit as $+\infty$ and obtain

$$g(t)=\int_{-\infty}^{\infty} f(\tau)h(t-\tau)\,d\tau \tag{6-3}$$

which is known as the convolution integral [in German, *Faltungsintegral*, "folding integral"—expressing the fact that $h(t)$ is "folded over" $f(t)$ before the ordinates of f and h are cross-multiplied and the results added together]. We say that $f(t)$ and $h(t)$ are *convolved* (NOT "convoluted"). A simpler and more convenient notation for Equation 6-3 is

$$g(t)=f(t)*h(t)=h(t)*f(t) \tag{6-4}$$

(A simple change of variables in Equation 6-3 will prove the last equality.) The convolution integral also expresses the fact that the output is a weighted integral of the past input. After the time interval t_e, the "memory" of the past input vanishes.

PROBLEM

6-1 Prove that $f(t)*h(t)=h(t)*f(t)$.

B. The z-Transform Method

An alternative, and perhaps even simpler, way to demonstrate the input-output relationship is to postulate that *both* $f(t)$ and $h(t)$ consist of distinct impulses of various heights. Accordingly, $f(t)$ consists of $(f_0, f_1, f_2 \ldots)$ at $(t_0, t_1, t_2 \ldots)$ separated by time interval δt (Figure 6-4), and $h(t)$ is $(h_0, h_1, h_2 \ldots)$ at the same time interval δt.

Impulse f_0 gives rise to output $f_0h_0, f_0h_1, f_0h_2, f_0h_3$ at times t_0, t_1, t_2, t_3, respectively. Similarly, f_1 generates $(f_1h_0, f_1h_1, f_1h_2, f_1h_3)$ at times t_1, t_2, t_3, t_4: that is, at an interval δt behind the output resulting from f_0. The same reasoning applies to f_2, etc. Hence, at time t_4 the output, g_4, is

$$\begin{aligned} g_4 &= f_4h_0+f_3h_1+f_2h_2+f_1h_3 \\ &= \sum_{i=0}^{3} f_{4-i}h_i \end{aligned}$$

f_0	f_1	f_2	f_3	f_4	f_5			*e.g.*,	2	1	−1	0.5	−2	0.3			*a*
h_0	h_1	h_2	h_3					*e.g.*,	1.8	2	0.4	−0.2					*b*
f_0h_0	f_0h_1	f_0h_2	f_0h_3					Output from f_0	3.6	4	0.8	−0.4					*c*
	f_1h_0	f_1h_1	f_1h_2	f_1h_3				Output from f_1		1.8	2	0.4	−0.2				*d*
		f_2h_0	f_2h_1	f_2h_2	f_2h_3			Output from f_2			−1.8	−2	−0.4	0.2			*e*
			f_3h_0	f_3h_1	f_3f_2	f_3h_3		Output from f_3				0.9	1	0.2	−0.1		*f*
				f_4h_0	f_4h_1	f_4h_2	f_4h_3	Output from f_4					−3.6	−4	0.8	0.4	*g*

FIG. 6-4 *(a) The input f_0, f_1, f_5 (left) and possible numerical values (right). (b) The unit impulse response h_0, h_1, h_2, h_3 and possible numerical values. (c) The output from f_0, with resulting numerical values. (d) to (g) The outputs from f_1 to f_4, with resulting numerical values.*

Or, at time t_j:

$$g_j = \sum_{i=0}^{3} f_{j-i} h_i$$

More generally, if h consists of p impulses, then

$$g_j = \sum_{i=0}^{p-1} f_{j-i} h_i \tag{6-5}$$

which is the discrete convolution operation. By comparing this equation with Equation 6-3, we see that $g(t) \approx g_j \times \delta t$. Thus, if a comparison of the results of these two operations is needed, g_j should be multiplied by δt. In cases in which the scale is unimportant, Equation 6-5 may be used directly. We can also obtain g_j as follows. First, we define $F(z)$, the z-transform of a series of impulses (f_i), as

$$F(z) = f_0 + f_1 z + f_2 z^2 + \cdots \tag{6-6}$$

where z is (for this purpose) an ordinary variable. $G(z)$ and $H(z)$ then are similarly defined. By direct polynomial multiplication and equating equal powers of z, we find

$$G(z) = F(z) \times H(z) \tag{6-7}$$

which is the z-transform equivalent of Equation 6-5.

PROBLEMS

6-2 Prove Equation 6-7.

6-3 Using the data given below, convolve f_i with $h_i^{(1)}$.
Plot $g_i^{(1)} = f_i * h_i^{(1)}$.
Do the same with $h_i^{(2)}$. Obtain $g_i^{(2)}$.
What difference between $g_i^{(1)}$ and $g_i^{(2)}$ do you observe? Can you guess, roughly, why the difference exists?

$$f_i = 1, 0.6, 0, -1, -1, -0.5, 0.5, 0.5, 0, 1, 0, 0, -1$$

$$h_i^{(1)} = 1, 0.8, 0.4, 0.1$$

$$h_i^{(2)} = 1, 0, -0.5$$

C. System Response

We know how to compute the amplitude spectrum $F(\nu)$ and the phase spectrum $\phi(\nu)$ for any function of time (see Equations 4-13 and 4-14). Computing the amplitude and the phase of $f(t)$, $g(t)$, and $h(t)$, we obtain, for the amplitude and the phase spectra, respectively,

$$f(t) \leftrightarrow F(\nu), \qquad \phi(\nu)$$

$$g(t) \leftrightarrow G(\nu), \qquad \gamma(\nu)$$

$$h(t) \leftarrow H(\nu), \qquad \eta(\nu)$$

It is easy to show that the following relations are true. If

$$g(t) = f(t) * h(t) \tag{6-8}$$

then

$$\gamma(\nu) = \phi(\nu) + \eta(\nu) \tag{6-9}$$

$$G(\nu) = F(\nu) \times H(\nu) \tag{6-10}$$

The addition of phase angles is self-evident. The multiplication of amplitude spectra means that the amplitude $F(\nu_o)$ at any specific frequency ν_o must be multiplied by $H(\nu_o)$ to find $G(\nu_o)$. This is what we would intuitively expect. If a loudspeaker responds well to low frequencies and poorly to high frequencies, then its $H(\nu)$ is relatively large when ν is small and relatively small when ν is large. Such a loudspeaker will attenuate (or distort) the high frequencies that it receives.

Whereas $h(t)$ is called the (unit) impulse response, $H(\nu)$ is the transfer function. Equations 6-8, 6-9, and 6-10 express the fact that convolution in the time domain corresponds to multiplication in the frequency domain. If f, g, and h are functions of space rather than of time, then multiplication in the wave number domain corresponds to convolution in the space domain.

D. Filtering

Any instrument by its very existence acts as a filter, but special filters can also be built. With the advent of digital computers it has become common to think of $f(t)$ not as a continuous function but as a series of closely spaced samples f_i, known as a time series; and the same is true for $h(t)$. Normally this concept involves the sampling of a continuous voltage $f(t)$ at a regular interval δt, called the sampling interval. Once this operation has been performed, f_i can be filtered by being convolved with h_i, see Equation (6-1). Here h_i often is *not* constrained to being causal—for example, if $f(t)$ is recorded on magnetic tape, the relative future is known at any instant, since it is already recorded. This relative time is known as *fiducial time*. In such a case a filter may use the relative future as well as the relative past to obtain the output at any instant. This is, of course, impossible in *real time*.

The design of causal filters is a difficult matter. In contrast, the design of "even" filters—those for which $h(-t)=h(t)$ or $h_{-i}=h_i$—is very simple, for such filters introduce no phase shift between the input and the output (they are said to be zero-phase). To design an even filter of a given frequency response $H(\nu)$ with $\eta(\nu)=0$, we need merely compute (see Equation 4-15).

$$h(t)=2\int_0^{\infty} H(\nu)\times\cos(2\pi\nu t)\times d\nu \tag{6-11}$$

Because there is no sine term $[B(\nu)=0]$, $h(t)=h(-t)$, which proves that $h(t)$ is even. Since $B(\nu)=0$, $\eta(\nu)=0$ (see Equation 4-5).

PROBLEMS

6-4 Faced with data (say, grain size) obtained every 0.5 cm that appear to be badly scattered, you decide to take moving averages over seven points. This, of course, is a very simple convolution, so it affects your results—but how? Specifically, what does it do to the frequency spectrum? Or, if you prefer, what is the amplitude response of the filter? (For simplicity, compute only A and use the limits $-L/2$, $+L/2$.) *Note*: To take a moving average over seven points, you first take the average value of the first seven points and attribute this value to the center point. Next, you take the average value of points 2–8 and attribute it to the center point, and so on.

6-5 Compute analytically $h(t)$ for the even filter whose amplitude response is shown in Figure 6-5. Plot $h(t)$ from $t=0$ to $t=20$ s, every 1 s. Sketch $h(t)$ for $t=0$ to $t=-20$ s.

6-6 Compute analytically or numerically $h(t)$ for $H(\nu)$ shown in Figure 6-6, if $\eta(\nu)=0$. Plot $h(t)$ for $t=0$ to $t=10$ s, every 0.5 s. Sketch $h(t)$ for negative values of t.

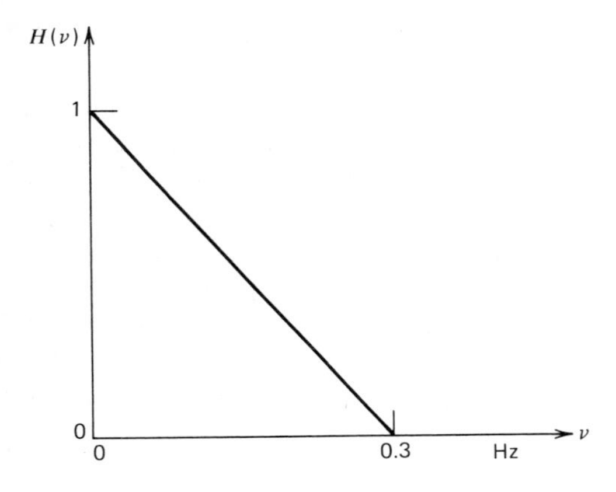

FIG. 6-5 *Amplitude response for the even filter of Problem 6-4.*

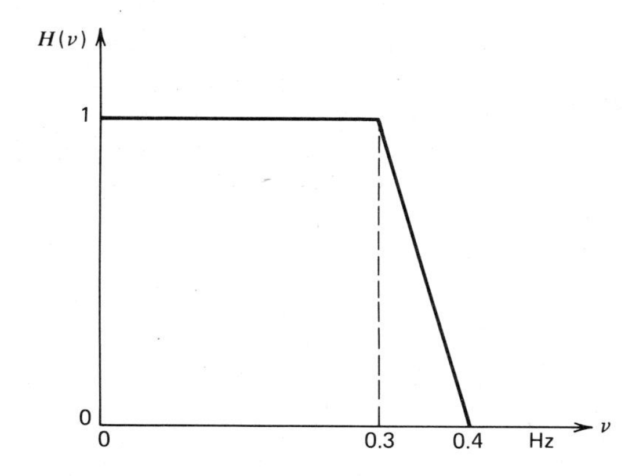

FIG. 6-6 *Amplitude response for the even filter of Problem 6-5. The frequency at which the amplitude response is approximately zero is the cutoff frequency.*

E. Geological and Geophysical Applications

As has already been mentioned, every instrument may be considered a system and not all systems are linear. Our sense of hearing, for example, has a response that is logarithmic in amplitude. Take, as the reference, a note of a single frequency and of amplitude 10 (on a given scale): the same note of amplitude 100 (on the same scale) will *not* sound 10 times louder, but only twice as loud. This enables us to hear very loud and very soft sounds, but it makes clear the fact that our hearing is not a linear system.

Somewhat similarly, before the advent of digital recording in seismic reflection prospecting it was necessary to increase the gain as time passed in order to compensate for the much smaller amplitudes of later arrivals (see Chapter 3, Section A). (In digital recording this procedure usually is not necessary, because the dynamic range of the instruments is sufficiently large for both early and late arrivals to be properly recorded.) Sometimes the gain was automatically adjusted (automatic gain control) to yield legible records over the whole record length. In such cases filtering after AGC would be meaningless, because it would result in the addition of "impulses" representing different amplitudes at the input (Equations 6-1 and 6-5). However, if the memory of the filter (t_e; see Figure 6-2) is much shorter than the time over which the gain is changed, filtering after AGC is still meaningful.

Another example of filtering in the seismic reflection method can be found in the low-pass filters used between the geophone (seismograph) and the digital recorders. These filters are designed to pass frequencies below a limiting frequency and greatly to attenuate higher frequencies. (The response shown in Figure 6-6 is a highly idealized version of such a filter, but the frequencies used in seismic reflection might be nearer 40 and 50 Hz than 0.3 and 0.4 Hz. Furthermore, these low-pass filters operate in real time.)

Maps provide another example of the result of filtering. Any map is an idealized representation of reality. Contour levels, for instance, cannot be drawn with the same "wiggliness" they display in the field. In terms of Fourier analysis, this means that the high spatial frequencies (or high wave numbers) are eliminated in the map-drawing process, somewhat as the high seismic frequencies are eliminated before digital recording. Furthermore, as the scale of the map increases, larger and larger features become obliterated—that is, the cut-off frequency decreases. (It is common to use the term *frequency* rather than *wave number* when dealing with space-dependent phenomena.)

With the advent of digital recording in earthquake seismology, it has become possible to filter the record of one instrument to extract both the short-period and the long-period waves. Finally, it is possible to use the idea of system response to relate relief and gravity, and to examine the isostatic adjustment of a region (see Chapter 9, Section C).

These examples suffice to show that the concept of system response has a wide application in geology and geophysics.

F. Deconvolution

In many cases it is desirable to retrieve the input of an instrument from its output and its impulse response—for example, when the input consists of a series of impulses that cannot be distinguished on the record. Equation 6-7 suggests that in order to obtain F from G and H, one need merely perform the operation

$$F(z)=\frac{G(z)}{H(z)} \tag{6-12}$$

Such an operation corresponds to a polynomial division of the z-transforms, also called a deconvolution. Such a division is, however, not always possible, for the coefficients of F may increase rapidly (diverge). To be convinced of this fact, one merely needs to attempt a division by $H_1(z)=1\pm 2.5\,z$. In contrast, a division by $H_2(z)=1\pm 0.4\,z$ presents no problem. The root of H_1 is $z=\mp 0.4$; that of H_2 is $z=\mp 2.5$. Since any polynomial may be written as a product of such "doublets" (H_1 or H_2), we induce that a division by H converges if all of the roots of H are greater than unity in absolute value, and that it diverges (is unstable) if any root is less than unity in absolute value. Unfortunately, most polynomials have "complex" roots. But because of the need to use complex numbers in examining this problem, we will not consider it further.

7

Quantification, Prediction, Noise, and Instruments

A. Intensity

It is useful to have a measure of the intensity of the "shaking" experienced at various places during an earthquake. So simple a measure as the amount of damage figured in constant dollars would fail to take into account the differences in damage totals stemming from variations in the concentration of population. Therefore, researchers have developed a few standard scales that attempt to measure the acceleration at any given point by relying on how strongly the earthquake was felt and on how much damage it did to what kind of building. In the United States, the most commonly used intensity scale is the modified Mercalli scale, which ranges from intensity 1 (felt only by a very few people, under especially favorable circumstances) to intensity 12 (damage total; objects thrown upward). Byerly (1942, p. 58), among others, gives the exact definition of the values. Questionnaires are often distributed after a major earthquake, and intensity maps are prepared on which the contours, or isoseismal lines, define zones of equal intensity. The recent emphasis on sophisticated instrumentation has raised the danger that such sources of information will be disregarded in the future. The following should be noted however, that

- A survey taken after the 1906 earthquake conclusively showed that accelerations were much larger (hence the damage much greater) in alluvial valleys than on the bedrock. (This increase in damage is quite independent of slumping.) The theoretical explanation came only some 55 years later (Haskell 1960, 1962; Hannon, 1964). Essentially, the great impedance contrast between the bedrock and the sediments results in the trapping of the seismic energy in the latter, and this causes much longer shaking. Additionally, the lower density and velocity in the sediments result in a compensatory increase in amplitude (see Chapter 3, Section A).
- Elongated isoseismal lines suggest the presence of a fault. The isoseismal lines of the 1906 shock were extremely elongated.
- A very large area of almost constant intensity suggests a deep focus.

Since the intensity varies from place to place, however, it does not provide a measure of the total energy released by the shock.

B. Magnitude

The necessity for some measure of the energy released by local shocks was first recognized by Charles F. Richter in 1935 in Southern California (see Richter 1958, Ch. 22). The procedure was later extended to measure the energy of distant earthquakes. At the present time there are a number of slightly different magnitude scales. One of the most commonly used (e.g., by the U.S. Geological Survey) determines the magnitude of shallow earthquakes from surface waves according to the equation

$$M = \log_{10} \frac{A}{T} + 1.66 \log \Delta + 3.3 \tag{7-1}$$

where T is the period in seconds, with $18\text{ s} < T < 22\text{ s}$; A is the maximum horizontal ground motion, in μm; and Δ is the epicentral distance in degrees. The value of M thus obtained is sometimes referred to as the surface wave magnitude, M_s. Since the energy at any given distance is proportional to the square of the amplitude, one might expect to find that

$$\log_{10} E = 2M + \text{constant} \tag{7-2}$$

That is, the energy increases by a factor of 100 for each unit increase in magnitude. In fact, mostly because of dispersion, this is not exactly the case. The generally accepted relation is

$$\log_{10} E = 1.5M_s + 4.4 \text{ (for } E \text{ in joules)} \tag{7-3}$$

An increase of one magnitude unit thus corresponds *approximately* to an energy increase by a factor of $30 = 10^{1.5}$, but according to some authors, this factor may be as high as 60.

In Chapter 1, we noted that most of the energy released by earthquakes is radiated away as seismic waves. This statement can now be refined: for magnitudes greater than 6 it is correct, but for smaller shocks the efficiency appears to drop rapidly, perhaps to as low as 1% for $M = 2$ (Stacey 1977, p. 129).

For deep earthquakes, from which the surface waves are small, a body wave magnitude, m_b, can be similarly defined, but the formula takes into account both the geometrical spreading and the depth of focus.

PROBLEMS

7-1 How long would a 1000 MW power plant have to operate to produce the same energy as a $M_s = 8.5$ earthquake? Give the time in the largest possible unit —years, days, or hours (rather than, say, "10^3" days, use " ~ 3 years").

7-2 Derive Equation 7-2.

C. Seismic Moment

In the 1960s, seismologists realized that some earthquakes that generate extremely large surface waves of periods as long as 300 s, and that are accompanied by faults up to 1000 km in length, do not generate correspondingly large surface waves of periods near 20 s. The physical reason for this phenomenon has been alluded to before. The volume involved in generating a wave of a given period T_o is on the order of the cube of the wavelength. When the volume greatly increases, then, the amplitude of the wave of period T_o does not increase correspondingly, and hence the magnitude scale reaches saturation. On the other hand, by using these very long surface waves, geophysicists realized that the "correct" magnitude of these "great" earthquakes (so-called to distinguish them from "large" earthquakes) was as high as 9.5 for the 1960 Chile earthquake and 9.2 for the 1964 Alaska earthquake.

As a result of these observations, the energy of an earthquake is now often given by its moment, M_o, which can be defined as

$$M_o = \mu \times \text{average slip} \times \text{fault area} \tag{7-4}$$

where μ is the rigidity of the rocks adjacent to the fault (Aki and Richards 1980, p. 49). M_o can be computed from surface waves; it ranges from $\sim 10^5$ Nm for very small earthquakes (micro-earthquakes) up to 10^{23} Nm for the largest ones.

It may be noted that the slip is a vector, and the fault can be idealized as a plane. This combination is thus very similar to the one encountered in the discussion of the stress tensor in Chapter 1. The only difference is that the slip on both sides of the fault constitutes a couple, which is why M_o is a moment. What we now have is therefore a couple, or a moment, associated with a plane (the fault plane). We conclude that M_o is properly a moment tensor.

The seismic energy released every year fluctuates by more than an order of magnitude (see Figure 7-1). On the order of 10^{18} J from 1950 to 1968, it reached 5×10^{18} J in 1960 but has dropped to about 10^{17} J since 1970. The reason for these fluctuations is not known.

D. Prediction

Prediction has three different aspects: place, time, and magnitude.

Place

Although it is statistically true that areas that have experienced earthquakes in the past can expect them in the future, significant variations occur over periods of several centuries or more. In parts of the Middle East (Lebanon, Israel, Syria), there were many destructive earthquakes from biblical times up to approximately 1000 A.D., but at present the area experiences only relatively small shocks.

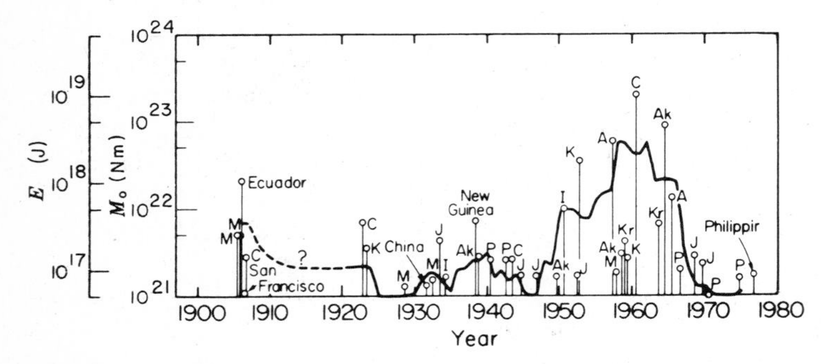

FIG. 7-1 *Seismic moment M_0 and seismic energy E, for large earthquakes, from 1900 to 1980. Solid curve: annual average seismic energy released (averaged over five years). The locations of the individual earthquakes: A, Aleutian Islands; Ak, Alaska; C, Chile; I, India; J, Japan; K, Kamchatka; Kr, Kurile Islands; M, Mongolia; and P, Peru. (After Kanamori 1978. Reprinted by permission from Nature 271:411–414. Copyright © 1978 by Macmillan Journals Limited.)*

Somewhat similarly, northern China suffered enormously destructive earthquakes ($M \approx 8$) from the fourteenth to the seventeenth centuries (1303, 1556, 1668, and 1679), then enjoyed a period of quiet lasting until the early nineteenth century. Earthquake activity then resumed, culminating in 1976 in a large earthquake ($M \approx 7\frac{3}{4}$) near Tangshan, 150 km east of Beijing (Peking) that killed about a quarter of a million people by official count. Other areas seem free from earthquake activity but then experience sudden major earthquakes (for instance, New Madrid, in southern Missouri, at the head of the Mississippi embayment, in 1812). However, seismographs installed in these areas record fairly numerous if small shocks (see Chapter 3, Section E).

In attempting to predict the location of large earthquakes, seismologists often use the concept of "seismic gap." This term refers to a segment of an active plate boundary (transform fault or subduction zone) where either (a) large earthquakes have occurred in the past, but not within the preceding 30 years, or (b) the adjacent plate boundaries have either experienced major earthquakes during the preceding 30 years or are slipping aseismically past each other. The 30-year period is somewhat arbitrary and has been debated. The approximately 10 such seismic gaps that have been identified are shown in Figure 7-2. One of the advantages of locating such gaps is that earth scientists can then concentrate observational efforts there, which may lead to improved prediction capabilities.

In summary, it is relatively easy to define areas that will definitely experience large earthquakes, but it is more difficult to identify with certainty areas that definitely will not do so. (It should be noted that for insurance purposes the place is more important than the time.)

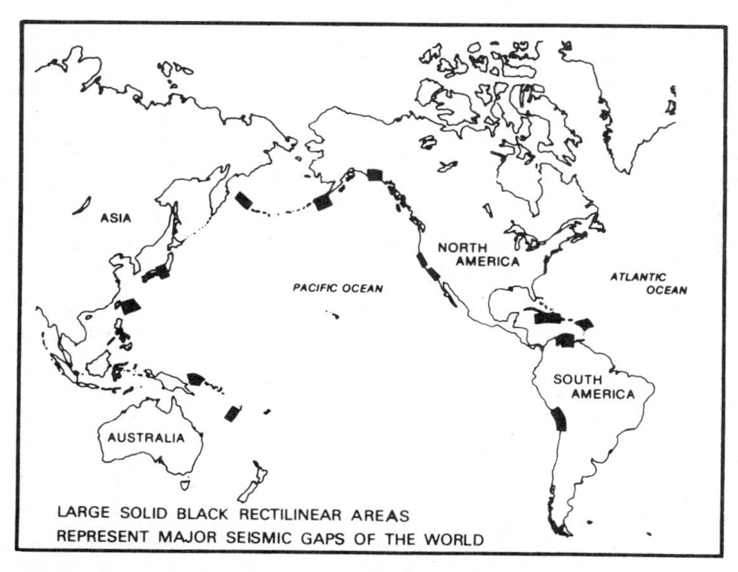

FIG. 7-2 *Major seismic gaps of the world (Earthquake Information Bulletin 1979).*

Time

It is much more difficult to predict the time of an earthquake than to predict its place. This problem has several aspects. From a sociological point of view:

1. A prediction must define a short "time window"—a very few days in which the event will occur.
2. A prediction must be made with a high level of confidence. A false prediction may lead to mass evacuation, disruption, and grave economic losses (as happened when a false prediction of a volcanic eruption was made recently for an island in the Carribbean), but a "missed" prediction may lead to many casualties (as happened with the Tangshan earthquake in 1976).

And, from a scientific point of view:

1. There are many precursory phenomena. So far, no single phenomenon has been shown to occur before every earthquake or to occur *only* before an earthquake.
2. There is no model that can successfully predict the precursory phenomena and their variability.
3. It is possible that some phenomena that are reliable precursors in one tectonic setting (say, a subduction zone) are not precursors (or are rarely so) in another setting (say, a transform fault).

Some investigators believe that there may be a certain pattern of events that reliably indicates the coming of an earthquake—that is to say, that a definite number of phenomena *may* precede an earthquake and may occur in a more or less fixed sequence. According to this view, if at least 90% of these phenomena occur, the earthquake is highly probable; if only 80%, it is less probable, and so forth. The entire scheme is susceptible to quantification (Barenblatt et al. 1981), but so far it appears to have been more successful in accurately predicting the locations of epicenters in one area (Gelfand et al. 1976) than in predicting the times of earthquakes, and, as previously noted, there is some doubt that the pattern that characterizes one tectonic setting also characterizes another one or, indeed, that a time pattern exists at all (Kanamori 1981).

Pattern recognition has become a field with its own scientific journal and has such varied applications as the automatic reading of seismograms, earthquake prediction, and even the prediction of the outcome of presidential elections (Lichtman and Keilis-Borok 1981).

Magnitude

It is a general rule that precursor times increase as the magnitude increases, but it is not always easy to decide beforehand what constitutes a precursor. If an event precedes a magnitude-8 earthquake by 10 years or more, for example, its significance may not be recognized until much later.

Many papers on earthquake prediction are published annually; the most recent book by a single author is that written by Rikitake (1976). It should also be mentioned that the United States has bilateral agreements with China, Japan, and the Soviet Union for research on this subject.

E. Noise

The part of the input that is of interest to the observer is termed the signal; the rest is "noise." In earthquake seismology, such noise is caused largely by two sources.

Traffic

In or near a major city or highway, this noise usually reaches a maximum near 2 Hz. The reason for this frequency is not clear. Since the *P*-waves of local earthquakes may reach their maximum near this frequency, seismographic stations should be located well away from cities and highways.

Waves at Sea

As is clear from Equation 3-5, the very large difference in density between the air and the ground makes for very poor transmission of energy from one to the other. The direct effect of the wind on the ground is thus small, although it is somewhat enhanced by the presence of trees, buildings, and other structures, because of nonlinear effects (turbulence). The best coupling from air to ground, however, occurs via the ocean. The exact mechanism is still debated, but the best such coupling occurs when a storm approaches the coast, possibly because of the rapid decrease in the depth of the ocean there. Waves thus generated are called microseisms (not "micro-earthquakes," which are earthquakes with $M < 1$). Their typical period ranges from approximately 6 to 8 s and more. It is generally thought that this is one-half the dominant period of waves at sea, and Figure 7-3 tends to support this view. The most characteristic feature of microseisms is that they propagate with little loss of amplitude across a continent (presumably as surface waves). They are present everywhere and constantly, but there are occasionally large "microseismic storms" that may last for a few days. Some seismographic stations are therefore equipped with filters that attenuate waves in this frequency band; unfortunately, these filters also attenuate waves of interest in the same band. Alternatively, microseisms can be "filtered out" if the station records on magnetic tape.

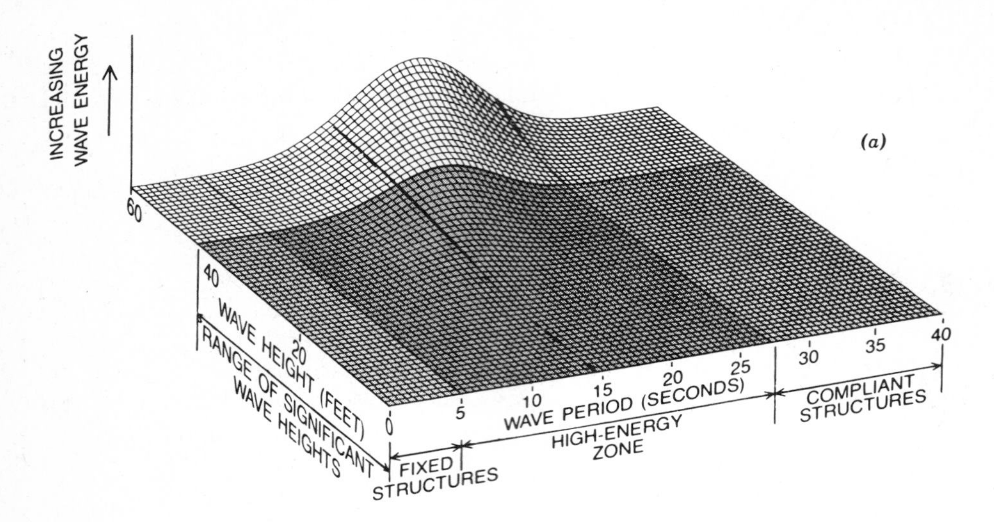

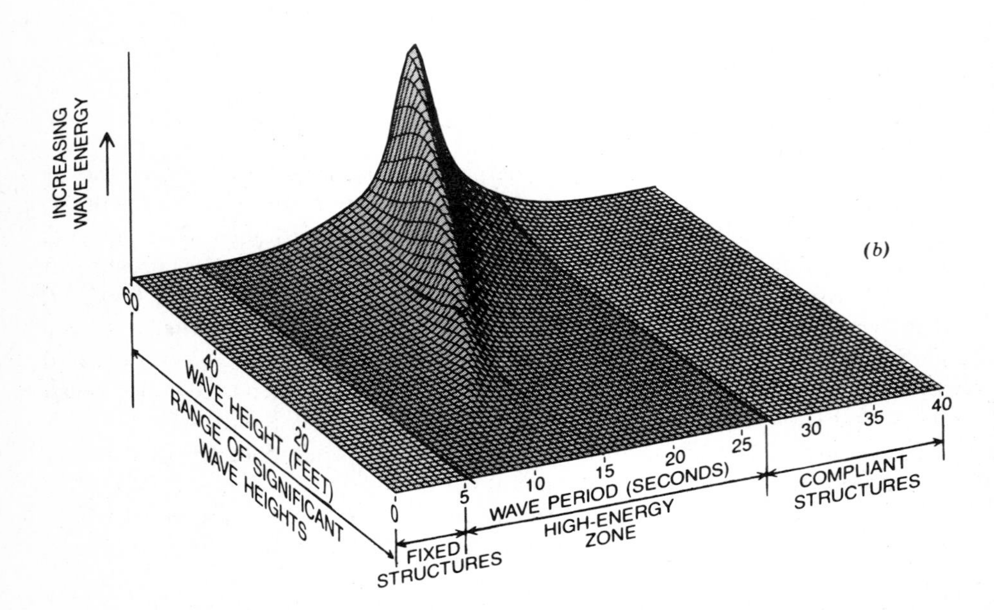

FIG. 7-3 *Sea-spectrum diagrams (a) for the Gulf of Mexico, (b) for the North Sea. "Significant" waves are approximately 60% as high as the maximum height to be expected during a period of a few hours. (Ellers 1982. Copyright © 1982 by Scientific American Inc. All rights reserved. By permission of Ross Cowan, International Software Services Corp., Mission Viejo, CA, 92691).*

F. Seismographs

We have repeatedly mentioned earthquake records, or seismograms, and we should examine briefly how they are obtained. A conventional seismograph has three parts: a mobile mass, a transducer, and a recorder. (The mass plus the transducer constitute the seismometer.).

The Mobile Mass

The instrument is generally designed so that its mass can only move in one direction, usually N–S, E–W, or up–down. The mobile mass thus consists in principle of a pendulum (Figure 7-4) constrained to move in a plane. The vertical instrument is schematized in Figure 7-5. To lengthen the period of the horizontal instrument and save space, a "horizontal pendulum" (Figure 7-6) whose axis is near-vertical (Θ is small) is often used; such instruments are sometimes characterized as being of the "swinging door" type. To prevent the instrument from continuing to oscillate once set in motion, damping is always introduced. (It is puzzling to note that the earliest seismographs lacked this feature.) The undamped period is known as the free period of the instrument, T_o.

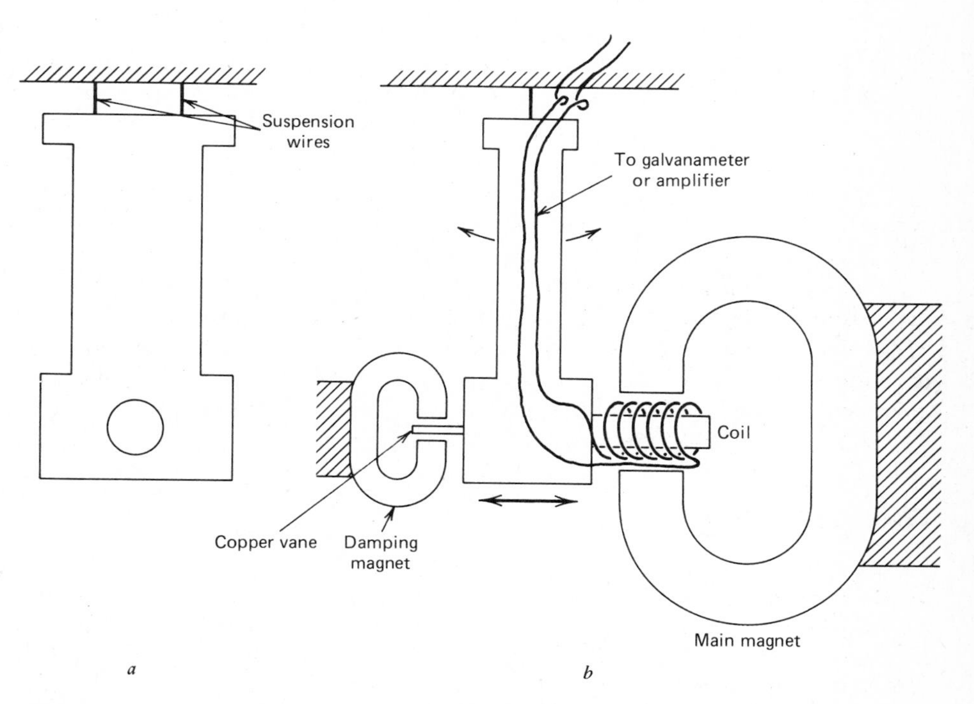

FIG. 7-4 *Principle of the horizontal seismograph. (a) Front view. (b) Side view.*

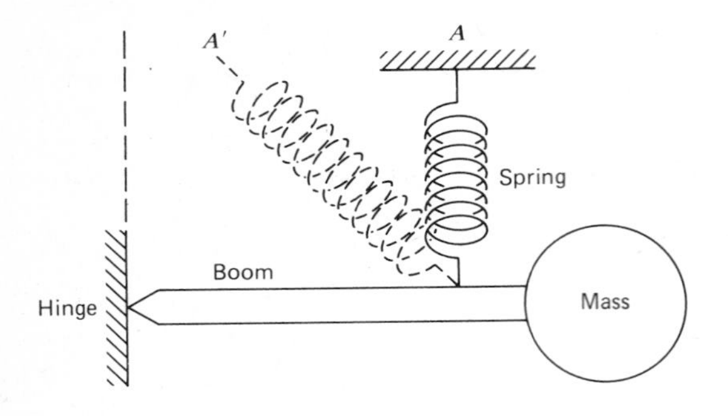

FIG. 7-5 *Principle of the vertical seismograph (transducer and damping system not shown). If the spring is attached at A′ rather than at A, the free period is increased; A′ must be to the right (i.e., toward the mass) of the vertical drawn through the hinge point to prevent instability.*

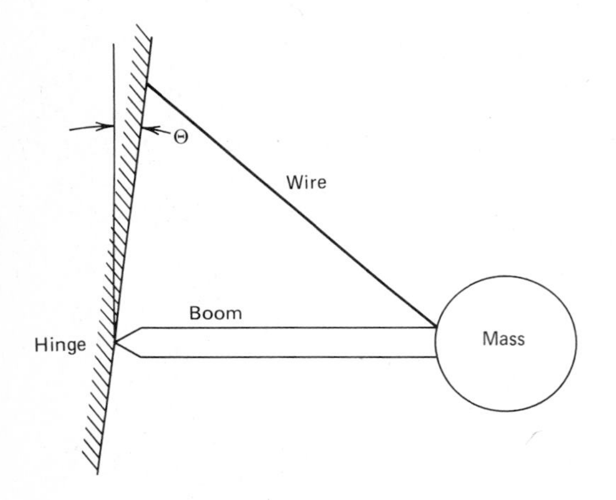

FIG. 7-6 *Principle of the "horizontal pendulum": Θ is small.*

If the period, T, of the ground motion is much shorter than T_o, the ground will, so to speak, move under a stationary mass. Discounting the effect of the transducer, the instrument is then a displacement meter.

Similarly if $T \gg T_o$, the instrument is an accelerometer, for reasons particularly easy to understand in a vertical pendulum (Figure 7-5). Placing a small mass on the mass of the pendulum is equivalent to exerting on it an additional force or an additional acceleration with an infinite period; clearly, this results in a "permanent" displacement of the mass, which proves the proposition.

Finally, if $T \approx T_o$, the instrument is a velocity meter.

The Transducer

Displacement Transducer A system that records (with amplification) the *position* of the mass is called a displacement transducer. Such transducers were actually the first to be used. Originally, they consisted of long, thin levers. In such a form, however, they could not avoid friction, and to minimize its effect, extremely heavy masses were used (up to 60 tons!). In the 1960s, electronic displacement transducers became somewhat more common, although they introduced a new difficulty. Most seismometers are subject to very long period noise known as drift, which may be several orders of magnitude larger than the signal. On the vertical seismograph, drift may result from thermal effects on the spring or from the barometric effect (Archimedian effect) on the mass; on horizontal instruments, it mainly stems from the unwanted tilt (in and out of the plane of the paper in Figure 7-6) of the nearly vertical axis. In such cases displacement transducers are not useful without some drift compensation.

Velocity Transducer A velocity transducer consists in general of a coil attached to the mass of the seismometer; the coil moves in and out of a magnetic field produced by a magnet attached to the frame (Figure 7-4). The current generated by this motion is proportional to the velocity of the mass. This current may be electronically amplified or sent directly to a sensitive galvanometer.

Other Types of Seismograph

The Extensometer This instrument, also called a strain meter, consists of a tube a few meters or tens of meters in length attached to one pier and equipped with a transducer at the end. The displacement between the ground and the other end is recorded; it is, of course, proportional to the normal strain in the direction of the tube. The latter is usually made of fused quartz, so as to be thermally insensitive.

The Dilatometer This instrument, like the preceding one, is also called a strain meter. It measures the dilatation, θ. The most successful instrument, the Sacks-Evertson strain meter, accomplishes this by measuring the change in volume in a sealed cavity lowered into a well, in which it is carefully sealed (Sacks et al. 1971). Because of the depth to which it is lowered, the instrument is protected from noise resulting from rain, temperature fluctuations, and so forth. This enables it to measure strains occurring over periods ranging from a few seconds to several weeks or more, and to measure values of θ down to 10^{-11}, at least (Sacks et al. 1982).

The Tilt-Meter As previously mentioned, a horizontal pendulum is sensitive to the tilt of its frame. It has been shown, however, that such tilts are poorly correlated over short distances (on the order of 1 m). For this reason, tilt-meters now generally consist of a liquid-filled horizontal tube at least 10 m long that connects two vertical cylinders. The relative displacement of the liquid in the cylinders is measured. The first such an instrument was built in the 1920s by Michelson, who used his well-known interferometer to record the displacement of the liquid in the cylinders and compute the large-scale elastic properties of the earth.

Brief Review of the Contributions of Seismology to Plate Tectonics

In the previous chapters we noted at various points the contributions made by seismology to the theory of plate tectonics. These contributions provided the subject of a classic paper, "Seismology and the new global tectonics" (Isacks et al. 1968), which helped usher in a new view of the earth. Because they have been previously described, a simple enumeration of the discoveries made possible by seismology will suffice.

- The Appalachians are a huge thrust sheet (Chapter 3, Section A).
- The motion on transform faults is opposite to that expected from the offset of the ridges (Chapter 3, Section C).
- The pattern of motions near a subduction zone is that expected from a heavy plate being subducted (Chapter 3, Section C).
- Earthquakes occur mostly along plane boundaries (Chapter 3, Section E).
- The existence of the LVZ and its variations agree with the expectations from plate tectonics (Chapter 5, Section B).
- The gradual thickening of the oceanic lithosphere with age is further confirmation of the plate tectonic model (Chapter 5, Section B).

PART TWO

Gravity

8

Gravitational Attraction

A. Introduction

In 1666, Isaac Newton (1642–1727), then 24 years old, apparently conceived the idea of universal gravitational attraction when seeing an apple fall. At the time, the idea that the same force that makes an apple fall extends all the way to the moon and beyond was unheard of. Something fell because it was attempting to go to "its proper place." Newton's leap of imagination was to surmise that the force that attracts the apple downward also holds the moon in orbit. There and then he proceeded to verify his hypothesis. Knowing the value of the centripetal force ($\omega^2 r$), and the moon's distance as well as the earth's radius, he "thereby compared the force requisite to keep the moon in her orb with the force of gravity at the surface of the earth, and found them answer pretty nearly" (Newton 1695?).* As it happens, Newton knew the correct distance between the earth and the moon in terms of the earth's radius; however, possibly following Galileo, he took the degree of latitude to be 60 Italian miles (1 it. mile = 5000 ft $\approx$ 1.544 km),

*(An interesting comparison of the role of creativity in science and in poetry can be found in Bronowski 1965, pp. 3–25).

rather than 60 nautical miles (1 n. mile = 1.853 km). This caused him to make a 20% error in the earth's radius and caused a 20% discrepancy between the two values of g. He then abandoned the subject until 1684.

Newton assumed that gravity is transmitted instantaneously and that there is no need for a "transmitting medium." For a long time the idea of instantaneous "action at a distance" was accepted without discussion. It is now generally believed that nothing can travel faster than light, including gravity. Furthermore, there is some indication that gravity does not act "at a distance" but either by means of waves or by warping the space-time continuum. To be sure, these considerations will not figure importantly in what follows, but they may help relate the following discussion to the rest of science.*

Geologists are primarily interested in gravity for the information that it provides concerning density variations in the earth's interior. These variations may be as shallow as 1 km or less, or as deep as 100 km or more. Although an infinity of density distributions yield the same gravity field at the surface, only a small number of such distributions are plausible and meaningfully different (see Chapter 9, Section C). For this reason, and because they are relatively easy to conduct, gravity surveys remain an important tool in geophysics. The subject is slightly more mathematical than seismology; the examples and the problems should help make it clear. Two well-known references for the following discussion are MacMillan (1958) and Ramsey (1961).

B. Gravitation as the Gradient of a Potential

Gravitational Field

Consider a point mass (that is, a mass concentrated at a point) m at $P(x, y, z)$ and another one, m_1, at $R_1(a_1, b_1, c_1)$. (See Figure 8-1.) Let the vector R_1P be $\mathbf{r}_1$ (from R_1 to P). The magnitude of the force of attraction at P due to m_1 is

$$A_1 = -G\frac{mm_1}{r_1^{\,2}} \tag{8-1}$$

It is directed opposite $\mathbf{r}_1$ and thus is negative if $\mathbf{r}_1$ is taken as the positive direction.

Let us now assume that m is unity. Then, by definition, $\mathbf{A}_1$ becomes $\mathbf{F}_1$, the *gravitational field* at P due to m_1. Hence,

$$F_1 = -G\frac{m_1}{r_1^{\,2}} \tag{8-2}$$

*An easy-to-read book on the subject is *Gravity* by G. Gamow, 1962.

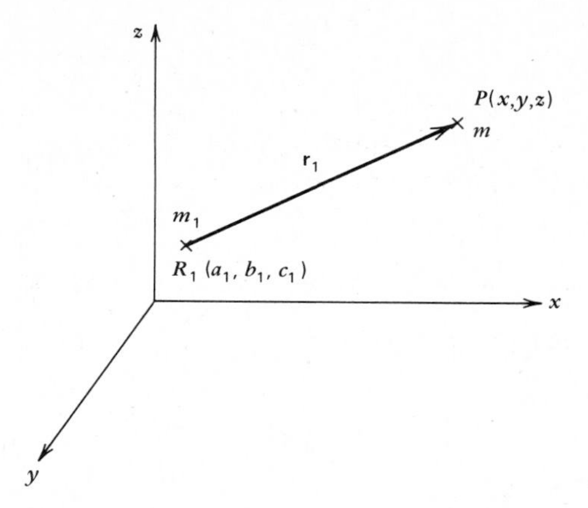

FIG. 8-1 Point masses m at P and m_1 at R_1.

which is oriented toward R_1. Its direction cosines are thus

$$-\frac{x-a_1}{r_1}=\frac{a_1-x}{r_1}, \quad \frac{b_1-y}{r_1}, \quad \frac{c_1-z}{r_1} \tag{8-3}$$

(obtained by subtracting the coordinates of the end of the vector from those of origin, dividing by the distance, and adding the minus sign in front), while

$$r_1^{\,2}=(x-a_1)^2+(y-b_1)^2+(z-c_1)^2 \tag{8-4}$$

The x-component of r_1, written r_{1x}, is

$$r_{1x}=x-a_1 \qquad (\textit{not } a_1-x) \tag{8-5}$$

From Equations 8-2 and 8-3, we find the x-component of F_1 by multiplying the absolute value of $F_1 = Gm_1/r_1^{\,2}$ by the direction cosine:

$$F_{1x}=G\frac{m_1}{r_1^{\,2}}\times\frac{a_1-x}{r_1}=G\frac{m_1}{r_1^{\,3}}(a_1-x) \tag{8-6}$$

with similar expressions for F_{1y} and F_{1z}. (This result is independent of the orientation of $\mathbf{r}_1$.)

G (sometimes γ or k^2) is the universal constant of gravitation; its value is 6.67×10^{-11} m^3/kg s^2. There has been some discussion over whether the value of G has changed in the past; the consensus appears to be that its change since the formation of the earth has not been significant ($<1\%$).

Gravitational Potential

The function U_1, called the gravitational potential at P due to m_1, can be defined as

$$U_1=G\frac{m_1}{r_1} \tag{8-7}$$

Here, U_1 is a scalar. Let us compute how it varies as the x-coordinate of P varies. (The coordinates of R_1 are kept fixed). To this effect, we first treat y and z as if they were constant, letting only x vary. We write this change of U_1 with respect to x only as $\partial U_1/\partial x$, using the symbol ∂ to indicate that y and z are kept constant (see also Appendix E). We thus need to compute $(\partial/\partial x)(1/r_1)$. By the chain rule, we obtain

$$\frac{\partial U_1}{\partial x} = \frac{\partial U_1}{\partial r_1} \times \frac{\partial r_1}{\partial x} = -Gm_1 \frac{1}{r_1^2} \times \frac{\partial r_1}{\partial x} \tag{8-8}$$

Then, from Equation 8-4, taking the square root of each side,

$$\frac{\partial U_1}{\partial x} = -\frac{Gm_1}{r_1^2} \times \frac{1}{2}\left[(x-a_1)^2 + (y-b_1)^2 + (z-c_1)^2\right]^{-1/2} \times 2(x-a_1) \tag{8-9}$$

$$= Gm_1(a_1 - x)r_1^{-3}$$

Since the RHS of this equation is identical to that of Equation 8-6, we obtain

$$F_{1x} = \frac{\partial U_1}{\partial x}$$

By induction, we immediately have

$$F_{1x} = \frac{\partial U_1}{\partial x}, \quad F_{1y} = \frac{\partial U_1}{\partial y}, \quad F_{1z} = \frac{\partial U_1}{\partial z} \tag{8-10}$$

In order to avoid writing these three equations separately, we can agree to write all three as *one* equation (see also Appendix E):

$$\mathbf{F}_1 = \nabla \mathbf{U}_1 = \operatorname{grad} \mathbf{U}_1 = -Gm_1 \frac{1}{r_1^2}\hat{\mathbf{r}}_1 \tag{8-11}$$

(The term ∇ can be read "del" or "gradient of" or "grad." It is important to recognize that equations 8-10 and 8-11 have *exactly* the same meaning; the latter is just shorthand for the former. Since the components of a vector **F** are F_x, F_y, and F_z,

$$\mathbf{F} = \hat{\mathbf{i}}F_x + \hat{\mathbf{j}}F_y + \hat{\mathbf{k}}F_z \tag{8-12}$$

It follows that

$$\nabla \mathbf{U}_1 = \hat{\mathbf{i}}\frac{\partial U_1}{\partial x} + \hat{\mathbf{j}}\frac{\partial U_1}{\partial y} + \hat{\mathbf{k}}\frac{\partial U_1}{\partial z} \tag{8-13}$$

The gradient operation can be applied to any scalar field—for example, the temperature or elevation—to yield a vector field. (A scalar field associates a scalar to any point in space inside a given domain; a vector field associates a vector to such a point. The temperature in a room is thus a scalar field, the motion of the water particles in a stream a vector field.) At any point, the gradient of the elevation (a scalar) is the slope, which is a vector. The gradient of the temperature is a vector that points in the direction of the maximum rate of change of temperature and whose magnitude is that rate of change. The gradient of U_1 may be defined in exactly the same terms as that of the temperature.

Comparing this operation on a scalar with the computation of the strain (Chapter 2, Section B), we see that the strain (a tensor) is essentially the gradient of the displacement (a vector). We conclude that the gradient of a vector is a tensor, just as the gradient of a scalar is a vector.

The gradient can often be approximately measured. To measure $\nabla \mathbf{T}$ we could measure

$$\frac{\delta T}{\delta x} \approx \frac{\partial T}{\partial x}, \quad \frac{\delta T}{\delta y} \approx \frac{\partial T}{\partial y}, \quad \text{etc.}$$

It is physically evident that the value and direction of the gradient are independent of the direction of the coordinate axes.

Reverting to Equation 8-11, we note that the equation would still be correct if U_1 was increased by a constant value rather than given by Equation 8-7. This is not done because we want to have $U_1 = 0$ for $r_1 = \infty$.

It is easy to generalize the preceding discussion when many (say, s) masses are present: we need only replace Equation 8-7 by

$$U = \sum_{i=1}^{s} U_i = G \sum_{i=1}^{s} \frac{m_i}{r_i} \tag{8-14}$$

Taking an increasing number of increasingly smaller masses, we replace m_i by $\rho\, dV$ and r_i by r. In that case,

$$U = G \int_V \frac{\rho\, dV}{r} \tag{8-15}$$

where ρ may be a function of the coordinates. To repeat: U is the potential at a point P due to a distribution of mass in the volume V external to the point—that is, there is no mass *at* P. Equation 8-11 now becomes

$$\mathbf{F} = \nabla \mathbf{U} = -\left(G \int_V \frac{\rho\, dV}{r^2} \right) \hat{\mathbf{r}} \tag{8-16}$$

and thus

$$F = -G \int_V \frac{\rho\, dV}{r^2}$$

PROBLEM

8-1 Compute the three components of the gravitational attraction at a point of coordinates ($x = 5$ m, $y = 2$ m, $z = 10$ cm) with respect to a point mass ($M =$ 5 kg) at the origin. Compute the resultant attraction. (Beware of signs!)

Work

The work dW performed by a force **F** acting over an infinitesimal distance **ds** (Figure 8-2) is

$$dW = F\cos\Theta\, ds$$

If we orient the x-axis along **ds**, this becomes

$$dW = F_x\, dx \tag{8-17}$$

In the general case in which the axes are arbitrarily oriented, we similarly can obtain

$$dW = F_x\, dx + F_y\, dy + F_z\, dz \tag{8-18}$$

If $\mathbf{F} = \nabla \mathbf{U}$, we thus obtain

$$dW = \frac{\partial U}{\partial x}\, dx + \frac{\partial U}{\partial y}\, dy + \frac{\partial U}{\partial z}\, dz \tag{8-19}$$

The calculus of more than one variable shows that the total derivative of any function $f = f(x, y, z)$ is

$$df = \frac{\partial f}{\partial x}\, dx + \frac{\partial f}{\partial y}\, dy + \frac{\partial f}{\partial z}\, dz$$

Equation 8-19 may thus be written

$$dW = dU \tag{8-20}$$

To compute the total work along an arbitrary path between A and B (Figure 8-3) *on which* **F** *is always the gradient of a potential*, we need only integrate Equation 8-20 along the path. But a definite integral of an exact

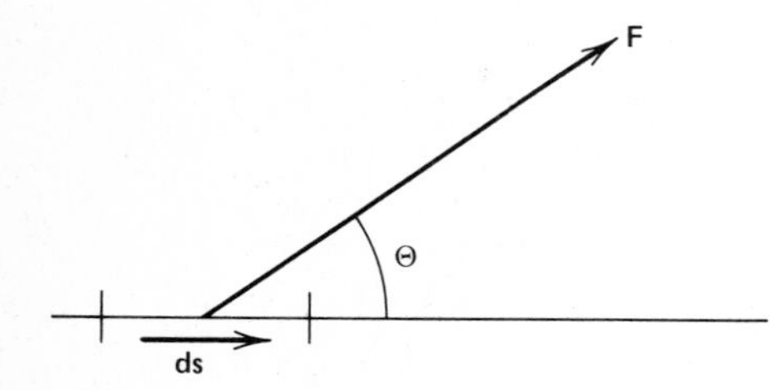

FIG. 8-2 *Force* **F** *acting over an infinitesimal path* **ds** *making an angle Θ with* **F**.

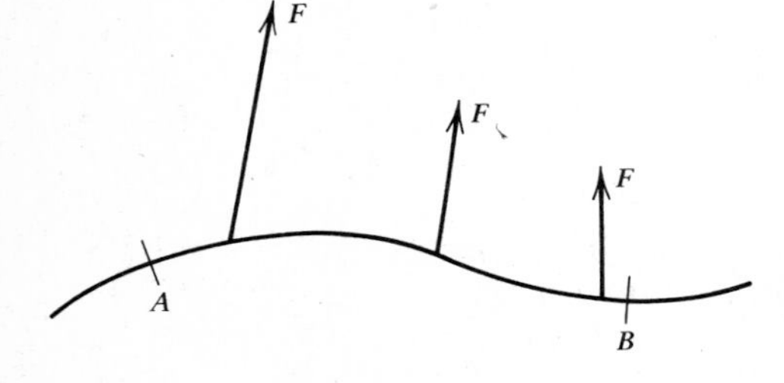

FIG. 8-3 *The force* F *varies (in magnitude and direction) along the path AB but is always the gradient of a potential.*

differential depends only on its limits, not on the path taken. Thus,

$$W = \int_A^B dW = \int_A^B dU = U_B - U_A \tag{8-21}$$

What is true for a point mass is easily generalized to an extended mass. It follows that the work performed in moving a mass parallel to itself between two points in a gravitational field is independent of the path (provided, of course, that there are no other forces). In particular, if points A and B coincide, W vanishes. A field of force that obeys Equation 8-21 is said to be conservative. Since this equation is true regardless of the form of the potential function [e.g., $U = -k/r^2$, $U = k\log(1/r)$, etc.], every force field that is the gradient of a potential is conservative. The magnetic field (Chapter 10, Section B) is another example of such a field.

A surface along which U is constant is an equipotential surface, or level surface. It takes no work to move a point on such a surface if the only force is $\nabla \mathbf{U}$, and it may be shown, on the one hand, that at any point the gravity field is perpendicular to the equipotential surface. On the other hand, the surface of a liquid at rest is normal to the force acting upon it, hence the surface of a liquid at rest is an equipotential surface. On the earth the direction of gravity is by definition vertical. The constancy of U on an equipotential does not imply the constancy of $\mathbf{F} = \nabla \mathbf{U}$ on it. Consider two equipotential surfaces, S_1 and S_2, close to each other: $U = U_1$ on S_1, $U = U_2$ on S_2. Hence, $\nabla \mathbf{U}$ is inversely proportional to the distance between S_1 and S_2. Because there is no reason for this distance to be constant, it follows that $\nabla \mathbf{U}$ may vary.

PROBLEMS

8-2 How much energy is dissipated by the earth in one orbit around the sun? Why?

8-3 Show that frictional forces are not the gradient of a potential.

8-4 Why is the surface of a liquid at rest horizontal?

Laplace's and Poisson's Equations

We already know how to compute $\partial U/\partial x$, $\partial U/\partial y$, and so forth. If we set, say, $f = \partial U/\partial x$ and compute $\partial f/\partial x = \partial^2 U/\partial x^2$, we obtain the second partial derivative with respect to x. It is easy to show (see Problem 8-5) that, outside of any mass,

$$\nabla^2 U \equiv \frac{\partial^2 U}{\partial x^2} + \frac{\partial^2 U}{\partial y^2} + \frac{\partial^2 U}{\partial z^2} = 0 \tag{8-22}$$

which is Laplace's equation. (The term $\nabla^2 U$ is read "del square U" or "Laplacian of U," after the French mathematician Pierre Simon de Laplace, 1749–1827.)

It plays a fundamental role in potential theory, heat conduction, and other fields of physics.

If P itself is an attracting mass, $U = \infty$, but nothing can immediately be asserted about $\nabla^2 U$.

To find the value of $\nabla^2 U$ at such a point, we will successively compute the potential of the following:

1. A spherical shell on an exterior point.
2. A sphere on an exterior point.
3. A spherical shell on an interior point.
4. A sphere on an interior point.

Attraction of a Spherical Shell on an Exterior Point Consider a thin homogeneous spherical shell of thickness t centered at O and attracting an exterior point P (see Figure 8-4).

If $OA = a$ and $OP = r$ $(r > a)$, the mass of the shell element shown in the figure is

$$\rho\, dV = \rho(2\pi a \sin\Theta) \times (a\, d\Theta) \times t$$

Its attraction at P is (in absolute value)

$$dF = 2\pi G\rho t(a^2 \sin\Theta \times d\Theta)x^{-2}\cos\alpha \tag{8-23}$$

since, by symmetry, it is directed along PO. Since x, α, and Θ are not independent variables, two of them must be eliminated. We decide to eliminate α

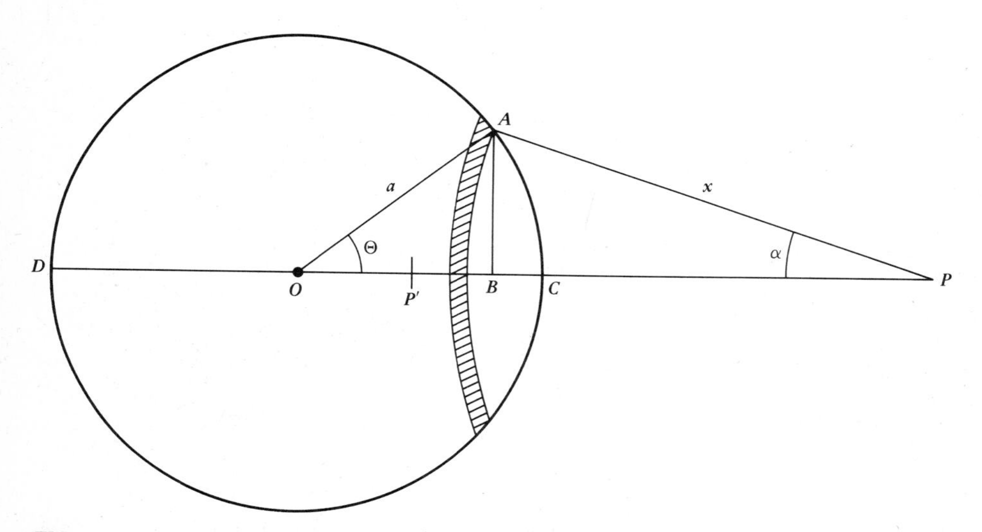

FIG. 8-4 A thin spherical shell with center O, a shell element (hachured), an arbitrary outside point P, and an arbitrary inside point P′.

and Θ. Since

$$x \cos \alpha = r - a \cos \Theta$$

we find

$$\cos \alpha = \frac{r - a \cos \Theta}{x} \tag{8-24}$$

Similarly,

$$x^2 = r^2 + a^2 - 2ra \cos \Theta \tag{8-25}$$

This equation eliminates Θ from Equation 8-24, which may then be substituted in Equation 8-23. Furthermore, by differentiation of Equation 8-25 (note that a and r are fixed),

$$\sin \Theta \, d\Theta = \frac{x \, dx}{ra} \tag{8-26}$$

This equation can also be substituted in Equation 8-23, yielding

$$dF = \pi G \rho t \frac{a}{r^2} \left(\frac{r^2 - a^2}{x^2} + 1 \right) dx \tag{8-27}$$

Thus,

$$F = \int_C^D dF = \pi G \rho t \int_{r-a}^{r+a} \frac{a}{r^2} \left(\frac{r^2 - a^2}{x^2} + 1 \right) dx$$

which yields

$$F = \frac{4\pi G \rho t a^2}{r^2} \tag{8-28}$$

But the mass of the shell, for small t, is

$$M = 4\pi \rho a^2 t \tag{8-29}$$

Hence,

$$F = \frac{GM}{r^2} \quad \text{and thus} \quad U = \frac{GM}{r} \tag{8-30}$$

Note that ρ must be constant: if it were variable, the variation would have to be taken into account in the integration process.

We thus conclude that a homogeneous thin spherical shell attracts an exterior point as if its mass were concentrated at its center.

Attraction of a Sphere on an Exterior Point Consider a sphere made up of a succession of thin homogeneous spherical shells. From the preceding discussion, we conclude that its attraction upon an exterior point is the same as if the mass were concentrated at the center. The density of the sphere may vary with depth (i.e., with the radius), but not with the longitude or the latitude. If the density of the sphere is constant, then its mass is $M = (4/3)\pi\rho a^3$.

$$F = \frac{4\pi G\rho a^3}{3r^2} \quad \text{and} \quad U = \frac{G}{r} \times \frac{4}{3}\pi\rho a^3 \tag{8-31}$$

All of this is valid only for an exterior point.

Today, these results can be obtained easily. Yet in order to demonstrate the law of gravitational attraction, Newton first had to invent calculus, which took quite a few years.

Potential of a Spherical Shell at an Interior Point Equation 8-27 is independent of the position of P, but when we integrate it for an interior point (say, P' in Figure 8-4), the limits become $a - r$ and $a + r$. That integral vanishes; the field of a homogeneous spherical shell is thus zero inside the shell.

But $\mathbf{F} = \nabla \mathbf{U}$, hence U is constant inside the shell. By symmetry, at the center, for small t, Equation 8-7 yields

$$U = \frac{GM}{a} = 4\pi G\rho a t \tag{8-32}$$

To obtain U inside a thick homogeneous spherical shell with inner radius r_1 (Figure 8-5) and outer radius r_2, we replace t by da in Equation 8-32 and integrate it.

$$U = \int_{r_1}^{r_2} 4\pi G\rho a\, da = 2\pi G\rho\left(r_2^2 - r_1^2\right) \tag{8-33}$$

which is constant.

Potential of a Sphere at an Interior Point To find the potential of a sphere at an interior point P, we subdivide the sphere into two regions: an inner sphere S_1 of radius $OP = r_1$ (Figure 8-5) with potential U_1 at P; and a spherical shell S_2. Then $U = U_1 + U_2$. Hence, supposing that ρ is constant, we obtain, from

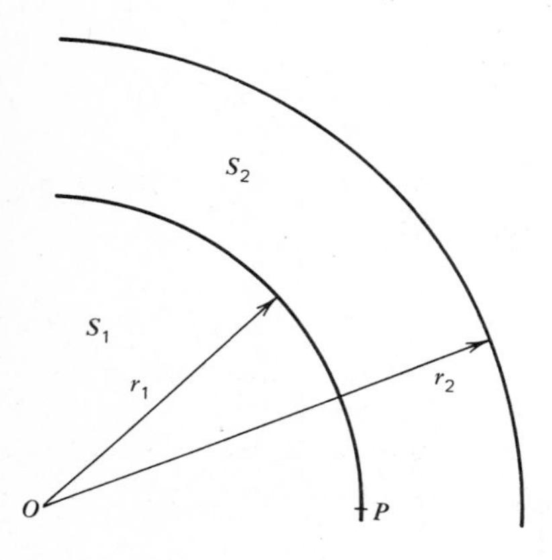

FIG. 8-5 A sphere with an interior point P may be divided into an inner sphere S_1 and a spherical shell S_2.

Equations 8-31 (with $a = r = r_1$) and 8-33,

$$U = \frac{4}{3}\pi G\rho r_1^2 + 2\pi G\rho(r_2^2 - r_1^2) = \frac{2}{3}\pi G\rho(3r_2^2 - r_1^2) \tag{8-34}$$

Poisson's Equation We are now able to compute $\nabla^2 U$ at P if P is "part of" the material. (Note that if P is not part of the material, but merely surrounded by it, $\nabla^2 U = 0$. This is true, of course, no matter how close the material is to P, provided it is not *at* P.) We divide the body (Figure 8-6) into two parts; a small sphere S_a of radius a centered at O (d, e, f) and containing P; and the rest of the body S_b. Let $OP = r$ and take a small enough that ρ is constant inside S_a; again $U = U_a + U_b$, and thus $\nabla^2 U = \nabla^2 U_a + \nabla^2 U_b$.

Since P is not in S_b, $\nabla^2 U_b = 0$ at P. U_a is given by Equation 8-34, in which $r_2 = a$, $r_1 = r$, and

$$r^2 = (x-d)^2 + (y-e)^2 + (z-f)^2 \tag{8-35}$$

Hence

$$\frac{\partial U_a}{\partial x} = -\frac{4}{3}\pi G\rho(x-d) \tag{8-36}$$

with similar expressions for $\partial U_a/\partial y$ and $\partial U_a/\partial z$. Hence,

$$\frac{\partial^2 U_a}{\partial x^2} = -\frac{4}{3}\pi G\rho$$

and

$$\nabla^2 U_a = -4\pi G\rho$$

Thus,

$$\nabla^2 U = -4\pi G\rho \tag{8-37}$$

This is Poisson's equation, which is valid *at* a point where the density is ρ. It is a generalization of Laplace's equation (so to speak: Poisson → Laplace when

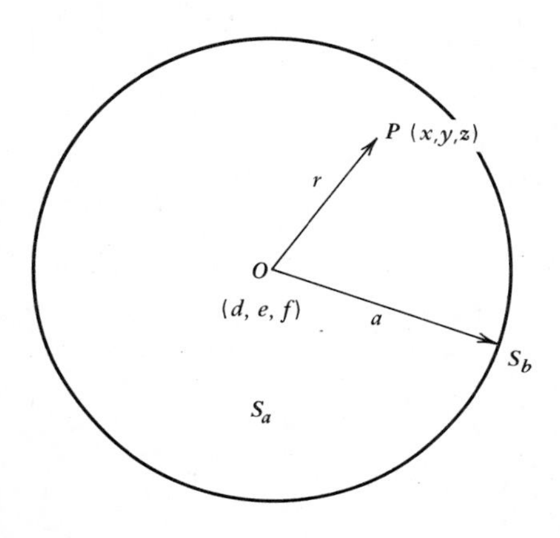

FIG. 8-6 *An arbitrary body containing a point P may be divided into a small sphere S_a centered at O and containing P, and the rest of the body S_b.*

$\rho \to 0$). The right-hand side of the equality is sometimes referred to as the source-term.

We will see that the Laplacian operator can be easily approximated, just as the partial derivatives can be.

PROBLEMS

8-5 Show that at P outside any mass,

$$\nabla^2 U \equiv \frac{\partial^2 U}{\partial x^2} + \frac{\partial^2 U}{\partial y^2} + \frac{\partial^2 U}{\partial z^2} = 0$$

if

$$U = \frac{1}{r} \left(\text{or } Gm\frac{1}{r} \right)$$

8-6 At a point, the potential U caused by two masses is the sum of the potentials caused by each mass ($U = U_1 + U_2$). Is this also true of the gravitational fields? That is, does $F = F_1 + F_2$? Why or why not?

9

The Earth's Gravity Field

Gravity measurements are made principally to determine their departure from the normal value, in order to infer variations of density at depth. It is thus necessary to define what this normal value is and where it is to be measured.

A. The Geoid

The first reference surface is the geoid, the equipotential surface that coincides with the surface of the oceans at rest. This, of course, is an idealized state: winds, salinity, temperature, and ocean currents produce important deviations (up to ~ 2 m in the mean sea level, much more over short periods) from the "perfect" geoid.

The geoid may be theoretically extended under the continents by "digging" imaginary, infinitesimally thin canals and by observing the water level in them. In practice, the geoid is determined by precision spirit leveling (orthometric leveling), in which the difference in height is measured along successive stations (Bomford 1952, p. 98; Vanicek et al. 1980). Because light rays may be bent by refraction caused by changes in temperature along the line of sight, such leveling is difficult to perform with a high degree of precision (say, better than 1 mm vertically for 1 km horizontal distance). The elevation shown on topographic maps is given with respect to the geoid. The determination of the shape of the geoid is the object of geodesy.

As stated above, the geoid does not coincide everywhere with the local mean sea level (MSL). Precision leveling across the Isthmus of Panama has shown that the mean sea level on the Pacific side is about 20 cm higher than that on the Atlantic side (mostly as a result of salinity differences); similarly, the Gulf Stream causes a difference in elevation of about 55 cm between Miami Beach, FL., and Cat Cay in the Bahamas, the latter being higher (Lisitzin 1965). Fortunately, the differences between the MSL and the geoid are small enough to be generally ignored. Thus, elevations can be given "relative to the MSL," rather than "relative to the MSL at a specified place." Obviously, a point along the Panama Canal has an elevation 20 cm higher relative to the Atlantic MSL than to the Pacific MSL.

B. The Reference Ellipsoid

One could, of course, define the shape of the geoid by reference to a sphere, but this would lead to large differences. Instead, the reference figure adopted is that of an ellipsoid of revolution, sometimes called the spheroid, with a flattening f determined from satellite orbits.

$$f = \frac{R_{eq} - R_{po}}{R_{eq}} \tag{9-1}$$

where R_{eq} and R_{po} are the equatorial radius and the polar radius, respectively. For many years f was computed to be $\sim 1/297$. It is now known to be 298.247. It is remarkable that one of the pioneers of geodesy, Hayford, gave a similar value (298.2) in 1910.

The radius vector of this ellipsoid (the distance from the center to a point on the surface) is given by

$$r = R_{eq}(1 - f\sin^2\phi) \tag{9-2}$$

where ϕ is the latitude and $R_{eq} = 6378.139$ km. On this ellipsoid the reference gravity, g_o, has been theoretically computed; a good derivation is given by Officer (1974, pp. 269–273). In SI units,

$$g_o = 9.7803185[1 + (5.278895 \times 10^{-3}\sin^2\phi) + (23.462 \times 10^{-6}\sin^4\phi)] \tag{9-3}$$

(for $\phi = 45°$, $g_{45°} = 9.8062$ m/s^2). This formula was internationally adopted in 1967, but from 1930 until that time (and sometimes even at present), another reference gravity formula was used:

$$g_o = 9.78049[1 + (5.2648 \times 10^{-3}\sin^2\phi) + (23.6 \times 10^{-6}\sin^4\phi)] \tag{9-4}$$

The variation of g_o with ϕ in these two equations is caused by a combination of two factors, the first of which is the direct effect of the earth's rotation. Figure 9-1 shows that g_ϕ at a point of latitude ϕ on the surface should be

$$g_\phi = g_{\text{pole}} - \omega^2 r \cos\phi \tag{9-5}$$

where $r = R_{\text{eq}} \cos\phi$, and ω is the radial frequency of rotation of the earth [$\omega = 2\pi/(24 \times 3600)$]. Hence, Equation 9-5 may be written

$$g_\phi = g_{\text{pole}} - \omega^2 R_{\text{eq}} \cos^2\phi$$

At the equator ($\phi = 0$), this becomes

$$g_{\text{eq}} = g_{\text{pole}} - \omega^2 R_{\text{eq}}$$

hence,

$$g_{\text{pole}} = g_{\text{eq}} + \omega^2 R_{\text{eq}}$$

Thus,

$$\begin{aligned} g_\phi &= g_{\text{eq}} + \omega^2 R_{\text{eq}} - \omega^2 R_{\text{eq}} \cos^2\phi \\ &= g_{\text{eq}} \left[1 + \frac{\omega^2 R_{\text{eq}}}{g_{\text{eq}}} (1 - \cos^2\phi) \right] \\ &= g_{\text{eq}} \left[1 + (3.446 \times 10^{-3} \sin^2\phi) \right] \end{aligned} \tag{9-6}$$

A comparison of Equations 9-3 and 9-6 shows that the direct influence of the earth's rotation accounts for about 60% (3.4 versus 5.3) of the variation of g with ϕ. The remaining ~ 40% stems in part from the difference between the polar and the equatorial radii (~ 21.4 km), and in part from the difference between the shape of the Earth and that of a sphere.

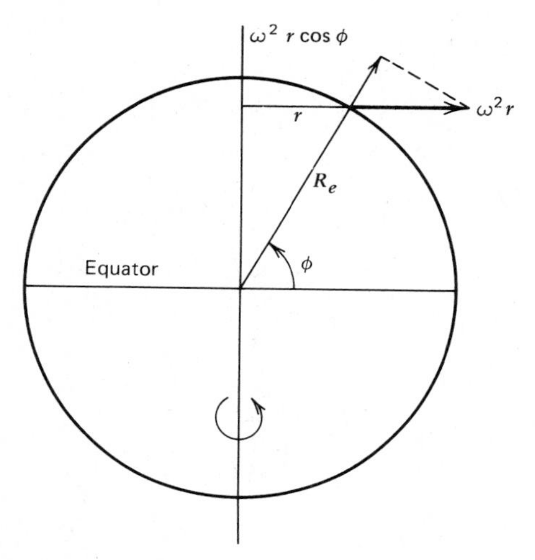

FIG. 9-1 *A spherical earth and the latitude ϕ.*

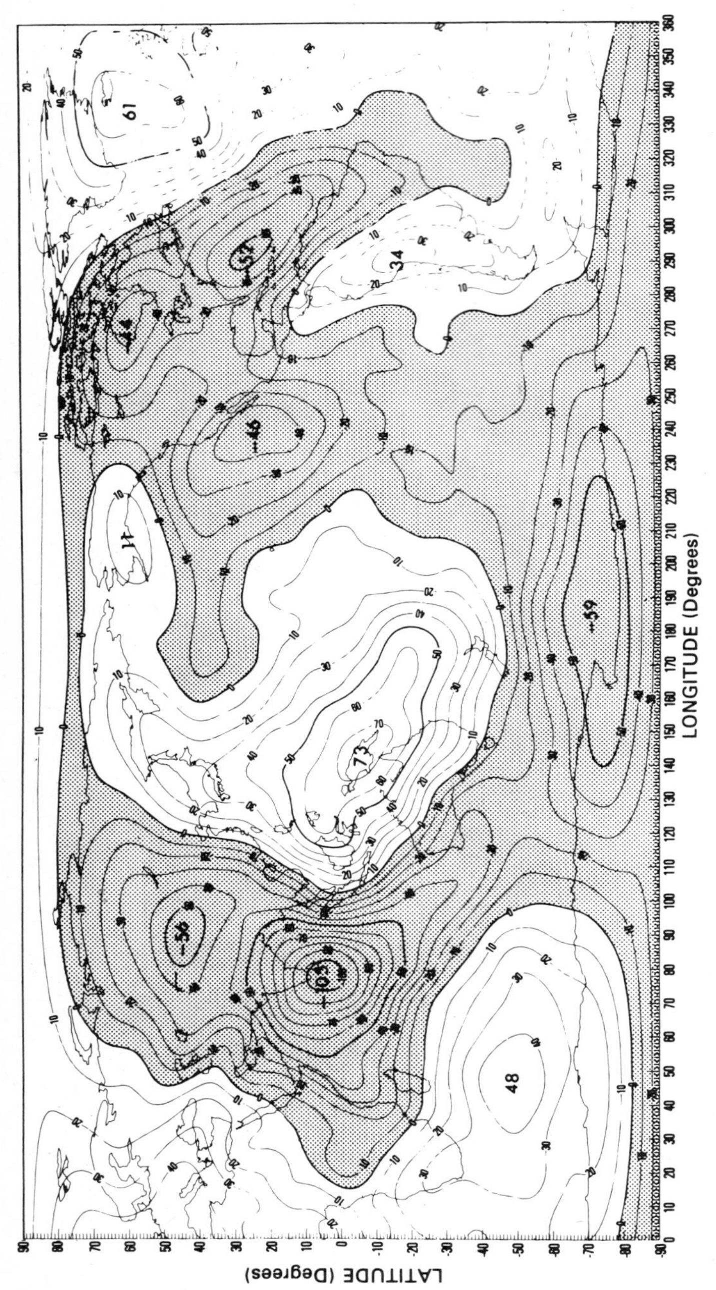

FIG. 9-2 *Height in meters of the geoid surface above— or below (in gray)— the reference ellipsoid. (From Lerch et al. 1979. Copyright © by the American Geophysical Union.)*

A recent map giving the height of the geoid above (white) and below (gray) the reference ellipsoid is shown in Figure 9-2. The two extremes are a "hole" 105 m deep south of India, and a "hill" 73 m high centered near New Guinea. The most remarkable feature of this map is the apparent absence of correlation between the contours of geoid height and other features on the earth's surface—continent-ocean boundaries, midocean ridges, mountain chains, and so on. The exact significance of these contour lines has so far resisted attempts at explanation. They are presumably correlated with large-scale heterogeneities in the deep mantle. Indeed, seismic waves such as *SS*, reflected at the earth's surface in the region of the Indian Ocean "hole," arrive about 3 s earlier than normal. The shear wave velocity in that area is thus abnormally high, but the causal relation between the geoid depression and abnormal seismic velocity remains speculative (Stark and Forsyth 1983).

If we consider only the very longest wavelengths of the geoid, a symmetric "tennisball" pattern seems to appear, conceivably related to the dispersal of Pangea (LePichon and Huchon 1984). If such a relation can be confirmed and extended, it may open the way for significant insights into the dynamics of the earth's interior. At short wavelengths (less than a few hundred kilometers), the correlation between geoid height and bathymetry is quite clear (see Section 9C).

Contrary to what is sometimes thought, in precision leveling there is no reference to stars. The reason is that leveling is aimed at the determination of the real shape of the geoid, whereas the direction of a star is known only with respect to the ellipsoid. The angular difference between the local vertical (perpendicular to the level surface as determined by the spirit level) and the normal to the reference ellipsoid (determined by star observations) is the deflection of the vertical.

C. Gravity Anomalies

In SI units, the gravity field g is measured in m/s^2 and its variations in gravity units (1 g.u. = 10^{-6} m/s^2), especially in Europe. In the United States the most commonly used unit is the gal, named after Galileo (1 gal = 1 cm/s^2) and the mgal (1 mgal = 10^{-5} m/s^2 = 10 g.u.). We always count g positive downward.

An anomaly γ_i of type i (see below) is defined as follows.

$$\gamma_i = g_{obs} - g_{exp} \qquad \text{or} \tag{9-7}$$

$$\gamma_i = g_{obs} \pm \delta g_i - g_{ref} = (g_{obs} \pm \delta g_i) - g_{ref}$$
$$= g_{obs} - (g_{ref} \mp \delta g_i)$$

where g_{obs} is the observed value; g_{exp} is the expected value; g_{ref} is the value at the same latitude on the reference ellipsoid; and δg_i is a correction or corrections of type i, always taken as positive (this convention is not universal). The terms of this equation will be made clear shortly. The types of anomalies and corrections are free air, Bouguer, and isostatic.

Since one often wishes to know γ_i to a precision better than 0.1 mgal (i.e., $10^{-6}\ g$), gravity must be measured in the field to a precision on the order of 10^{-7}. It is remarkable that such a precise measurement can be routinely attained in a few minutes and by purely mechanical means.

Free-Air Correction and Anomaly

The purpose of the free-air correction is to compensate for the decrease of gravity with the increase of elevation. For this purpose, we take *only* the elevation into account. We may thus say that we should either add a correction to the observed value and compare the result to the reference value (second line of Equation 9-7), or subtract a correction from the reference value before comparing it to the observed value (third line of Equation 9-7). (The $\pm$ and $\mp$ signs in that equation correspond to the fact that some corrections are added and others subtracted.)

As previously shown, a sphere possessing radial symmetry attracts an outside point P as if its mass were concentrated at its center, O: that is,

$$g = -\frac{GM}{r^2}$$

or, vectorially,

$$\mathbf{g} = -\left(\frac{GM}{r^2}\right)\hat{\mathbf{r}}$$

where $\mathbf{r} = \mathbf{OP}$, and

$$\frac{d\mathbf{g}}{dr} = \frac{2Gm}{r^3}(\hat{\mathbf{r}}) = \left[-\left(\frac{GM}{r^2}\right)\hat{\mathbf{r}}\right]\left[-\frac{2}{r}\right] \tag{9-8}$$

Except in the case of satellites, $r \cong R_e$. Writing $\mathbf{g}_o = -(GM/R_e^{\,2})\hat{\mathbf{r}}$ for the surface value, we obtain

$$\frac{dg}{dr} \cong -\frac{2g_o}{R_e} \cong -0.3086 \text{ mgal/m} \tag{9-9}$$

or, approximately,

$$\delta g \cong -\left(\frac{2g_o}{R_e}\right)\delta h$$

The value of gravity decreases by 0.3086 mgal/m as r or h (the elevation) increases. To compensate for this decrease, we must *add* a corresponding free-air correction

$$\delta g_{\mathrm{FA}} = +0.3086h \tag{9-10}$$

for δg_{FA} in milligals and h in meters. (This value is an average that has been used for many decades. It is based on the 1930 gravity formula and on a flattening of

1/297. The correct value ranges from $0.3072h$ at the equator to $0.3104h$ at the pole.) Below sea level ($h < 0$), of course, $\delta g_{FA} < 0$. The free-air correction is sometimes called the "elevation correction."

The free-air anomaly is thus

$$\gamma_{FA} = g_{obs} + \delta g_{FA} - g_{ref} \tag{9-11}$$

Partly for historical reasons, the free-air anomaly is not extensively used. Nevertheless, it is much preferable to the Bouguer anomaly, at least for correcting the effects of longer wavelengths (greater than approximately 200 km) in the topography. It is sometimes called the Faye anomaly.

The magnitude of the free-air correction explains the reason for the difficulty and cost of gravity surveys. If a final precision of 0.1 mgal is desired, all corrections must be exact to at least 0.03 mgal—which implies that the elevations must be known with an accuracy of 10 cm. Accurate leveling and precisely determined stations are thus necessary.

PROBLEM

9-1 What is the rate of change of g with depth as you go down a mine shaft that starts at sea level? Is it the same as the rate of change in free air? Why, or why not?

Bouguer Correction and Anomaly

In the eighteenth century, a controversy arose over whether the earth was flattened (as we now know to be the case) or elongated along its polar axis. Figure 9-3 shows that on a flattened ellipsoid the same angle Θ between two verticals corresponds to a greater length near the pole (AB) than near the equator (CD). In 1735 the French Academy of Sciences decided to settle the controversy by sending two expeditions to measure the length of 1° of meridian—one near the North Pole, in Lapland, and one near the equator, starting from Quito in present-day Ecuador, which was then part of the vice-royalty of Peru. Bouguer was one of the two leaders of the expedition to Peru, and he observed that the timekeeping of his pendulum clock was less affected than he had expected by the mass of the mountains. Nevertheless, he proceeded to take the attraction of the whole mountain mass into account.

In effect, the Bouguer correction assumes that mountains are just added onto the surface of the standard earth. Furthermore, in order to simplify the problem, a mountain at the elevation of the station is supposed to extend to infinity horizontally (Figure 9-4). To compute the attraction of this infinite slab at the station S (which is just above it, but not *exactly* on its surface), we observe that at S (outside the slab—index o).

$$\nabla^2 U_o = \frac{\partial^2 U_o}{\partial x^2} + \frac{\partial^2 U_o}{\partial y^2} + \frac{\partial^2 U_o}{\partial z^2} = 0$$

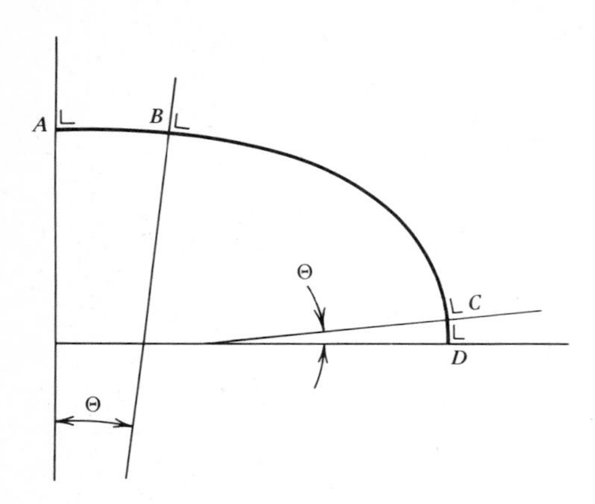

FIG. 9-3 On a flattened ellipsoid the same angle Θ between two verticals subtends a greater arc length near the pole (AB) than near the equator (CD).

But nothing changes along x and y, hence

$$\frac{\partial^2 U}{\partial x^2} = \frac{\partial^2 U}{\partial y^2} = 0 \tag{9-12}$$

Hence, $\partial^2 U_o/\partial z^2 = 0$. But $g = \partial U_o/\partial z$, since g is by definition along the z-axis. Hence, $\partial^2 U_o/\partial z^2 = \partial g_o/\partial z = 0$. Hence, the gravity g_o outside of the slab is constant. To find its value, we use Poisson's equation. Again, Equation 9-12 holds true, but inside the slab (index i),

$$\frac{\partial^2 U_i}{\partial z^2} = \frac{\partial g_i}{\partial z} = -4\pi G\rho$$

Hence,

$$g_i = -4\pi G\rho z + C \tag{9-13}$$

If z points upward and $z = 0$ at sea level, then in the middle of the slab $z = h/2$ where, by symmetry, $g_i = 0$. But there, from Equation 9-13,

$$g_i = 0 = -4\pi G\rho(h/2) + C$$

Using the resulting value of C in Equation 9-13, we find that at $S(z = h)$, where $g_i = g_o$,

$$g_o = g_i = -2\pi G\rho h \tag{9-14}$$

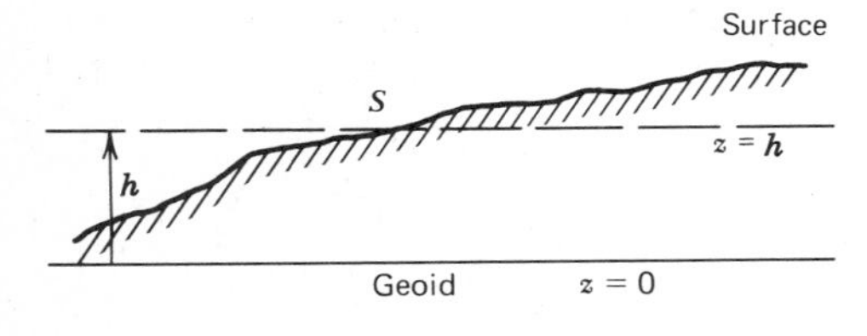

FIG. 9-4 Principle of the Bouguer correction for a station S.

which is the downward attraction caused by the slab. Hence,

$$\delta g_B = +2\pi G\rho h \text{ (in SI units)} \tag{9-15}$$
$$= 4.185 \times 10^{-5}\rho h \text{ (in mgal/m)}$$

when ρ and h are in SI units. To compensate for this downward attraction, we must subtract δg_B from the observed value for $h > 0$. Thus, the simple Bouguer anomaly may be defined as

$$\gamma'_B = g_{obs} + \delta g_{FA} - \delta g_B - g_{ref} \tag{9-16}$$

We observe that for $\rho = 2600$ kg/m^3, $\delta g_B = 0.1$ mgal/m, or one-third of δg_{FA}.

Note that the approximation of mountains by a horizontal slab is usually adequate: slopes are normally so small, and the vertical attraction falls off so rapidly with the horizontal distance, that a slab generally provides a sufficiently good representation of the topography.

In high-precision gravity work, in which the anomalies of interest are less than 0.1 mgal, or in mountainous areas where a 1-mgal precision is desired, an additional correction must be made in computing γ_B. It is the terrain or topographic correction. If the station is near a valley, δg_B will be too large, because it will include the attraction of a mass filling the valley. This mass is not there; thus, we must add a δg_T. If, however, the station is at the foot of a mountain (which is not taken into account in δg_B), the attraction of the mountain causes a decrease in g_{obs}, and we thus must also add a δg_T to compensate. Hence, δg_T is always added, and the complete Bouguer anomaly is defined as

$$\gamma_B = g_{obs} + \delta g_{FA} - \delta g_B + \delta g_T - g_{ref} \tag{9-17}$$

The terrain correction is determined either by hand, by reading the average elevation of various segments of circular zones from topographic maps at each station (Dobrin 1976, pp. 419–423), or from a digital terrain map, provided its scale is sufficiently large.

Over mountain ranges the Bouguer anomalies are much larger than the free-air ones. They are also strongly negative—on the order of 100 mgal for each km of elevation. As will be shown, this is a result of Equation 9-15 when $\rho \cong 2600$ kg/m^3. In general, Bouguer anomalies show a strong inverse correlation with the long-wavelength topography ($\Lambda \gtrsim 100 \ldots 200 \ldots$ km). In contrast, the free-air anomalies are small and positive over mountain ranges. We must conclude that δg_{FA} has added a little too much to g_{obs}, but that δg_B clearly has subtracted too much. It may seem odd that δg_{FA}, which takes *no* account of the mountains, would be slightly too large, but this fact will be clarified shortly. Yet the fact that δg_B is much too large immediately suggests that the density of mountains (or of something connected with them) is less than that of the average material.

From what has been said about the correlation between the Bouguer and free-air anomalies on the one hand, and the topography on the other, we may infer that the choice of the "best" anomaly depends on the scale of the feature of interest. For features of long wavelength (greater than approximately 200 km), it is better to use the free-air anomaly, because the Bouguer anomaly is essentially a mirror image of the topography (at an appropriate vertical scale). For short wavelengths, the Bouguer anomaly is preferable.

Both the Bouguer and the topographic corrections require a determination of the densities of rocks at various depths below the surface. There is no completely satisfactory method of determining these densities in all cases. For short wavelengths (less than approximately 10 km), the most satisfactory method is to rely on the fact that isostatic adjustments do not operate at this scale. Consequently, the best density is the one that yields the gravity anomaly least correlated with the topography. In practice, it generally suffices to run a single profile in an area of relatively high relief and to find the density that gives the flattest gravity profile.

For long wavelengths (e.g., a correction in the Rocky Mountains), one must resort to average values. Appendix C contains a table of minimum, average, and maximum densities for a few rock-types. All rock densities increase with depth, largely as a result of the decrease in porosity.

Isostatic Correction and Anomaly

From the preceding discussion it is clear that mountains are supported by a density difference somewhere—or, more precisely, that above a certain depth the masses of all columns are approximately equal. Although, as is now clear, no single explanation is valid everywhere, simple models remain useful as references. Of the two most common models, Pratt's assumes that above a certain depth, known as the depth of compensation or level of compensation (Figure 9-5*a*), the density of each column is inversely proportional to its height, so that the "lithostatic pressure," $\rho_i g h_i$, at that level is constant. (The existence of a lithostatic pressure implicitly assumes that the rocks are fluid.) The level of compensation is generally assumed to be ~100 km below the geoid (see the discussion below). There is no obvious geological basis for this scheme.

Airy's scheme supposes that the mountains have "roots" (Figure 9-5*b*). The equality of lithostatic pressure below the level of the deepest roots may then be written $\rho_1(h_1 + d_1) = \rho_2 d_1$. Hence,

$$\frac{d_1}{h_1} = \frac{\rho_1}{\rho_2 - \rho_1} \tag{9-18}$$

The depth of the roots is thus inversely proportional to the density contrast. For $\rho_2 = 3270\ \mathrm{kg/m^3}$ and $\rho_1 = 2840\ \mathrm{kg/m^3}$, the roots are thus about seven times larger than the topography. In Heiskanen's version of the Airy scheme, $\rho_1 = 2670$, $\rho_2 = 3270$; hence, $d_1/h_1 = 4.5$. In Airy's scheme the depth of compensation may be defined as the bottom of the deepest root.

Two somewhat related criticisms of the isostatic hypotheses have been made. First, it has been pointed out that both presume that the compensation is purely local, which is clearly absurd. However, the relief is estimated only for zones at least a few kilometers or more in radius, and so the matter is not quite so bad as it may seem. The second criticism is that both hypotheses assume that the

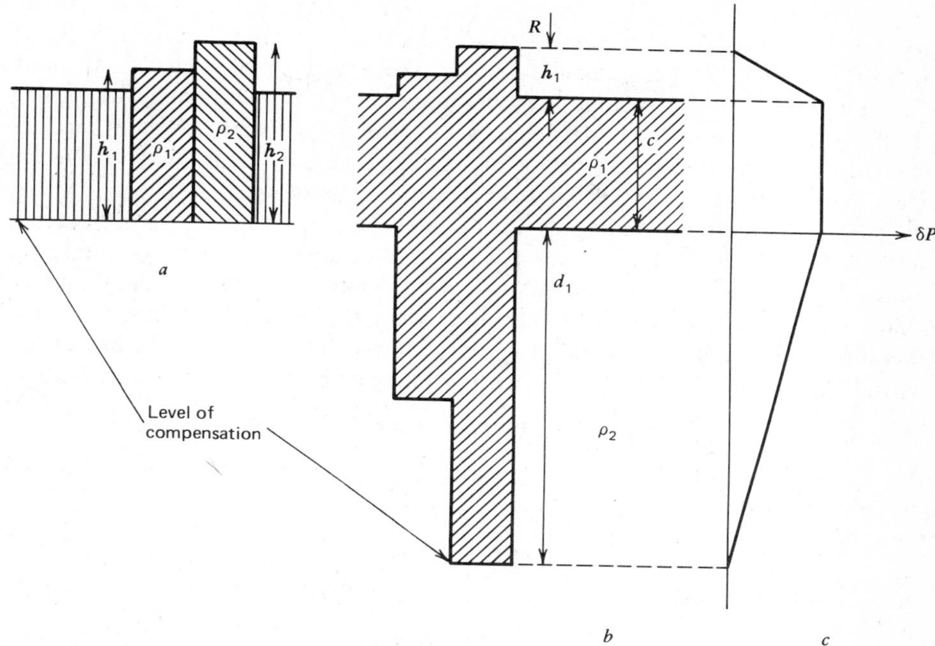

FIG. 9-5 *(a) Pratt's and (b) Airy's schemes of isostatic compensation. (c) Difference in lithostatic pressure versus depth below point R in the Airy case.*

adjustment is perfect vertically, which presumes that the material involved has no shear strength in that direction, whereas the inequality of densities (in Pratt's scheme) or of depth of roots (in Airy's) causes large shear stresses between adjacent columns. For Pratt's scheme the analogy would be to columns of, say, alcohol, water, and oil, which would quickly rearrange themselves into layers. For the same reason, deep, light roots would cause large stresses. These stresses are supposed to persist indefinitely. As a first approximation, it may thus be said that these isostatic hypotheses assume that the resistance of the material to horizontal deformation is much greater than its resistance to vertical deformation.

Further thought on the subject, taking into account Hooke's law, shows that we should really talk about resistance to strain rather than to deformation. Finally, it is probable that what is involved here is the rate of strain versus time, rather than the strain itself. Since a relation between stress and rate of strain characterizes viscous behavior (see Chapter 12, Section B), we may say that isostasy assumes that the material above the depth of compensation is anisotropic with respect to viscosity. Recall (Chapter 2, Section C) that shale is anisotropic with respect to elasticity. In both cases, all horizontal directions (or all those parallel to the bedding plane of the shale bed) are equivalent; such materials are thus said to be transversely isotropic.

The most remarkable aspect of this situation is that the need for this peculiar constitutive relation was not realized for more than a century. Credit for drawing attention to this problem goes to the Soviet geophysicist Artyushkov (1973). One may thus conclude that mountain roots are maintained by ongoing processes and/or are slowly decaying in the absence of these processes, or that such anisotropic behavior is real—for example, because nearly vertical faults make vertical displacements much easier than horizontal ones.

It is instructive to compute the difference in hydrostatic (or lithostatic) pressure between adjacent columns as a function of depth, but such computations are only approximate, since they assume that the material is an inviscid fluid (see Chapter 12) and is at rest. However, they yield the order of magnitude of the phenomenon. For the Airy case (Figure 9-5*b*) the difference in pressure versus depth below point R (shown in Figure 9-5*c*) may be computed from the equation

$$\delta P = \left[\rho_1(h_1 + c + z) - (\rho_1 c + \rho_2 z)\right] g \quad \text{for} \quad z \geq 0$$

where c is the normal thickness of the crust, and z is the depth below the base of the crust. Hence,

$$\delta P = \left[\rho_1 h_1 - (\rho_2 - \rho_1) z\right] g \tag{9-19}$$

which is maximum for $z = 0$: that is, at the base of the crust,

$$\delta P_{\mathrm{M}} = \rho_1 h_1 g$$

This is also the value at sea level, and δP can thus be immediately drawn.

For reasonable values of ρ_1 ($\sim 2.800\ \mathrm{kg/m^3}$) and h_1 (5 km),

$$\delta P_{\mathrm{M}} \approx 1.4 \times 10^8\ \mathrm{Pa} = 1.4\ \mathrm{kbar}$$

This result is clearly a maximum, as it supposes a 5 km vertical relief. With a more gradual relief, the differences in lithostatic pressure might be decreased by a factor between 3 and 5, to yield a maximum difference on the order of 0.4 kbar. Although this result is only a rough approximation, it agrees well with that of Jeffreys (1959, p. 202), who finds 0.45 kbar by a more sophisticated analysis.

A third scheme of isostatic compensation was proposed by the Dutch geophysicist Felix Andries Vening Meinesz, the first scientist to make gravity measurements at sea. (To minimize the accelerations caused by ship motions, he convinced the Dutch government to let him operate in a submarine. At the time —ca. 1930—submarines were very cramped; Vening Meinesz was very tall, and it is said that he could never stand up when at sea.) He suggested that the lithosphere acts approximately as an elastic plate floating on a liquid (the so-called Boussinesq problem). Under a point load (Figure 9-6), such a plate takes a shape that can be easily computed if the parameters of the system (densities, thickness of the plate, elastic constants) are known (Heiskanen and Vening Meinesz 1958, pp. 321–326). The distance from the center to the nearest point of zero deflection is the radius of regionality R_r. Although this regional compensation idea is probably valid in oceanic areas, its applicability to continents appears dubious (Forsyth 1979; Turcotte and Schubert 1982, pp. 112–133).

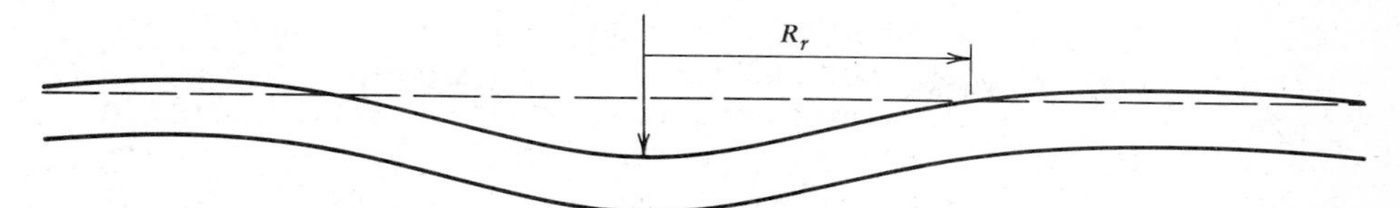

FIG. 9-6 *Regional isostatic compensation scheme of Vening Meinesz; R_r is the radius of regionality.*

Regardless of the type of isostatic compensation adopted, the isostatic correction is applied only in studies involving continent-size areas. In such a case, we are, in essence, taking into account the effect of the isostatic compensation—adding, for example, a correction for the presence of mountain roots. The difficulty then arises that this correction must take into account not only the vicinity of the station, but also the whole earth, which is very time-consuming. Needless to say, in such a case the effect of the sphericity of the earth must also be taken into account.

It is now clear that all isostatic compensation schemes are simplifications; consequently, they have lost much of their early appeal. Gravimetric interpretations now often attempt to account for free-air anomalies on the basis of the geologic structure (see below).

Correction for Drift and Tides

Most gravimeters (see Section 9D) exhibit a slow, more or less steady drift of their zero when left in place. In order to compensate for this drift, gravity measurements must be periodically repeated at one or more base stations (a procedure called looping); the drift is distributed among the stations under the assumption that it is linear with time. This correction is usually made before any other.

In addition, the luni-solar attraction causes a small, almost periodic variation (period ~ 12 h) of gravity amounting to ≤ 0.2 mgal. There are two ways of correcting for the luni-solar attraction if such precision is required. First, because the instrument normally returns to the base station much more frequently than every 6 h, the correction for drift automatically includes that for tides. Also, tables giving the luni-solar attraction are available, although they are now rarely used. The lunar component is the largest, because of the much greater proximity of the moon than of the sun.

Contrary to what might be thought, the tides are not caused solely by the attraction of the sun (or the moon) on the oceans, since the tidal bulge occurs both on the side of the earth facing the sun (or the moon) and on the opposite side: that is, the tides normally occur *twice* a day. Simple qualitative reasoning shows why. Consider first a point on the side of the earth facing the sun. If this point were not on the earth, it would need to orbit the sun more rapidly to stay in orbit, because it would be closer to the sun than is the center of the earth. (For the same reason, the space shuttle orbits around the earth in a period much shorter than that of more-distant satellites.) Because of the slower orbital velocity imposed upon it by the earth, this point tends to fall toward the sun—hence the tidal bulge. For a point on the opposite side of the earth the orbital velocity is too large, and the point tends to fly off into space. The lunar tides have the same explanation: one need only consider the earth-moon system as rotating around its center of mass.

Gravity Anomalies and Stresses

Gravity anomalies and stresses are inseparably linked. To see why, consider inviscid fluids in which shear stresses cannot exist. On a small scale, such fluids will arrange themselves in horizontal layers of increasing density; on a planetary scale, they will form a series of concentric shells. In neither case will any gravity anomaly appear anywhere, nor will there be any relief. We should not conclude, however, that the presence of gravity anomalies enables us to find the exact stresses, for a gravity distribution on a closed surface can result from an infinity of density distributions inside the surface. This is *the* fundamental indetermination of gravity interpretation, but too much can be read into this mathematical statement. Although there are an infinity of numbers between, say, 2400 and 2401, such numbers would not correspond to meaningful density differences in SI units. Similarly, reasonable densities are fairly limited in range. Furthermore, if we know the depth at which the anomaly occurs, the problem can be further narrowed.

A simplified example of a situation in which isostatic equilibrium is satisfied, but in which gravity anomalies and shear stresses exist, is shown in Figure 9-7. The reason for the gravity anomaly lies in the fact that the gravitational attraction varies as the inverse of the square of the distance, and hence the material closer to the station exerts a relatively more important influence than does the more distant one. This is sometimes referred to as the proximity effect.

Gravity Anomalies at Sea

Many gravity measurements are taken aboard surface vessels rather than in submarines. In such cases gravity corrections may be large, but the subject is

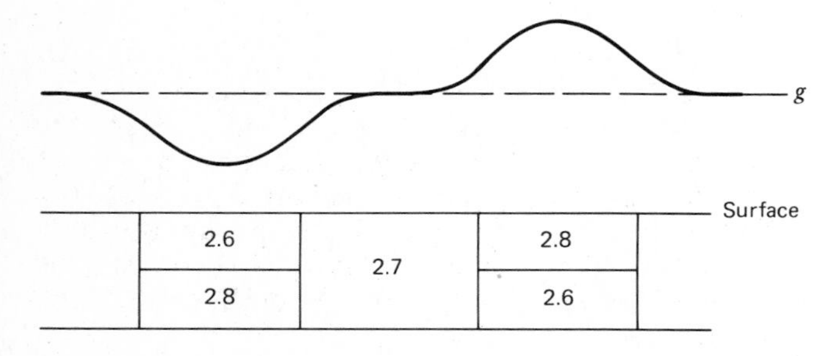

FIG. 9-7 *Structures that yield equal lithostatic pressures at depth may still give rise to gravity anomalies, because of the proximity effect. The numbers are the densities in g/cm³*

highly complex because of the many causes of acceleration. One simple correction —the Eötvös correction—is necessary because the vessel is in motion—or, more precisely, because the E–W motion of the vessel is algebraically added to the rotation of the earth. If the angular frequency of the motion of the vessels is ε, we have (see Equation 9-5)

$$(\omega+\varepsilon)^2 r \approx \omega^2 r + 2\omega\varepsilon r \tag{9-20}$$

since $\varepsilon \ll \omega$. The last term is the cause of the Eötvös correction; it is not negligible (Note that ε is caused only by the E–W component of the course).

To compute the Eötvös correction δg_{E} exactly, let the velocity of the vessel be $\mathbf{v}$ and its eastward component be v_E. Then, $v_E = \varepsilon r$, where r is the radius of the latitude circle of the observation point (see Figure 9-1). The resulting increase in acceleration is $2\omega\varepsilon r$; just as in Figure 9-1, it is directed normal to the earth's axis. Its vertical component is thus

$$\delta g = 2\omega\varepsilon r \cos\phi = 2\omega v_E \cos\phi$$

where ϕ is the latitude of the vessel. It is positive upward. To compensate for δg, we must thus add it to the observed value. Hence,

$$\delta g_{\mathrm{E}} = 2\omega v_E \cos\phi$$

To be exact, you should take $\omega = 2\pi/T_S$, where T_S is the length of the sidereal day (8.616×10^4 s)—the period of rotation of the earth with respect to the stars rather than to the sun (8.64×10^4 s).

The free-air correction at sea is, of course, very close to zero when the measurement is carried out on a surface vessel. The Bouguer correction is sometimes applied, but here it makes little physical sense, as it is equivalent to replacing the water of the oceans with an equal volume of rock. Since oceans are, on the average, close to isostatic equilibrium, the addition of this huge mass completely destroys the equilibrium and thus results in Bouguer gravity anomalies that are inversely correlated even more strongly with the "topography" (here, the bathymetry) than are those on land.

The fact that the free-air anomaly is very small over most midoceanic ridges, whereas the Bouguer anomaly there is large, immediately suggests that these features (like mountains on continents) are close to isostatic equilibrium. (For an interesting exception, see Liu et al. 1982.) This, of course, may not be strictly true because of the proximity effect. The opposite conclusion can also be tentatively reached for trenches; it is discussed further below.

PROBLEMS

9-2 A surface vessel making gravity measurements at 55°N latitude moves vertically with a period of 10 s and an amplitude of 50 cm. What is the corresponding maximum acceleration, in mgals?

9-3 The same vessel steers a course N 70°W at 10 knots (one knot = 1 nautical mile/h = 1.853 km/h). What is the magnitude and sign of the Eötvös correction?

9-4 What corrections must be applied when gravity observations are made in a submerged submarine, and why? (Neglect the Eötvös correction here.) Compare this result with that of Problem 9-1, and explain the similarity or difference.

Gravity Anomalies and the Geoid

Because the geoid is an equipotential surface, its shape is related to the gravity field. The form of this relation was first given by Stokes in 1849. He showed that, at any point P, the distance N from the geoid to the reference ellipsoid is given by

$$N = C \int_S D \, \gamma_{FA} \, dS \qquad (9\text{-}21)$$

where C is a constant, D is a function of the distance between P and the element dS of the geoid on which γ_{FA} is measured, γ_{FA} is the free-air gravity anomaly on dS, and S is the surface of the geoid. Equation 9-21 is known as the Stokes formula; it expresses the relation between gravimetry and geodesy. More precisely, it shows that the shape of the geoid (i.e., N) can be determined if the gravity anomalies over the whole earth are known. This method determines the gravimetric geoid; until about 1960, there was no other method of doing so.

Recently, it has become possible to determine the geoid through measurement of the altitudes of satellites above the oceans by radar altimetry. A radar pulse sent down by a satellite is reflected at the ocean surface, and the time of its return is measured. If the orbit of the satellite with respect to the reference ellipsoid is known, this method yields the "satellite geoid" (there is no standard term for it at present). Correction needs to be made for winds, ocean currents, tides, and so on.

These two geoids should be identical, but considering the very different methods on which they are based, differences are to be expected. In fact, these differences are remarkably small (Figure 9-8). As one might expect from the more global nature of the gravimetric geoid, the latter does not show local features as sharply as the satellite geoid does. Note that deep trenches (e.g., the Philippine trench) depress the geoid by as much as 20 m. The reason for this phenomenon can be most easily understood if there is a deficit of mass at the trench, which results in the gravity vector on either side of the trench axis being deflected as shown in Figure 9-9. Since the gravity is perpendicular to the geoid, the latter must be depressed.

Since at any point the height of the geoid above the ellipsoid is a weighted sum of the gravity values over the *whole* earth (Equation 9-21), the geoid shows long-wave features rather well. Short-wave features, in contrast, generally show up better on maps of the gravity field. Despite this restriction, and thanks to the progress in satellite altimetry, a recent map of the geoid (Figure 9-10) over

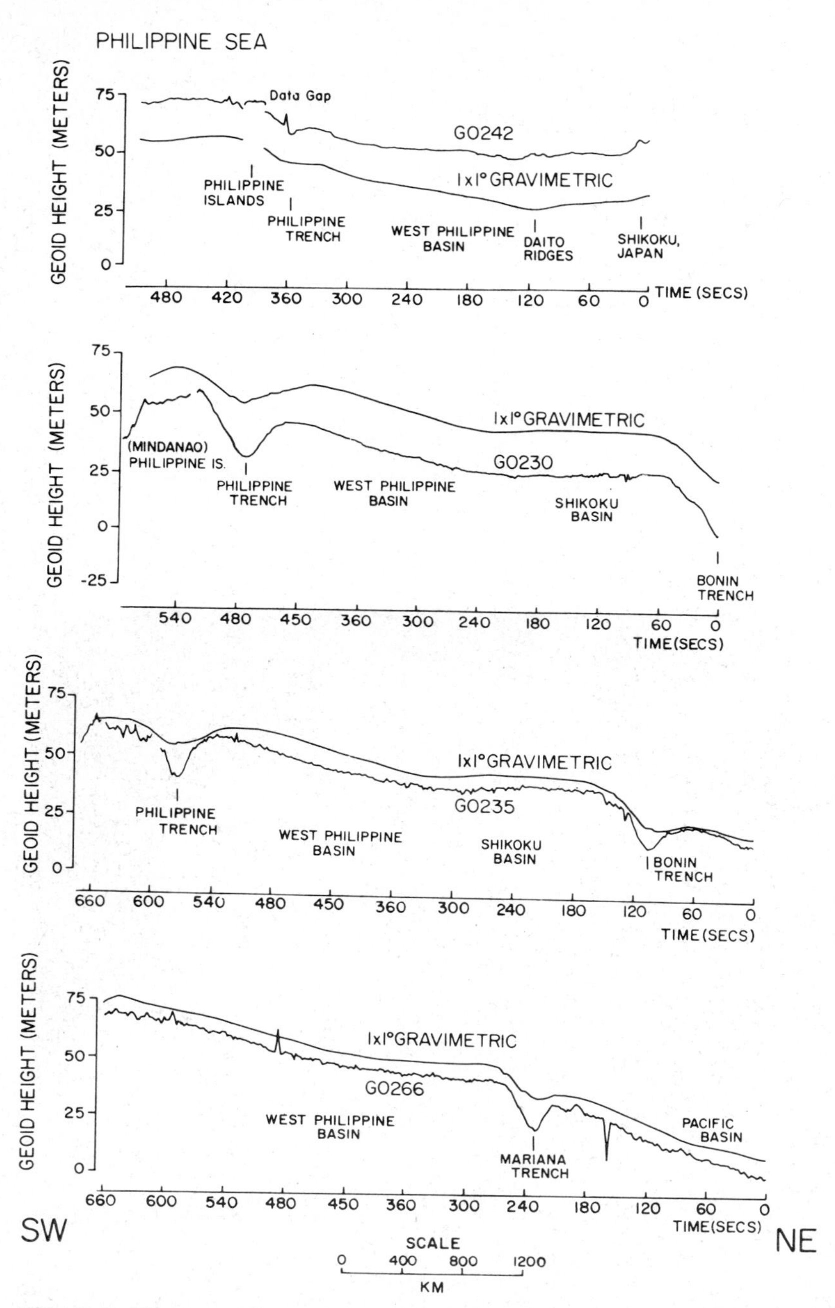

FIG. 9-8 *The gravimetric and the satellite geoid along four tracks in the northwestern Pacific Ocean. (From Chapman and Talwani, 1979. Copyright © by the American Geophysical Union.)*

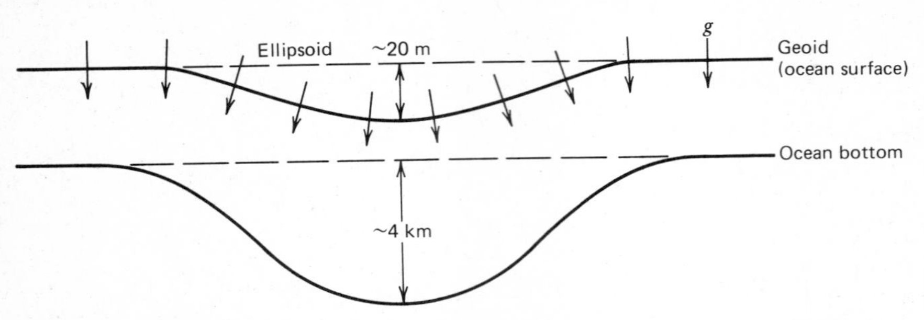

FIG. 9-9 *The depression of the geoid above a trench (or other mass deficiency).*

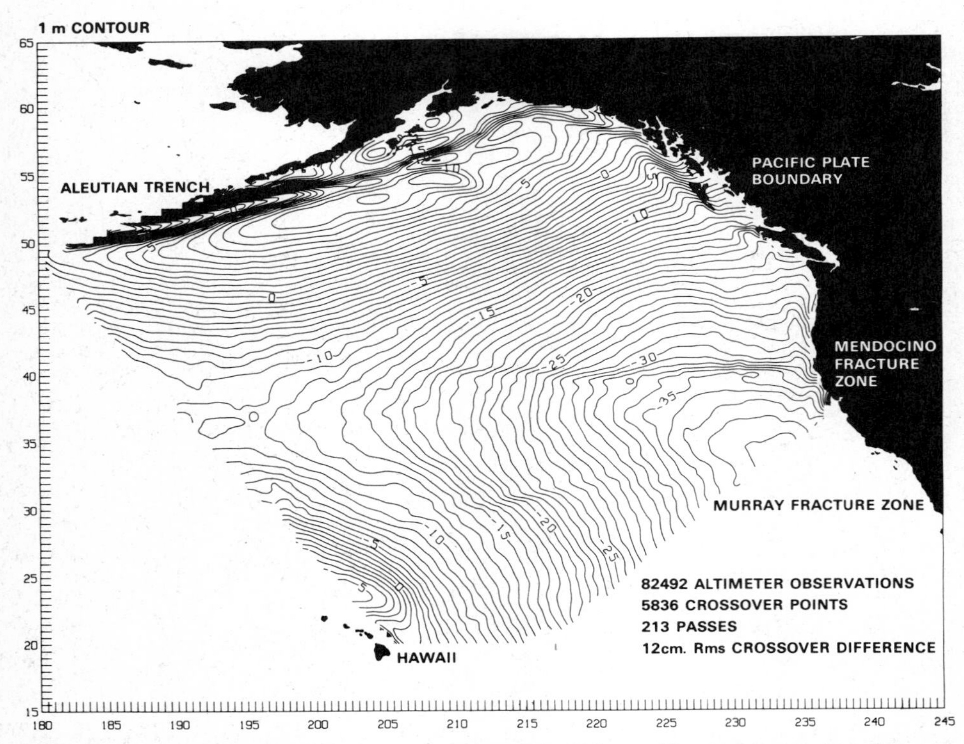

FIG. 9-10 *Satellite geoid over the northeastern Pacific Ocean. Fracture zones, trenches, and the boundary of the Pacific plate can be clearly seen. (Marsh et al. 1982, cover page. Copyright © by the American Geophysical Union.)*

the northeastern Pacific shows two fracture zones, the Aleutian trench, and the Pacific plate boundary (Marsh et al. 1982).

Finally, it may be shown that N at point P (see Equation 9-21) is approximately given by

$$N = -\frac{2\pi G}{g}\int_0^h z\,\delta\rho\,dz \tag{9-22}$$

where g is the gravity at P, $\delta\rho = \rho - \rho_o$ is the difference between the real density ρ and the normal density ρ_o, and h is the depth of compensation (30...100 km). This formula (Turcotte and Schubert 1982, pp. 223–225) assumes that (a) Isostasy prevails, (b) $\delta\rho$ is shallow (a few kilometers), and (c) $\delta\rho$ changes only slowly in the horizontal direction (wavelength of 100 km or more).

Some Major Gravity Anomalies

By far the most conspicuous anomalies on earth are those associated with deep-sea trenches. Figures 9-11*a* and 9-11*b* have frequently been reproduced. The gravity anomalies shown are of the regional isostatic type, with a radius of regionality of 174.3 km, but other types of isostatic compensation often give generally similar results. The free-air anomaly over other trenches is somewhat similar (see Figures 9-12 and 9-13).

It appears that a multiplicity of causes concur in the phenomenon shown in Figure 9-11.

- Because of the great depth of the ocean, the proximity effect is very large in oceanic trenches. However, it should be entirely compensated for by the isostatic correction, and thus the bathymetry should not be correlated with isostatic anomalies. In contrast, the bathymetry should—and does—show on a free-air anomaly map (see Figure 9-15). In a general way the minimum of the regional isostatic anomalies is closer to the continent than to the trench axis. Whether this is also the case for all other types of anomalies is not clear.
- The isostatic scheme used may bear little or no relation to the true situation. This is shown in a very simplified way in Figure 9-14. If the trench depicted in the figure is 6.5 km deep, the "antiroot" required by the regional isostatic scheme will be on the order of 25 km. But in fact the slab may be going down, as shown by the dashed lines; hence the assumed densities, especially those of the hachured area, are incorrect. The isostatic scheme thus correctly compensates for the difference in density between the water and the rocks below the ocean bottom, but then introduces an incorrect assumption whose exact effect is hard to evaluate.

Recent work in the eastern part of this area has shown that the real situation is very complex. The free-air gravity anomalies shown in Figure 9-15 indeed bear a resemblance to those shown by Vening Meinesz (Figure 9-11*a*), but there are some important differences, as well.

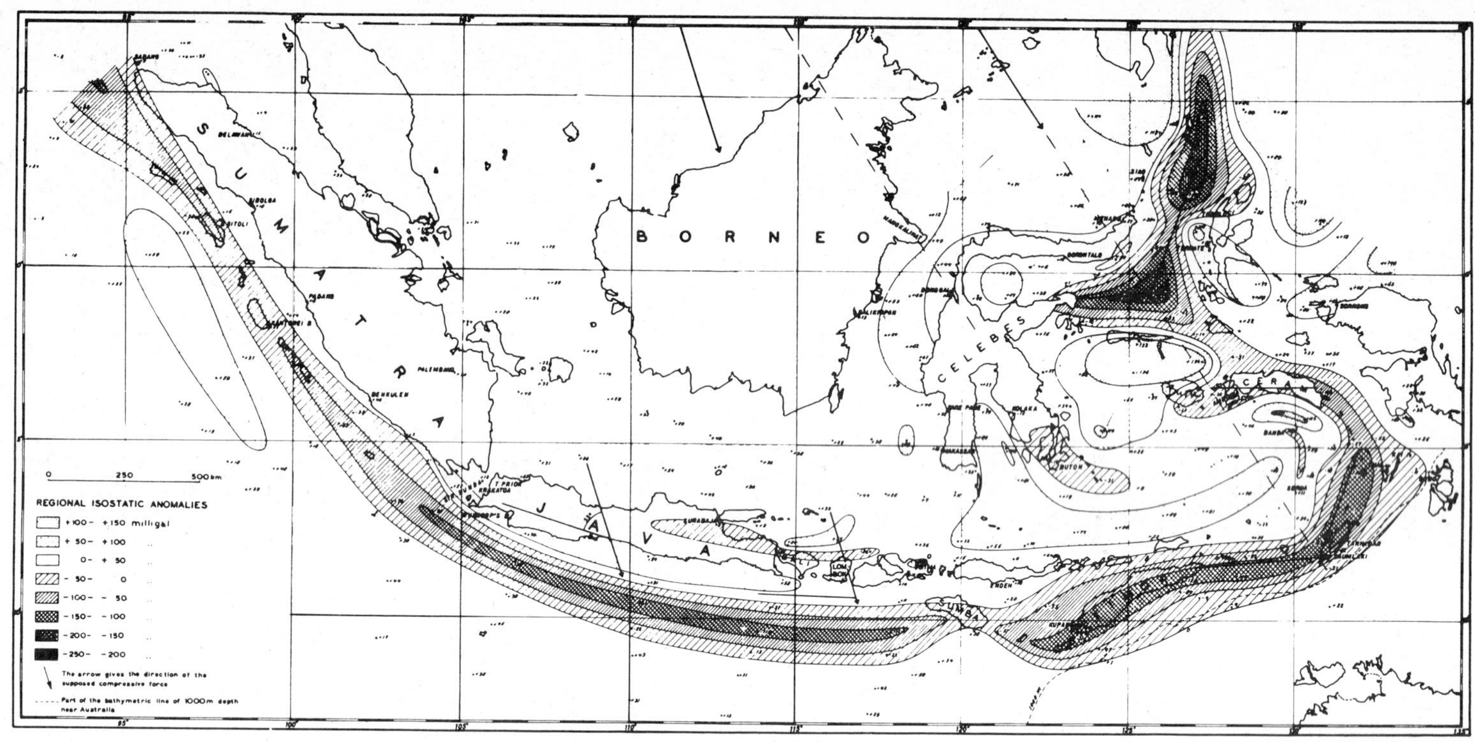

(a)

FIG. 9-11 *(a) Regional isostatic anomalies (depth of compensation, 30* km; *radius of regionality, 174.3* km*) in the Indonesian archipelago. (b) Bathymetric map of the same region (From Heiskanen and Vening Meinesz 1958, maps 10C1 and 10C2. Reprinted by permission from McGraw-Hill Book Co.)*

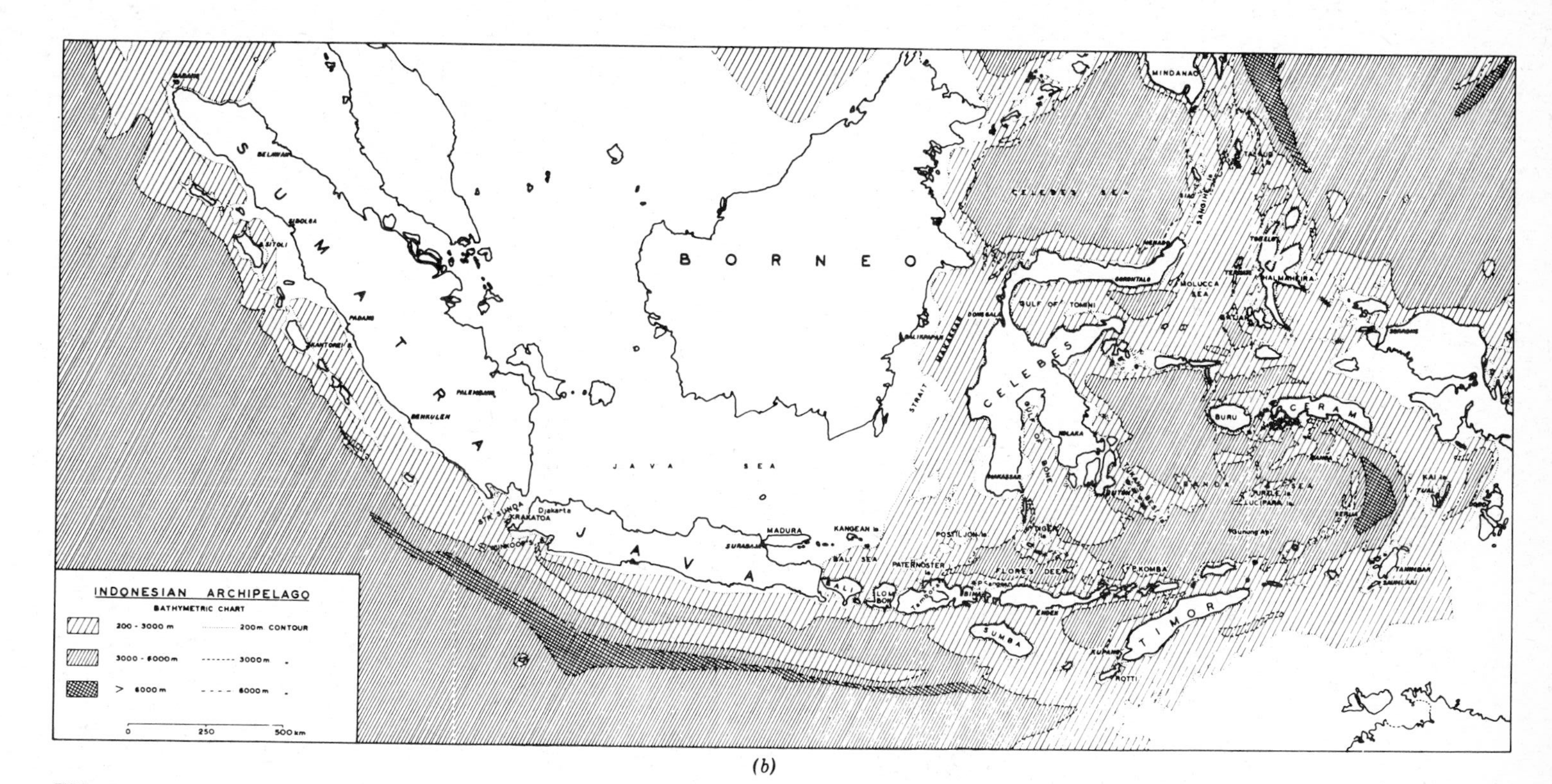

(b)

FIG. 9-11 (Continued)

- A strong positive (rather than negative) anomaly over Timor is accompanied by a negative anomaly to the south of it.
- There are to separate minima near Tanimbar, and so forth.
- The strong minimum east of the Northern Celebes is preserved.

The structure inferred along line *AA′* is shown in Figure 9-12, for which the densities were inferred from an empirical relation between density and seismic velocity. This structure bears so little relation to any simple isostatic model that nothing could be gained by comparing the two. It is far better to compute the mass anomaly, as is done at the bottom of the figure; this mass anomaly reaches almost 1000 kg/cm^2, or 10^7 kg/m^2. If the stress were purely hydrostatic, such a mass difference would correspond to a pressure difference of 10^8 Pa (1 kbar) at the bottom of the column; this figure is remarkably close to the one obtained previously from simple considerations of isostasy. Such a mass anomaly yields an immediately understandable picture that requires no additional geologic or geophysical assumptions.

Figure 9-13 shows the observed free-air anomalies and the structure computed on a profile across the Chile Trench and the Andes. Here again, the mass anomaly is on the order of 1000 kg/cm^2. We must conclude that the dynamics of

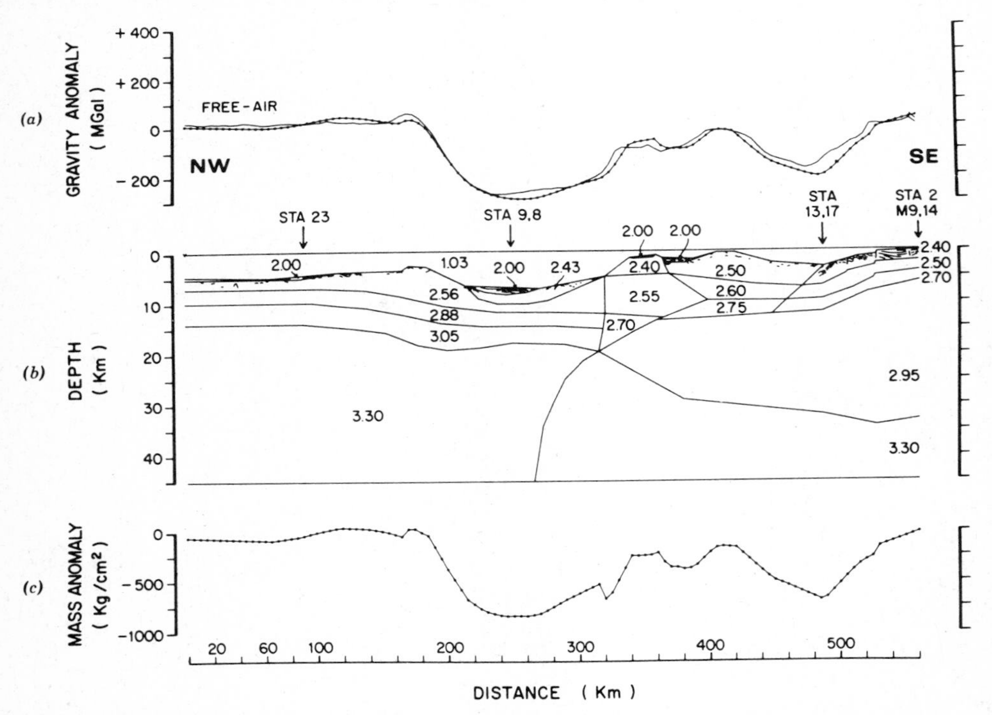

FIG. 9-12 *(a) Free-air gravity anomalies (thin line indicate observed values, dotted line indicates values computed from model), (b) structural model, and (c) resulting mass anomaly along profile AA′ in Figure 9-15 (From Bowin, Purdy, Johnston, Shor, Lawrer, Hartono, and Jezek 1980. Reprinted by permission.)*

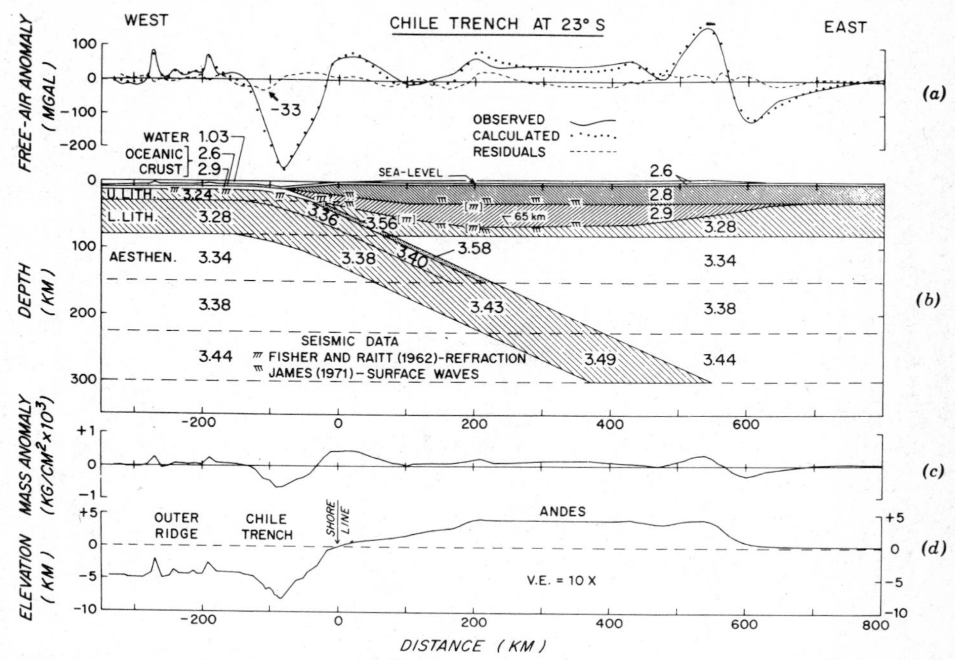

FIG. 9-13 *(a) Free-air gravity anomalies, (b) structural model, (c) mass anomaly, and (d) elevation for the Chile Trench. (From Grow and Bowin 1975. Copyright © by the American Georphysical Union.)*

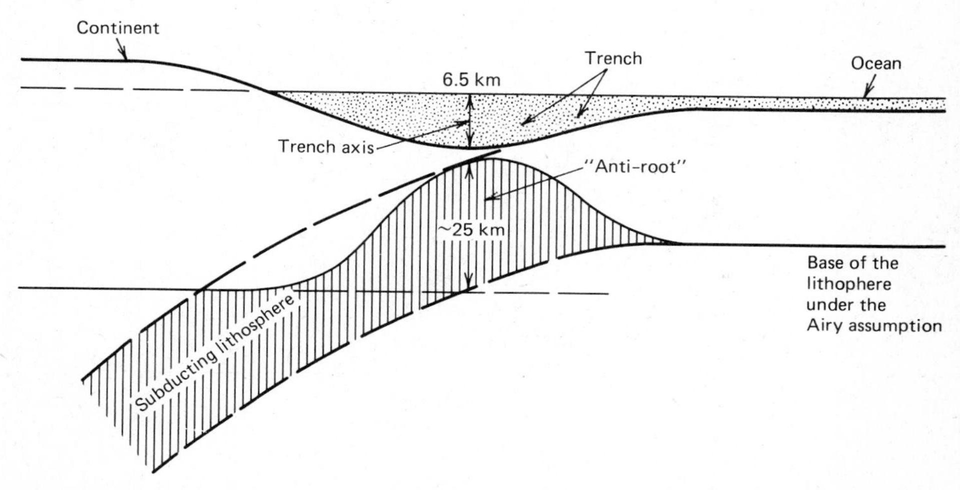

FIG. 9-14 *Comparison of a schematic isostatic model and a schematic plate tectonic model.*

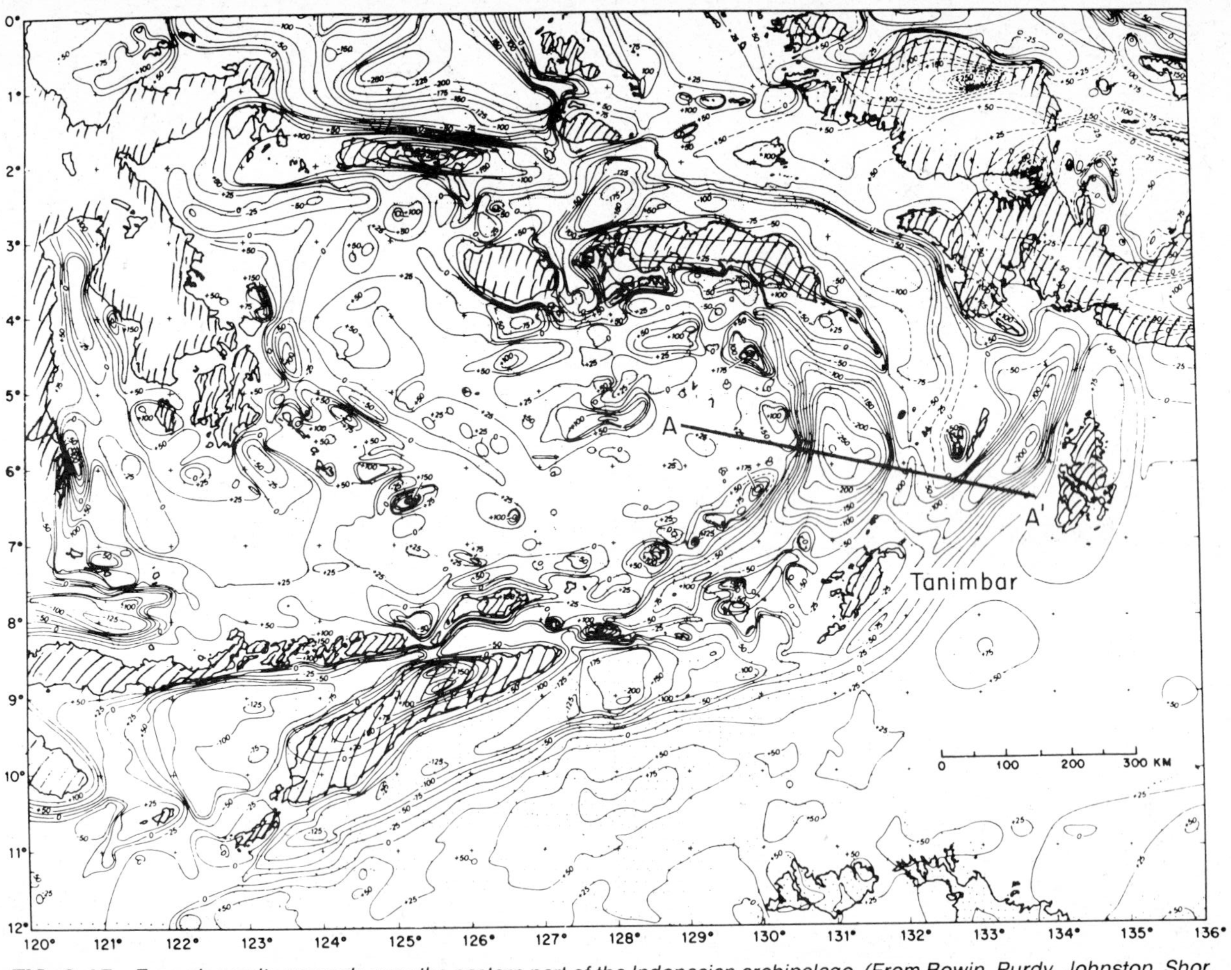

FIG. 9-15 *Free-air gravity anomaly over the eastern part of the Indonesian archipelago. (From Bowin, Purdy, Johnston, Shor, Lawrer, Hartono, and Jezek 1980. Reprinted by permission.)*

the system and/or the strength of some components are sufficient to maintain this disequilibrium, unless we assume the present situation to be strictly transient.

Although at short wavelengths the gravity field is strongly negative near the trenches, at long wavelengths there generally is a gentle positive anomaly, presumably related to the heavy downgoing slab.

In the continental United States the most prominent feature on a general gravity map is the midcontinent gravity high (Figure 9-16), which is caused by a Precambrian basic intrusion (i.e., a paleorift). The anomalies there reach 100 mgal.

An interesting example of the failure of the "pure" isostatic hypothesis is provided by the rift valleys of Africa. In 1936, Bullard showed that in the center of the rift is a negative isostatic anomaly flanked by positive ones on either side (Figure 9-17*a*). It was thought that such a negative anomaly, or mass deficiency, could not exist in a graben bounded by normal faults, because nothing would prevent the bottom from rising. The only way to explain the anomaly was to assume that the faults became thrust faults at depth, even though fairly spectacular normal faults are visible along many parts of the graben. Since thrust faults are caused by compression, graben were considered to be compressive features for almost 25 years, despite the evidence to the contrary. This idea probably would not have taken hold if geophysicists had known more geology, and geologists more geophysics. In fact, one of the causes of the anomaly is that the isostatic hypothesis incorrectly assumes that the adjustment is local and that the faults are vertical. If the fault planes dip toward the rift (Figure 9-17*b*), and if the crust behaves elastically, the smaller base of the rift block will tend to let the block drop downward; the opposite phenomenon occurs on the flanks (Heiskanen and Vening Meinesz 1958, pp. 389–395). Furthermore, the isostatic hypothesis presumes that the bottom of the crust is as shown in Figure 9-17*b* (in which the relief is greatly exaggerated). It is thus assumed that there is an excess of mass under the rift itself (upward arrows) and a deficit under the sides (downward arrows). These fictitious masses produce fictitious isostatic anomalies of about 37 mgal (Heiskanen and Vening Meinesz 1958, p. 395). The anomalies to be explained are on the order of 50 to 100 mgal. (The anomaly shown in Figure 9-17*a* is among the largest.) The remaining difference probably results from two factors: (a) the higher temperatures and thus lower densities prevailing under the rift, which are reflected in fairly widespread volcanism in the area; and (b) the relatively great thickness of low-density sediments that have accumulated in the rift. The relative importance of these three phenomena (isotasy, high temperature, and sedimentation) probably varies along the African rift. Similar anomalies have been found along other rifts, such as the Rhine graben; their cause is similar.

Determination of Minor Anomalies

In geophysical prospecting, the only important gravity anomalies are those caused by bodies located at relatively shallow depths (less than 10 km in any case, often shallower). It is physically evident that such near-surface density variations can

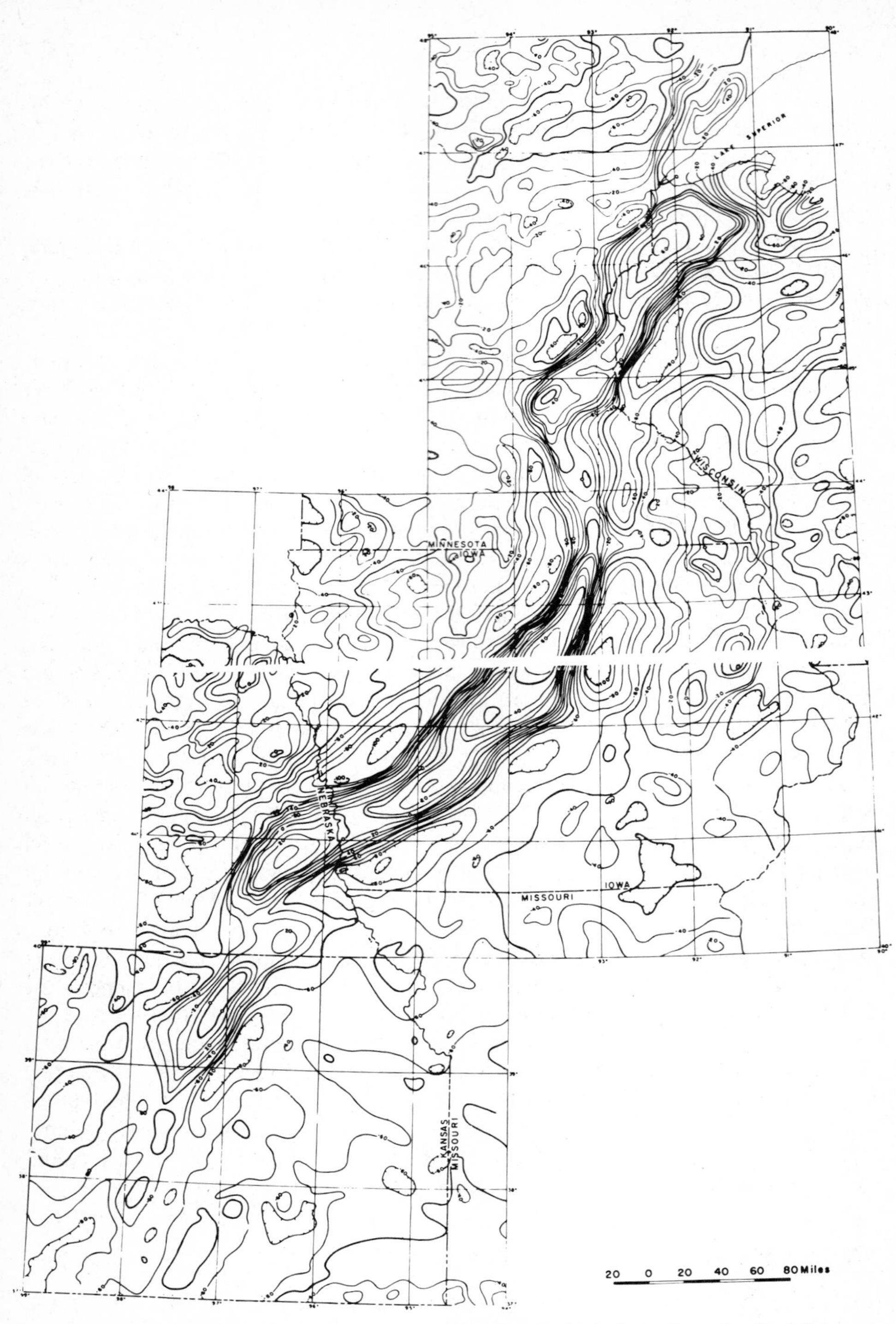

FIG. 9-16 *Bouguer anomaly of the midcontinent gravity high. Lake Superior is at the northeast corner of the area. (From Coons, Woollard, and Hershey, 1967.)*

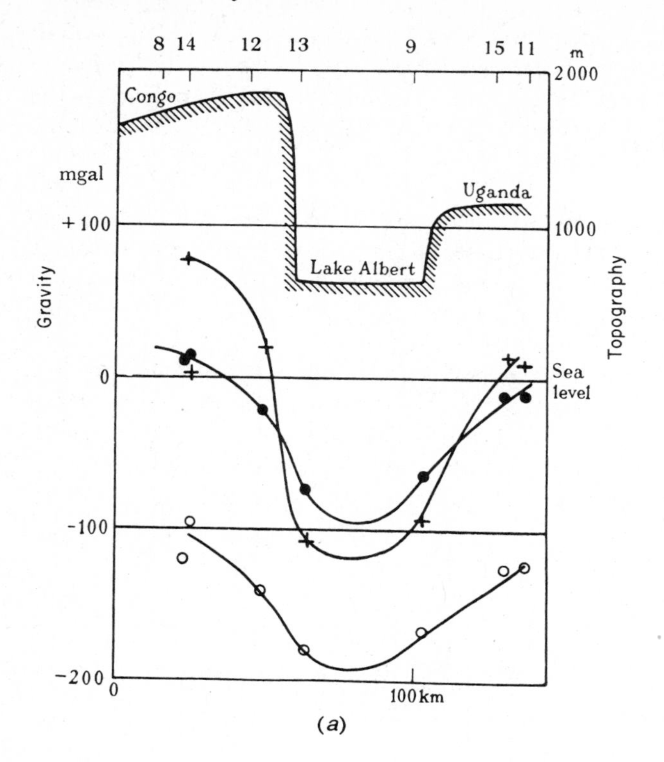

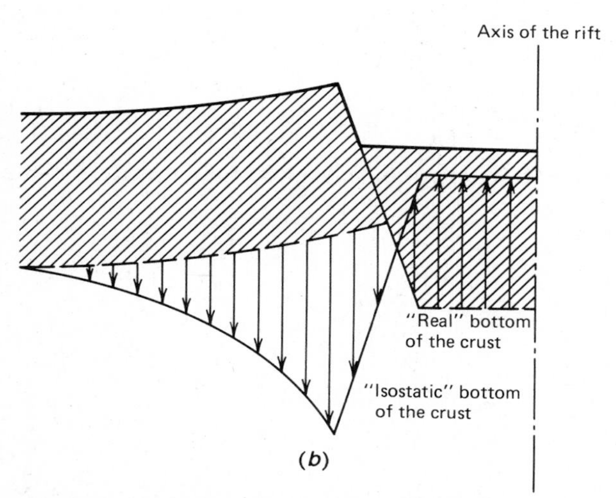

FIG. 9-17 *(a) Topography and gravity across the Lake Albert rift in East Africa. (From Bullard 1936. Reprinted by permission.) (b) Comparison of the isostatic structure and of a possible real structure near a rift zone.*

induce gravity variations of shorter wavelengths than those induced by deeper bodies. (This will also be proven later—see Equation 9-35.) The long-wavelength gravity field, or "regional field," is thus of no interest. The relevant part of the field consists of wavelengths of a few kilometers and is called the "local field," the "residual field," or, somewhat confusingly, the "anomaly." The question is how to isolate this local field.

The most straightforward method might appear to be the two-dimensional counterpart of the filtering process discussed during our treatment of seismology (Chapters 4 and 6). Although this method is sometimes used, one difficulty with it is that gravity measurements are almost never equispaced, which makes filtering appreciably more difficult, and its results less certain.

A simple method of emphasizing short wavelengths is to subtract from the measurement at a point the average of the measurements at the nearest four surrounding stations. This method also presumes that the stations are equispaced. It provides an approximate way of computing the second derivative of g in the vertical direction. To prove this, we note that since $\nabla^2 U = 0$ wherever $\rho = 0$, we also have there

$$\frac{\partial}{\partial z}\nabla^2 U = 0 = \frac{\partial}{\partial z}\left(\frac{\partial^2 U}{\partial x^2} + \frac{\partial^2 U}{\partial y^2} + \frac{\partial^2 U}{\partial z^2}\right) \tag{9-23}$$

$$= \frac{\partial^2}{\partial x^2}\frac{\partial U}{\partial z} + \frac{\partial^2}{\partial y^2}\frac{\partial U}{\partial z} + \frac{\partial^2}{\partial z^2}\frac{\partial U}{\partial z} = \nabla^2 g = 0$$

Hence,

$$\frac{\partial^2 g}{\partial z^2} = -\left(\frac{\partial^2 g}{\partial x^2} + \frac{\partial^2 g}{\partial y^2}\right) \tag{9-24}$$

But (see Figure 9-18)

$$\left.\frac{\partial g}{\partial x}\right]_A \approx \frac{1}{\delta x}(g_P - g_W)$$

where $]_A$ indicates the value at A. It is intuitively clear that the RHS best

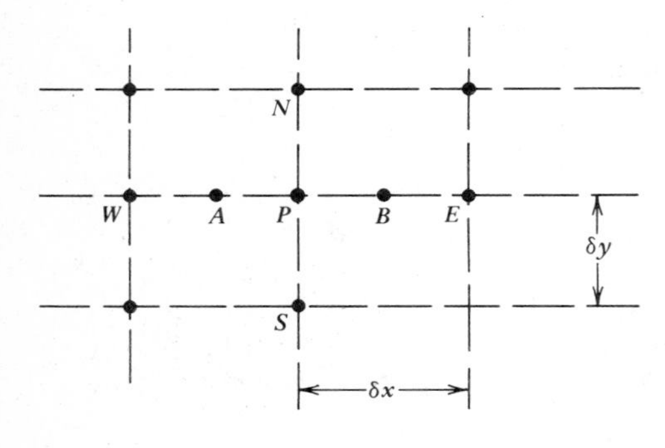

FIG. 9-18 *Point P, its four nearest neighbors, points A and B, and the intervals δx and δy.*

approximates $\partial g/\partial x$ at A, the midpoint between P and W. Hence, similarly,

$$\left.\frac{\partial g}{\partial x}\right]_B \approx \frac{1}{\delta x}(g_E - g_P)$$

Again similarly,

$$\left.\frac{\partial^2 g}{\partial x^2}\right]_P \approx \frac{1}{\delta x}\left(\left.\frac{\partial g}{\partial x}\right]_B - \left.\frac{\partial g}{\partial x}\right]_A\right)$$

and, by substitution,

$$\left.\frac{\partial^2 g}{\partial x^2}\right]_P \approx \frac{1}{\delta x^2}(g_E - 2g_P + g_W)$$

Thus,

$$\left.\frac{\partial^2 g}{\partial z^2}\right]_P \approx -\left[\frac{1}{\delta x^2}(g_E - 2g_P + g_W) + \frac{1}{\delta y^2}(g_N - 2g_P + g_S)\right] \tag{9-25}$$

If $\delta x = \delta y$, the RHS is the difference between the center value and the average of the nearest four, except for a scale factor. If $\delta x \neq \delta y$, Equation 9-25 should be used. In any case, the result will be in, say, mgal/m²: that is, $10^{-5}\ \mathrm{m}^{-1}\ \mathrm{s}^{-1}$.

With the advent of computers the least-squares method discovered by the German mathematician and physicist Karl Friedrich Gauss (1777–1855) has become very popular. This method was devised and is properly used to minimize random errors of observation. The present situation presents only a misleading similarity to that resulting from random errors. Failure to appreciate the differences between the two situations may lead to absurd conclusions (see below).

Let us first take a very simple, unrealistic example. Suppose that there are only four gravity measurements, g_i, in a given area. We assume that the regional field, g_r, can be represented by a plane

$$g_r = a + bx + cy \tag{9-26}$$

Hence, at each point the local field ε_i is

$$\varepsilon_i = g_i - (a + bx_i + cy_i) \tag{9-27}$$

We must choose a, b, and c in some definite way. In the one-dimensional case (Figure 9-19) we would want to draw a line that passes, in some way, "in between" the observations. One solution might be to minimize $\Sigma_{i=1}^4|\varepsilon_i|$. (If we do not take absolute values, large positive and negative errors might balance each other to some extent.) For mathematical convenience, let us decide instead to minimize $R = \Sigma_{i=1}^4 \varepsilon_i^2$. This implies, from the calculus of more than one variable, that $\partial R/\partial a = \partial R/\partial b = \partial R/\partial c = 0$. But

$$\frac{1}{2}\frac{\partial R}{\partial a} = g_1 - (a + bx_1 + cy_1) + g_2 - (a + bx_2 + cy_2) + g_3 + \cdots = 0$$

or

$$4a + b \sum_{i=1}^{4} x_i + c \sum_{i=1}^{4} y_i = \sum_{i=1}^{4} g_i$$

Similarly,

$$\frac{1}{2}\frac{\partial R}{\partial b} = b\left(x_1^2 + x_2^2 + \cdots x_4^2\right) + a\left(x_1 + x_2 + \cdots x_4\right)$$
$$+ c(x_1 y_1 + \cdots x_4 y_4) - (g_1 x_1 + \cdots g_4 x_4) = 0$$

or

$$a \sum_{i=1}^{4} x_i + b \sum_{i=1}^{4} x_i^2 + c \sum_{i=1}^{4} x_i y_i = \sum_{i=1}^{4} g_i x_i$$

$\frac{1}{2}(\partial R/\partial c) = 0$ gives a similar equation. If there are n observations, rather than four, we obtain

$$na + b \sum_{i=1}^{n} x_i + c \sum_{i=1}^{n} y_i = \sum_{i=1}^{n} g_i \tag{9-28}$$
$$a \sum_{i=1}^{n} x_i + b \sum_{i=1}^{n} x_i^2 + c \sum_{i=1}^{n} x_i y_i = \sum_{i=1}^{n} g_i x_i$$
$$a \sum_{i=1}^{n} y_i + b \sum_{i=1}^{n} x_i y_i + c \sum_{i=1}^{n} y_i^2 = \sum_{i=1}^{n} g_i y_i$$

which is a system of three equations (the "normal equations") with three unknowns, a, b, and c, which can be thus found. We can then determine the local field by substituting for a, b, and c in Equation 9-27.

This simple example shows the assumptions and the dangers involved. To begin with, we have *assumed* the shape of the regional field, but there is no theoretical reason (other than simplicity) to choose one regional field over another. As the formula becomes more complicated, the question becomes more

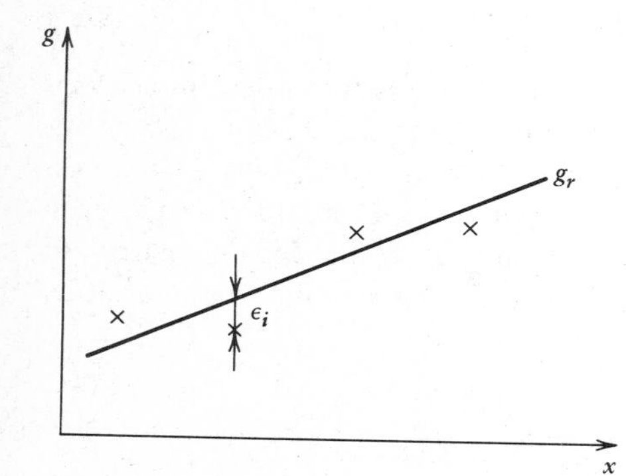

FIG. 9-19 *Plot of g versus x and the "error" ε_i. The crosses show the data.*

serious. If we want to fit an equation of the type, say,

$$g_r = a + bx + cy + dx^2 + exy + fy^2$$

we will obtain six normal equations of the same general type as Equation 9-28, the first two of which are

$$na + b\sum x_i + c\sum y_i + d\sum x_i^2 + e\sum x_i y_i + f\sum y_i^2 = \sum g_i \tag{9-29}$$

$$a\sum x_i + b\sum x_i^2 + c\sum x_i y_i + d\sum x_i^3 + e\sum x^2 y_i + f\sum x_i y_i^2 = \sum g_i x_i$$

Similar equations can be easily derived for, say, 20 coefficients. Although they are very easy to solve on a computer, they may lead to meaningless results. For one thing, high-order polynomials tend to get very "wiggly" in areas where they are poorly constrained (i.e., areas with few data points), and this may produce entirely fictitious anomalies. Since a Fourier representation gives a much better-behaved (less "wiggly") surface, it would seem preferable to choose a two-dimensional counterpart of the scheme discussed in Chapter 4, such as

$$g_r = \left(A + B\cos\frac{2\pi x}{w} + C\sin\frac{2\pi x}{w}\right)\left(D + E\cos\frac{2\pi y}{h} + F\sin\frac{2\pi y}{h}\right)$$

That is,

$$\begin{aligned} g_r = a + b\cos\frac{2\pi x}{w} + c\cos\frac{2\pi y}{h} + d\sin\frac{2\pi x}{w} \\ + e\sin\frac{2\pi y}{h} + f\cos\frac{2\pi x}{w}\cos\frac{2\pi y}{h} + g\cos\frac{2\pi x}{w}\sin\frac{2\pi y}{h} \\ + j\sin\frac{2\pi x}{w}\cos\frac{2\pi y}{h} + k\sin\frac{2\pi x}{w}\sin\frac{2\pi y}{h} \end{aligned} \tag{9-30}$$

where w is the length of the area along the x-axis and h is the length along the y-axis. For higher-order formulae, we could simply use higher harmonics in x and y—that is, terms in $2\pi x/w$, $4\pi x/w$, $6\pi x/w$, and similarly for y. We would then form the normal equations according to Equation 9-29, by replacing x_i by $\cos(2\pi x_i/w)$, y_i by $\cos(2\pi y_i/h)$, x_i^2 by $\sin(2\pi x_i/w)$, and so forth—. Unfortunately there is no guarantee that the real regional field is closer to Equation 9-30 than to Equation 9-26 or to any other.

The method just described amounts to a determination of Fourier coefficients by the least-squares method, rather than by the much simpler method given in Chapter 4. The reason that the two-dimensional equivalent of this simple method generally cannot be used here is that it supposes that the points are equispaced, which is generally not the case in gravity measurements.

Another difficulty stems from "bunching": if the points are very unevenly distributed over the domain, the area of maximum concentration may unduly influence the results. We can resolve this difficulty either by weighing more heavily the observations in areas where they are sparse—that is, by having every observation there count as N elsewhere ($N = 2, 3$, etc.)—or, equivalently, by averaging the observations in the areas of greatest concentration.

Finally, there is no theoretical justification for minimizing the sum of the squares of the residuals. This minimization has a sound theoretical basis when the residuals are errors of observations; in the present case, it is merely a mathematical convenience.

Because of these difficulties, there is in general no "best" way of selecting a "regional" field and thus of finding the local anomalies. A prudent way to do so may be to use a low-order polynomial surface combined with a low-order Fourier expansion, that is a combination of Equations 9-29 and 9-30, but the inherent difficulty of the problem cannot be circumvented.

PROBLEMS

9-5 Let the values of g_i at x_i, y_i, $z_i = 0$ be those given in the first three columns below. (x and y in km, g in mgal).

x_i	y_i	g_i	z_i	g_i
3.5	0.0	4.5	1.5	5.0
21.0	0.0	4.5	1.2	4.9
48.0	1.2	4.5	0.6	4.7
29.0	2.1	4.0	2.0	4.6
37.0	8.0	3.0	1.3	3.4
3.0	10.0	4.0	0.9	4.3
9.5	12.0	3.5	0.5	3.6
57.5	12.0	3.2	0.8	3.4
16.0	14.0	2.5	1.7	3.0
21.0	16.0	1.5	2.7	2.3
2.0	21.0	4.0	2.6	4.8
11.0	30.0	2.5	1.5	3.0
25.0	28.0	0.5	1.0	0.8
56.0	22.0	1.7	0.2	1.8
38.0	29.0	0.8	0.9	1.1
32.0	23.0	1.0	1.6	1.5
27.0	38.0	0.8	1.2	1.2
3.0	46.0	2.5	0.6	2.7
49.0	37.0	2.0	0.1	2.0
41.0	43.0	2.0	0.0	2.0
30.0	49.0	1.5	0.9	1.8
4.0	54.0	2.5	2.1	3.1
17.0	58.0	2.8	2.5	3.6
49.0	54.0	3.5	2.0	4.1
52.0	58.0	3.8	1.2	4.2
37.0	16.0	2.8	1.1	2.8

First contour the data, then

(a) Use a regional given by

$$g_r = a + bx + cy + dxy + ex^2 + fy^2$$

to determine the anomalies. Contour the anomalies, then interpret them.

(b) Use g_r as given by Equation 9-30 to determine the anomalies. Contour and interpret them. Assume h = 60 km, w = 61 km.

9-6 **(a)** Assume that x and y are given in the first two columns, z (in m) in the fourth one, and g in the fifth one. Take g_r as in Problems 9-5a and 9-5b, but add a linear term in z:

$$g_r = a + bx + \cdots fy^2 + Az$$

and repeat Problems 9-5a and 9-5b.

(b) Assume that the area has the size given in Problem 9-5b. Assume also that the relief is moderate and that corrections have been made only for drift and tides. What is the value of A in the present problem, and why?

Attraction of Simple Two-Dimensional Bodies

The attraction of a few three-dimensional bodies and of many more two-dimensional ones can be computed exactly—that is, the resulting integral can be computed analytically. In a two-dimensional case a point in the (x, z) plane represents a line parallel to the y-axis (Figure 9-20). Although no body is strictly two-dimensional, the inverse-square law shows that for bodies whose third dimension exceeds 10 times the other two dimensions, the error is generally less than ~ 1% (see Problem 9-7).

At P the attraction along the z-axis of an element dy of a line parallel to the y-axis is, for a density m per unit length,

$$dg = Gm\frac{dy}{r^2}\sin\Theta$$

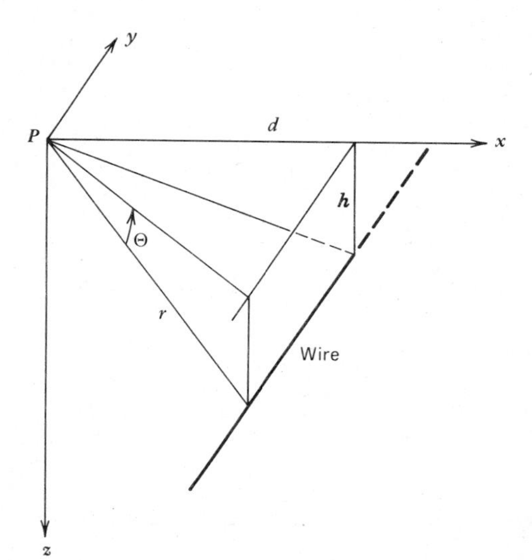

FIG. 9-20 An infinite wire parallel to the y-axis and point P where g is calculated.

But

$$\sin\Theta = \frac{h}{r}$$

Hence,

$$g = G\int_{-\infty}^{\infty} \frac{mh\,dy}{r^3}$$

where $r = (d^2 + h^2 + y^2)^{1/2}$. Thus, by integration, $g = 2G\sigma h/(d^2 + h^2)$, for a point of density σ per unit surface. For a body of arbitrary cross section but constant surface density, $d \to x$, $h \to z$, and an integral along x and z must be taken, resulting in

$$g = 2G\sigma \iint_S \frac{z\,dx\,dz}{z^2 + x^2}$$

For most two-dimensional bodies, however, the line integral method given below is by far the most practical one.

PROBLEM

9-7 Compute the attraction of a line of finite length along the y-axis extending from $-b$ to $+b$. Compare it to the result for an infinite line, plotting the ratio R of these values for $b/\sqrt{d^2 + h^2}$: 1, 3, 10, 30, 100. The value $1 - R$ is called the end effect.

Line Integral Method Because of the wide availability of computers, this method offers by far the most convenient way of computing the attraction of two-dimensional bodies of arbitrary shape. It relies on the fact that the attraction at P of an infinitely long prism, say $ABCDE$ (Figure 9-21) of constant density ρ

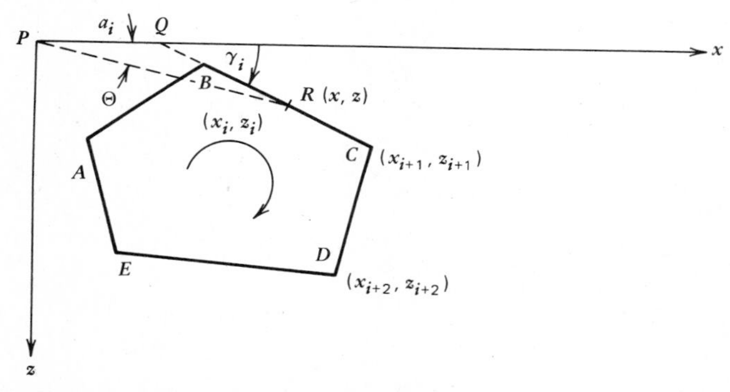

FIG. 9-21 *Coordinates of a simple two-dimensional body. Note that the points are numbered clockwise around the perimeter.*

can be expressed as a line integral that can be evaluated analytically (Talwani et al. 1959). This simply means that the integral is evaluated first along line AB, then along line BC, and so forth.

The derivation of the formula is lengthy but straightforward. It is based on the fact that the vertical component of the attraction of a prism of constant density, extending to infinity in the y-direction and bounded by the planes $\Theta = \Theta_1$, $\Theta = \Theta_2$, $z = z_1$, and $z = z_2$ (Figure 9-22), is

$$g = 2G\rho(z_2 - z_1)(\Theta_2 - \Theta_1)$$

To demonstrate this formula, we will use the elementary volume shown (shaded area) in Figure 9-22. Its mass is

$$dm = \rho\, dx\, dy\, dz$$

But $dx = R\,d\Theta/\sin\Theta$ and $R = r\cos\phi$, whereas $dy = r\,d\phi/\cos\phi$. Hence,

$$dm = \frac{\rho r^2\, d\Theta\, d\phi\, dz}{\sin\Theta}$$

The vertical component of attraction at P is thus

$$d^3g = G\frac{dm}{r^2}\sin\Theta\cos\phi = G\rho\cos\phi\, d\phi\, d\Theta\, dz$$

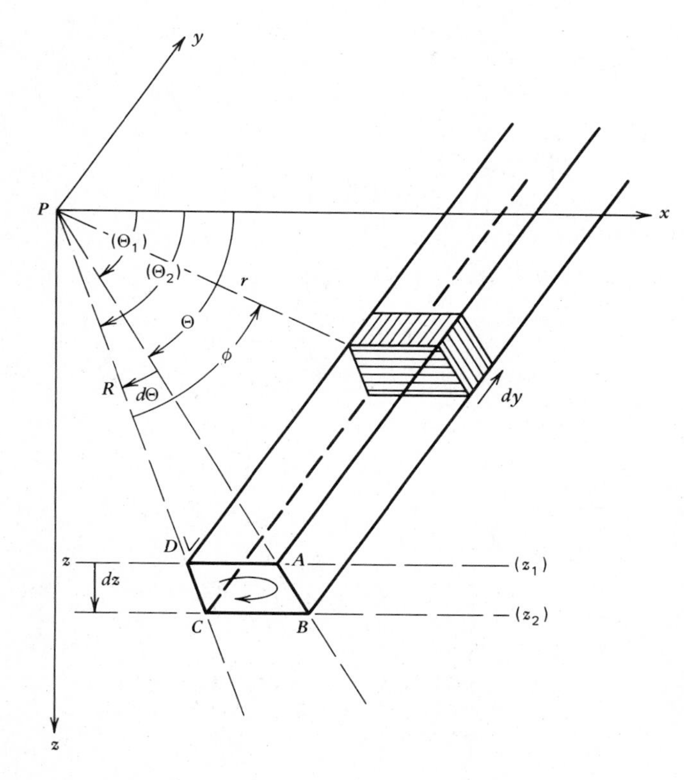

FIG. 9-22 *An elementary prism limited by the planes Θ, $\Theta + d\Theta$, z, and $z + dz$ and extending to infinity in the y-direction. Alternatively, this may be viewed as a infinite prism limited by the planes (Θ_1), (Θ_2), (z_1) and (z_2).*

For the whole elementary infinite prism, we thus obtain

$$d^2g = G\rho\, d\Theta\, dz \int_{-\pi/2}^{\pi/2} \cos\phi\, d\phi = 2G\rho\, d\Theta\, dz$$

For a prism limited by the planes Θ_1, Θ_2, z_1, and z_2, we then obtain

$$g = 2G\rho \int_{\Theta_1}^{\Theta_2} d\Theta \int_{z_1}^{z_2} dz = 2G\rho(\Theta_2 - \Theta_1)(z_2 - z_1)$$

Such a prism may be taken as the one shown in Figure 9-22, in which the planes Θ_1, Θ_2, z_1, and z_2 are shown in parentheses.

This formula may be obtained as a line integral,

$$g = G\rho \oint z\, d\Theta$$

(Note that although a contour normally is turned around counterclockwise, here we turn clockwise.)

Let us evaluate this integral around the contour $ABCD$ (Figures 9-22 and 9-23). Along AB, $\Theta = \Theta_1$ is constant. Hence,

$$\int_A^B z\, d\Theta = 0$$

Along BC, $z = z_2$. Hence,

$$\int_B^C z\, d\Theta = z_2 \int_B^C d\Theta = z_2(\Theta_2 - \Theta_1)$$

Along CD, $\Theta = \Theta_2$. Hence,

$$\int_C^D z\, d\Theta = 0$$

Finally, along DA, $z = z_1$. Hence,

$$\int_D^A z\, d\Theta = z_1 \int_D^A d\Theta = z_1(\Theta_1 - \Theta_2)$$

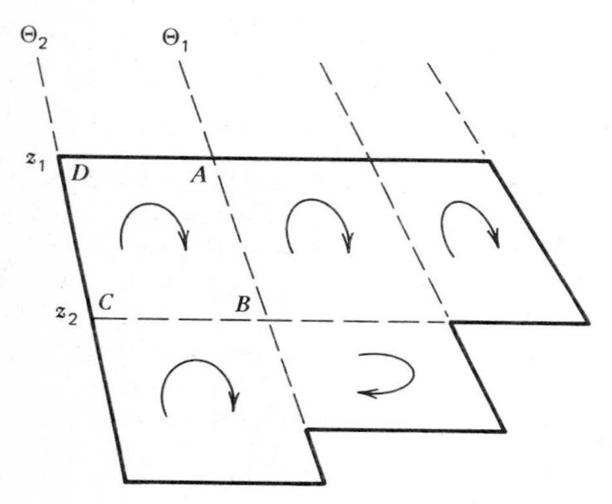

FIG. 9-23 *Five contours adjacent to one another, showing why the line integrals cancel along the inner lines.*

Finally,

$$\oint z\,d\Theta = z_2(\Theta_2 - \Theta_1) + z_1(\Theta_1 - \Theta_2)$$
$$= (\Theta_2 - \Theta_1)(z_2 - z_1)$$

which demonstrates the proposition.

Let us now consider, say, five such contours next to each other, as shown in Figure 9-23. Note that all the inner lines are followed twice, but in opposite directions. Since

$$\int_A^B f(x)\,dx = -\int_B^A f(x)\,dx$$

these integrals cancel out, and we obtain a line integral taken along the periphery.

The final step is to realize that a contour of any shape can be viewed as made up of an infinity of infinitely small contours, and that the cancellation of the integrals along the inner lines produces a line integral around the periphery. Hence,

$$g = 2G\rho \oint z\,d\Theta$$

Referring to Figure 9-21, we wish to evaluate $\oint z\,d\Theta$ along the contour $ABCDE$. To evaluate it, we need only show how to evaluate it along one side—say, BC. Defining Θ, γ_i, and $a_i = PQ$ as shown on the figure, we find at an arbitrary point R on BC,

$$z = x\tan\Theta = (x - a_i)\tan\gamma_i$$

Hence,

$$z = \frac{a_i \tan\Theta \tan\gamma_i}{\tan\gamma_i - \tan\Theta}$$

and

$$\int_{BC} z\,d\Theta = a_i \tan\gamma_i \int_B^C \frac{\tan\Theta\,d\Theta}{\tan\gamma_i - \tan\Theta} \equiv Z_i$$

This is a classical integral (Peirce 1929, Equation 335), and to find the expression given in Equation 9-31, we need only note that at B, $\Theta = \Theta_i$, and at C, $\Theta = \Theta_{i+1}$.

The definition of Θ_i shows that $\Theta_i = \pi/2$ for $x_i = 0$ (i.e., B on the z-axis). The definition of $\int_A^B z\,d\Theta$ shows that if $z_i = z_{i+1}$, then $Z_i = z_i(\Theta_{i+1} - \Theta_i)$. If $x_i = x_{i+1}$, then $\gamma_i = \pi/2$, $\tan\gamma_i = \infty$, and $a_i = x_i$. We rewrite Z_i by dividing numerator and denominator by $\tan\gamma_i$, and obtain

$$Z_i = x_i \int_B^C \tan\Theta\,d\Theta = x_i\left[\log_e(\cos\Theta_i) - \log_e(\cos\Theta_{i+1})\right]$$

Finally, $Z_i = 0$ for $\Theta_i = \Theta_{i+1}$ or $x_i = z_i = 0$ or $x_{i+1} = z_{i+1} = 0$.

The attraction of the prism is thus given by

$$g = 2G\rho \sum_{i=1}^{n} (a_i \sin\gamma_i \cos\gamma_i) \tag{9-31}$$
$$\times \left[(\Theta_i - \Theta_{i+1}) + (\tan\gamma_i)\{\log_e[(\cos\Theta_i)(\tan\Theta_i - \tan\gamma_i)]\}\right.$$
$$\left. - \log_e[(\cos\Theta_{i+1})(\tan\Theta_{i+1} - \tan\gamma_i)]\right]$$

where n is the number of points that define the contour ($n = 5$ in Figure 9-21) and

$$a_i = x_{i+1} + z_{i+1}\frac{(x_{i+1} - x_i)}{(z_i - z_{i+1})}$$

$$\gamma_i = \tan^{-1}\left(\frac{z_{i+1} - z_i}{x_{i+1} - x_i}\right)$$

$$\Theta_i = \tan^{-1}\left(\frac{z_i}{x_i}\right)$$

and, lastly, x_i and z_i are the coordinates of the corners that define the contour.

This formula lends itself readily to evaluation on a computer. The program for this method has been widely distributed. Through division of any geological cross section in subdomains of constant ρ, the attraction of any combination of bodies can readily be found at any point. The method is often referred to as Talwani's method.

Sheet Method The purpose of this method is not to arrive at an exact expression for the attraction of a body, but rather to find the approximate depth at which the cause of an anomaly lies if the density contrast $\delta\rho$ is known. For this purpose we approximate the body causing the anomaly by a sheet of surface density σ.

We assume that the anomaly at the surface is approximately one-dimensional and sinusoidal in character: that is,

$$g_o = A\cos bx \qquad (b = 2\pi k) \tag{9-32}$$

where A and b are thus known.

We now can guess that at a depth z, the potential of a sheet of variable surface density $\sigma = \sigma_o \cos bx$ lying at a depth d will be

$$U = C\sigma_o(\cos bx)\exp[(z - d)b] \tag{9-33}$$

where C and σ_o remain to be determined. It is easy to verify that $\nabla^2 U = 0$; thus, U satisfies a necessary condition of a gravitational potential. It remains to be shown how C and σ_o may be computed. Equation 9-33 gives

$$g = \frac{\partial U}{\partial z} = Cb\sigma_o(\cos bx)\exp[(z - d)b] \tag{9-34}$$

At depth $z=d$, $g=g_d$ can be computed in one of two ways: (a) from the preceding equation, or (b) from the slab formula (Equation 9-14), in which we set $\rho h = \sigma_o \cos bx$. This is valid despite the fact that the surface density of the sheet is not uniform, because the point at which we are computing g is just above the sheet. Thus, only the density immediately below the point is important. The sign has to be taken as positive, since z points downward. Thus,

$$g_d = Cb\sigma_o \cos bx = 2\pi G\sigma_o \cos bx$$

Hence, $C = 2\pi G/b$.

By introducing this value in Equation 9-34 at $z=0$, and equating it to Equation 9-32, we obtain

$$2\pi G\sigma_o(\cos bx)e^{-bd} = A\cos bx$$

which gives

$$\sigma_o = \frac{Ae^{bd}}{2\pi G} \tag{9-35}$$

The amplitude of the density variation necessary to cause a given gravity anomaly at the surface thus increases exponentially with the depth at which this density occurs and with the wave number of this anomaly. When the observed gravity is approximately sinusoidal, this method yields a simple way to determine the maximum possible depth.

Another way to look at the preceding results is to write the gravity at a depth d as

$$g_d = Cb\sigma_o \cos bx = 2\pi G\sigma_o \cos bx = Ae^{bd}\cos bx$$

whereas the gravity at the surface is

$$g_o = A\cos bx = g_d e^{-bd} = g_d \exp(-2\pi kd) = g_d \exp\left(\frac{-2\pi d}{\Lambda}\right) \tag{9-36}$$

A comparison of this result with the one obtained for the attenuation of body waves (Equation 3-6) shows that in both cases the amplitude of the fluctuation (respectively, g_d and A) decreases exponentially with distance, and that in both cases the exponent is inversely proportional to the wavelength (i.e., directly proportional to the frequency or wave number).

PROBLEM

9-8 A gravity anomaly of 20 mgal amplitude and $\Lambda = 30$ km is presumed to be caused by a variation of density of 0.5 g/cm^3 in a layer 3 km thick. What is the approximate depth of this layer? Is this approximately the depth of the top or of the bottom of the layer?

Attraction of Three-Dimensional Bodies

In many cases the attracting body cannot be considered even approximately two-dimensional (seamounts, volcanoes, etc.). In such cases, we must resort to three-dimensional computations, which are much lengthier. Three different methods have been proposed.

The Line Method In this method the body is divided into rectangular vertical prisms, and each prism is represented by a vertical line through its center. The accuracy of this representation increases with the distance to the prism. The attraction of a vertical line at a point can be easily computed. In Figure 9-24, we wish to compute at P the vertical component of a segment BA of line, of mass m per unit length. Let $MP = p$, and R be an arbitrary point on BA. Since $PR = p/\cos\Theta$ and $RR' = (PR) \times d\Theta/\cos\Theta$, we find $RR' = dz = p\,d\Theta/\cos^2\Theta$. Hence the mass of RR' is

$$m\,dz = \frac{mp\,d\Theta}{\cos^2\Theta}$$

The attraction at P of this element is

$$dF = G\frac{mp\,d\Theta}{\cos^2\Theta} \times \frac{\cos^2\Theta}{p^2} = G\frac{m\,d\Theta}{p}$$

It is directed along PR. Hence, its vertical component is $dg = (\sin\Theta) \times dF$, and the total vertical component of attraction at P is

$$d = G\int_{\beta}^{\alpha} \frac{m}{p}\sin\Theta\,d\Theta = G\frac{m}{p}(\cos\beta - \cos\alpha) \tag{9-37}$$

if B is below P.

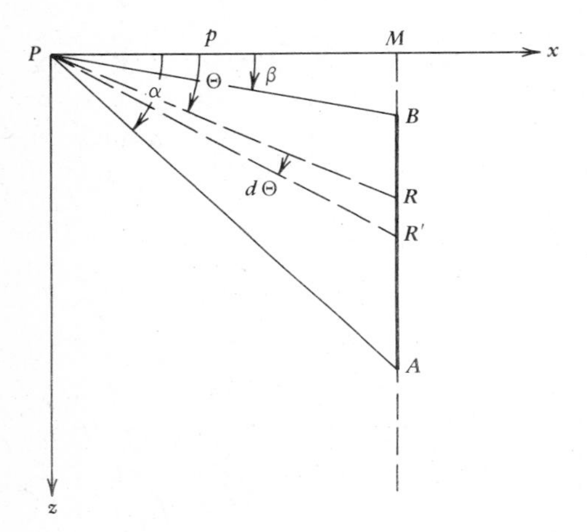

FIG. 9-24 Pertaining to the vertical attraction at P of a vertical segment BA of a straight line.

The Rectangular Parallelepiped Method The attraction of a rectangular parallelepiped can be computed exactly (MacMillan 1958, pp. 72–79), but the formula is very lengthy. This method and the line method are advantageous when gravity corrections are computed, because they can be almost completely automated. They require only that the elevations be read at a regular grid of points. They may also be combined—for example, by use of the line method for more-distant parts of the body, and the other method for a few close ones.

The Right Polygonal Prism Method (Plouff's Method) The attraction of a right polygonal prism can be computed exactly by a method somewhat similar to Talwani's (Plouff 1976). This method is probably less advantageous than the preceding ones in the computation of terrain corrections, for it does not lend itself so easily to automation; however, it is the preferred method when the shape of an attracting body must be determined. It will not be described in detail.

Determination of the Depth of Compensation

The depth of compensation over a continental area of the size of the United States varies greatly. There is thus no "best" value for it. The same is probably true for the compensation mechanism. However, it is possible to use gravity anomalies to determine the *maximum* depth of compensation, especially if an important geological boundary (e.g., continent-ocean) is located nearby. Such a boundary is normally accompanied by a rapid change in the value of the gravity anomaly (i.e., a change of short wavelength). But the density contrast necessary to cause an anomaly increases exponentially with the depth of the body and the wave number of the anomaly (see Equations 9-35 and 9-36). Since the range of possible densities is quite restricted, it follows that the maximum depth of compensation can be determined. For example, the maximum depth of compensation from the Pacific through the Sierra Nevada (near the latitude of San Francisco) does not exceed 60 km (Thompson and Talwani 1964).

Over a more restricted area, on the other hand—and especially over an oceanic region, where the lithosphere is fairly homogeneous—there is a justification for searching for the depth or for the mechanism of compensation that best fits the data.

The method used for this purpose can be most easily understood if we first examine a highly schematic problem. Knowing the bathymetry, $r(x)$, we want to find the theoretical gravity profile $g_t(x)$, if the following assumptions obtain.

1. The problem is purely two-dimensional.
2. The isostatic compensation is of the local Airy type.
3. All densities are known.
4. The normal crustal thickness is known.

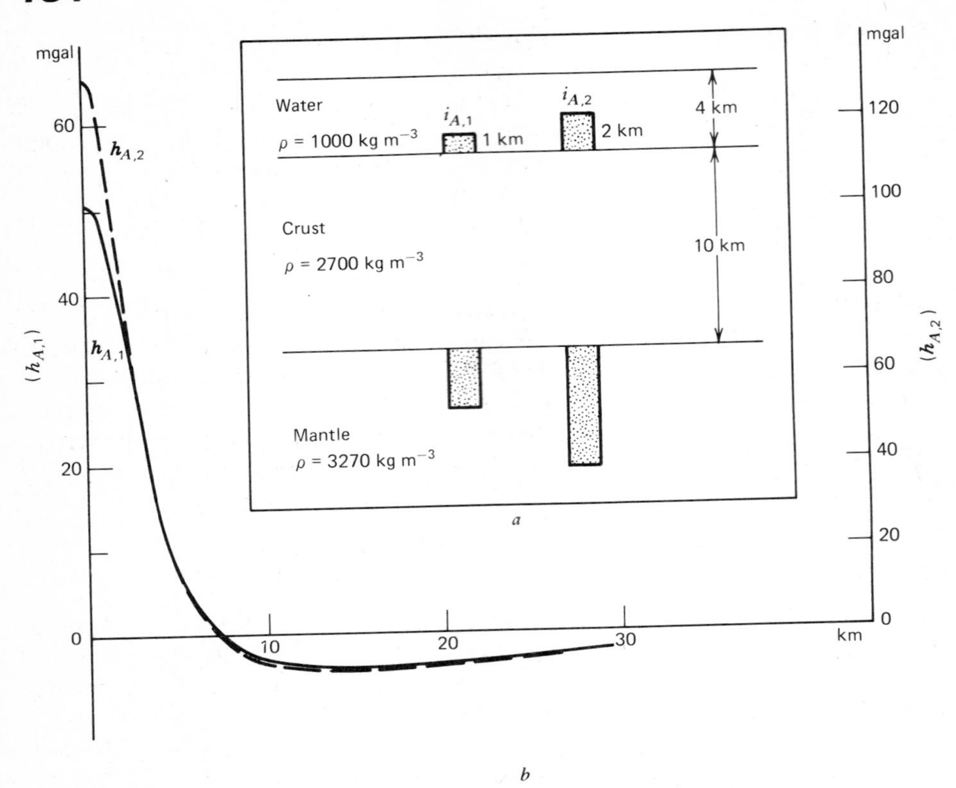

FIG. 9-25 *(a) A schematic oceanic area with impulses $i_{A,1}$ and $i_{A,2}$—1 and 2 km high, respectively—and their roots. (b) The gravity profiles at the ocean surface due to these two structures; only the right half is shown. The scale on the left is for $h_{A,1}$ and that on the right for $h_{A,2}$ for a 1-km-wide "impulse," as shown.) (Two-dimensional case.)*

Figure 9-25*a* shows a crust of normal thickness c overlain by water and perfectly horizontal except for a 1-km-high, infinitely thin vertical mountain, $i_{A,1}$, and its root. (The reason why it is shown 1 km wide will be explained later). The resulting gravity profile, $h_{A,1}$, at the surface of the ocean may be computed analytically (see Problem 9-9); it is shown in Figure 9-25*b*. Under the terminology of linear systems (Chapter 6, Section A), the "mountain" and its root constitute the impulse and the computed gravity becomes the impulse response. But the relief of the ocean bottom (bathymetry) may be considered to consist of an infinity of such impulses of various heights and next to one another. Each results in (approximately—see below) the same impulse response, except for a scale factor introduced by the varying heights. The theoretical gravity profile, $g_t(x)$, may thus be obtained by the addition of all these impulse responses properly scaled and displaced. Each impulse response is symmetric with respect to the impulse. It follows that "flipping it over" does not alter it. Hence, we can obtain $g_t(x)$ by convolving the impulse response with the bathymetry, $r(x)$.

$$g_t(x) = r(x) * h(x) \tag{9-38}$$

In a slightly more realistic case,

- We have an observed gravity profile, $g_o(x)$, and the bathymetry, $r(x)$.
- We compute $h(x)$, $h'(x)$, and so forth for various normal crustal thicknesses —c, c', and so on.
- We use Equation 9-38 to obtain theoretical gravity profiles—g_t, g_t', and so on.
- We compare g_t, g_t', and so on with $g_o(x)$. The one that best fits, determines the normal crustal thickness.

Rather than determining the crustal thickness, we can use this method to compare the various schemes of isostatic compensation. Use of the method for three-dimensional structures necessitates convolution along both the x- and y-axes, but nothing is changed in principle. (Of course, the impulse becomes a vertical line rather than a vertical plane.) Usually all quantities are first transformed from the space domain to the wave-number domain, but the principle is not thereby altered. This method is often called the transfer function, or the admittance method (Dorman and Lewis 1970; Cochran and Talwani 1979).

The gravity profile $h_{A,1}$ shown in Figure 9-25*b* was computed for a vertical plane 1 km high with a surface density, σ, of $(2.7-1.0)\times 10^3$ kg/m^2 for the impulse itself and of $(3.27-2.7)\times 10^3$ kg/m^2 for its root. This is equivalent to a sheet 1 m thick of density $\rho = 1700$ kg/m^3 for the impulse and 570 kg/m^3 for the root. The resulting maximum for $h_{A,1}$ is ~ 50 μgal. If the sheet were 1 km thick, the maximum would be ~ 50 mgal (as shown), except for a negligible difference caused by the finite thickness of the sheet.

Turning briefly to the approximation involved in the Airy case, we note that the gravity $h_{A,2}$, resulting from a 2-km-high impulse, $i_{A,2}$, and its root is not exactly twice as large as that resulting from a 1-km-high impulse (see Figure 9-29*b*). Because of the proximity effect the gravity maximum is multiplied by ~ 2.6 rather than by 2.0, and the gravity is very slightly more negative at large distances. The maximum nonlinearity with respect to the height of the impulse, and thus to the bathymetry, is thus 30%. Within a distance of 2 km of the impulse, the nonlinearity has practically vanished ($h_{A,1} \approx \frac{1}{2}h_{A,2}$). Strictly speaking, then, the gravity profile resulting from the bathymetry cannot be obtained exactly by the addition of impulse responses—that is, Equation 9-38 is not *strictly* correct. However, many other causes of errors are usually more important, especially the lateral variation in density of the crust and mantle and the effect of poorly known bathymetry.

The method outlined above was recently used to determine the thickness of the lithosphere in the Atlantic Ocean (McKenzie and Bowin 1976). Even in such an area the scatter of the data is such that it is not possible to decide on the type of compensation—Airy or flexure of an elastic plate. Geological considerations favor the latter. By taking reasonable values for the densities, the authors determined that a normal plate thickness of 10 km best fits the data.

PROBLEMS

9-9 Compute $h_{A,1}$ and $h_{A,2}$ (shown in Figure 9-25) analytically.

9-10 Do the same but for Pratt's hypothesis, then plot the results and compare with Figure 9-25. Assume a depth of compensation of 100 km below sea level and an average density of 3.27 g/cm^3.

Inverse Problem

Just as in seismology, it is rather easy to solve the direct problem (given the structure, find the gravity anomaly). The inverse problem is much more difficult, and highly nonlinear. (For a discussion of such problems, see Tarantola and Valette 1982*a*). In some cases, however, additional geophysical or geological information can be used to restrict the range of densities or fix the depth of discontinuities. Under these circumstances the method outlined below may give satisfactory results, if it is used with care (Corbato 1965). This method is a generalization of the least-squares method; its principle is applicable to many other physical situations, such as the determination of the distribution of β versus depth from dispersion curves (Chapter 5, Section B.)

We will first examine a very simple case. In Figure 9-26, the local gravity field $d^{(1)}, d^{(2)}, d^{(3)}$ ($d=$"data") is known at points (1), (2), and (3). The field is caused by two separate bodies, X and Y, of known geometry but of unknown densities ρ_X and ρ_Y. These must be determined as closely as possible.

We start by giving ρ_X and ρ_Y reasonable initial values $\rho_{X,0}$ and $\rho_{Y,0}$. We then compute the gravity at the three points (1), (2), and (3): that is, $g_o^{(1)}$, $g_o^{(2)}$, and $g_o^{(3)}$ (see Table 9-1). This computation may be done by Talwani's method, for example. Next, we slightly change the value of ρ_X to $\rho_{X,1}$ but keep ρ_Y at $\rho_{Y,0}$. The values of gravity are now $g_{X,1}{}^{(1)}$, $g_{X,1}{}^{(2)}$ and $g_{X,1}{}^{(3)}$. Clearly,

$$\frac{\partial g_o{}^{(1)}}{\partial \rho_X} \cong \frac{\delta g_o{}^{(1)}}{\delta \rho_X} = \frac{g_{X,1}{}^{(1)} - g_o{}^{(1)}}{\rho_{X,1} - \rho_{X,0}}$$

$$\frac{\partial g_o{}^{(2)}}{\partial \rho_X} \cong \frac{\delta g_o{}^{(2)}}{\delta \rho_X} = \frac{g_{X,1}{}^{(2)} - g_o{}^{(2)}}{\rho_{X,1} - \rho_{X,0}}$$

and similarly for point (3). Thus,

$$\frac{\partial g_o{}^{(i)}}{\partial \rho_X} \cong \frac{g_{X,1}{}^{(i)} - g_o{}^{(i)}}{\rho_{X,1} - \rho_{X,0}} \qquad (i=1,2,3) \tag{9-39}$$

We now return ρ_X to the value $\rho_{X,0}$ but change ρ_Y to $\rho_{Y,1}$ (see Table 9-1). We compute $g_{Y,1}{}^{(1)}$, $g_{Y,1}{}^{(2)}$, $g_{Y,1}{}^{(3)}$, and again find

$$\frac{\partial g_o{}^{(1)}}{\partial \rho_Y} \cong \frac{\delta g_o{}^{(1)}}{\delta \rho_Y} = \frac{g_{Y,1}{}^{(1)} - g_o{}^{(1)}}{\rho_{Y,1} - \rho_{Y,0}}$$

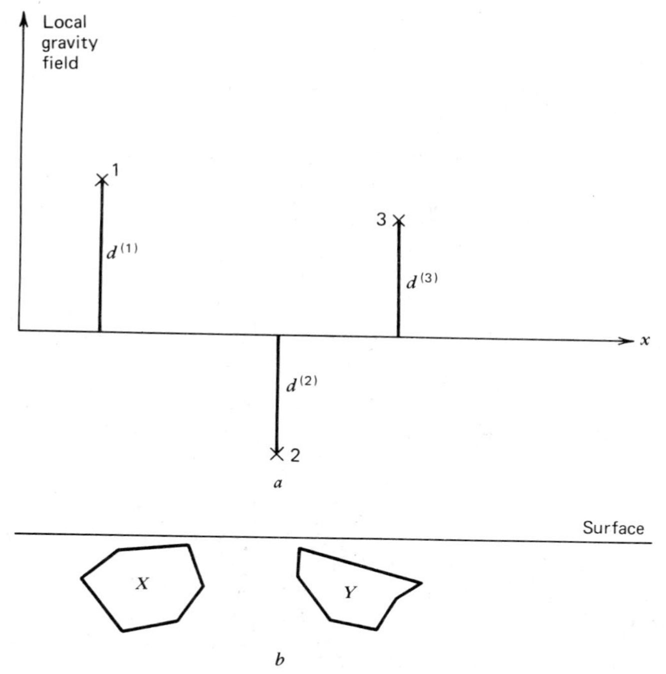

FIG. 9-26 *(a) Local gravity values (data) $d^{(1)}$, $d^{(2)}$, and $d^{(3)}$ at points 1, 2, and 3 due to prisms X and Y of known geometry but unknown densities ρ_X and ρ_Y. (b) Cross section showing the assumed prisms X and Y.*

with similar equations for points (2) and (3). Thus,

$$\frac{\partial g_o^{(i)}}{\partial \rho_Y} \cong \frac{g_{Y,1}^{(i)} - g_o^{(i)}}{\rho_{Y,1} - \rho_{Y,0}} \qquad (i = 1, 2, 3) \tag{9-40}$$

Equations 9-39 and 9-40 enable us to approximate the partial derivatives, as they are needed in what follows. Now, as a first approximation, by Taylor series

TABLE 9-1

Points →		(1)	(2)	(3)
$\rho_{X,0}$, $\rho_{Y,0}$	Initial	$g_o^{(1)}$	$g_o^{(2)}$	$g_o^{(3)}$
$\rho_{X,1}$, $\rho_{Y,0}$		$g_{X,1}^{(1)}$	$g_{X,1}^{(2)}$	$g_{X,1}^{(3)}$
$\rho_{X,0}$, $\rho_{Y,1}$		$g_{Y,1}^{(1)}$	$g_{Y,1}^{(2)}$	$g_{Y,1}^{(3)}$

expansion, omitting higher-order terms in $\delta\rho_X$ and $\delta\rho_Y$,

$$g_1^{(i)} \approx g_o^{(i)} + \frac{\partial g_o^{(i)}}{\partial \rho_X}\,\delta\rho_X + \frac{\partial g_o^{(i)}}{\partial \rho_Y}\,\delta\rho_Y \qquad (i=1,2,3) \tag{9-41}$$

where $\partial g_o^{(i)}/\partial \rho_X$ is given approximately by Equation 9-39 and $\partial g_o^{(i)}/\partial \rho_Y$ is given by Equation 9-40.

Thus, at every point i the difference $\varepsilon^{(i)}$ between the data $d^{(i)}$ and the values of gravity $g^{(i)}$ computed with $\rho_{X,1}$ and $\rho_{Y,1}$ may be written

$$\varepsilon^{(i)} = d^{(i)} - g_1^{(i)} = d^{(i)} - \left[g_o^{(i)} + \frac{\partial g_o^{(i)}}{\partial \rho_X}\,\delta\rho_X + \frac{\partial g_o^{(i)}}{\partial \rho_Y}\,\delta\rho_Y \right] \qquad (i=1,2,3) \tag{9-42}$$

We wish to minimize, as previously,

$$R = \sum_{i=1}^{3} [\varepsilon^{(i)}]^2 \tag{9-43}$$

by an appropriate choice of $\delta\rho_X$ and $\delta\rho_Y$. In doing so, we consider $\delta\rho_X$ and $\delta\rho_Y$ to be variables with no relation to $\partial g_o^{(i)}/\partial \rho_X$ or any other quantity. (This is only approximately correct.) In Equation 9-42 $\delta\rho_X$ and $\delta\rho_Y$ play exactly the same role as a, b, and c played in Equation 9-27. By exactly the same process as previously, we find

$$-\frac{1}{2}\,\frac{\partial R}{\partial(\delta\rho_X)} = \sum_{i=1}^{3} \frac{\partial g_o^{(i)}}{\partial \rho_X}\left[d^{(i)} - g_o^{(i)} - \frac{\partial g_o^{(i)}}{\partial \rho_X}\,\delta\rho_X - \frac{\partial g_o^{(i)}}{\partial \rho_Y}\,\delta\rho_Y \right] = 0$$

or

$$\delta\rho_X \sum_i \frac{\partial g_o^{(i)}}{\partial \rho_X}\,\frac{\partial g_o^{(i)}}{\partial \rho_X} + \delta\rho_Y \sum_i \frac{\partial g_o^{(i)}}{\partial \rho_X}\,\frac{\partial g_o^{(i)}}{\partial \rho_Y} = \sum_i \frac{\partial g_o^{(i)}}{\partial \rho_X}\left[d^{(i)} - g_o^{(i)}\right] \tag{9-44a}$$

and similarly,

$$\delta\rho_X \sum_i \frac{\partial g_o^{(i)}}{\partial \rho_Y}\,\frac{\partial g_o^{(i)}}{\partial \rho_X} + \delta\rho_Y \sum_i \frac{\partial g_o^{(i)}}{\partial \rho_Y}\,\frac{\partial g_o^{(i)}}{\partial \rho_Y} = \sum_i \frac{\partial g_o^{(i)}}{\partial \rho_Y}\left[d^{(i)} - g_o^{(i)}\right] \tag{9-44b}$$

This is a system of two equations in $\delta\rho_X$ and $\delta\rho_Y$, which can thus be determined. Hence, the optimum values of ρ_X and ρ_Y are

$$\rho_X^{\text{opt}} = \rho_{X,0} + \delta\rho_X, \qquad \rho_Y^{\text{opt}} = \rho_{Y,0} + \delta\rho_Y$$

The generalization of the method is obvious: if more densities are allowed to vary, there will be correspondingly more equations, and more $\delta\rho$s to be determined. If the position of one vertex, say, C (see Figure 9-21), is allowed to vary, we need to compute

$$\frac{\delta g^{(1)}}{\delta x_C} \approx \frac{\partial g^{(1)}}{\partial x_C} \quad \text{and} \quad \frac{\delta g^{(1)}}{\delta z_C} \approx \frac{\partial g^{(1)}}{\partial z_C}$$

If the bodies are partly superposed, a vertical change in their interface has the same effect as a variation in density, and so forth. Additionally, note that the values of all $\partial g/\partial \rho$ in Equations 9-44a and 9-44b may be computed exactly from Equation 9-31. For two bodies of different densities ρ_X and ρ_Y, this equation takes the form

$$g = \rho_X A + \rho_Y B$$

Hence,

$$\frac{\partial g}{\partial \rho_X} = A, \qquad \frac{\partial g}{\partial \rho_Y} = B$$

There are several difficulties with this method, as well as with all similar "simple" inverse methods, but they are not discussed here. Above all, one should restrict the number of quantities allowed to vary and use all other available information to constrain the solution. Because of the approximation made (that Equation 9-41 is linear is $\delta\rho_X$ and $\delta\rho_Y$), the present method is a linearization of the problem.

PROBLEM

9-11 This is a very simple example of least-squares inversion by linearization. It is not realistic, but it will help you to understand the method. It can be programmed, but it can also be done easily by hand.

Take the origin at O. The following four readings have been made in a straight line from O (the first number in each pair is the distance from O in meters, the second one is the anomaly in mgals): (200, −1.4), (700, −1.35), (1100, −0.32), (1700, −0.11).

These anomalies are caused by two spheres, X and Y, with centers under the profile.

(a) Sphere S_X has a radius of 600 m, center at $x = 600$, $z = 1000$, and sphere S_Y has a radius of 300 m, center at $x = 1300$, $z = 400$. Determine the densities of each sphere if the surrounding rock has a density of $\bar{\rho} = 2.4$ g/cm^3. *Suggestion*: Take $\rho_{X,0} = \rho_{Y,0} = \bar{\rho}$. What is $g_o^{(i)}$?

This problem can be solved exactly, but this is not true in general.

(b) You are given the x-coordinate of the center of S_X (600 m), along with its density (2.1 g/cm^3) and radius (600 m), but not its depth. The data for S_Y are the same as above. Determine the depth of the center of S_X and the density ρ_Y.

Density Distribution in the Mantle

In 1923, Adams and Williamson made the remarkable discovery that the density distribution in the mantle can be computed from seismic velocities. As will be shown, this method assumes homogeneity and isentropy (Chapter 13, Section B), neither of which obtains exactly in the earth. Nevertheless, the use of this method enabled its authors to show that the density at the bottom of the mantle is much smaller than at the top of the core (see Figure 3-15), thus confirming the presumed difference in chemical composition.

The Adams-Williamson equation may be obtained as follows. Assuming that the earth is radially symmetric, if the z-axis points downward and the r-axis points upward, then $dz = -dr$, and we have

$$\frac{d\rho}{dz} = -\frac{d\rho}{dr} = -\frac{d\rho}{dP}\frac{dP}{dr} \tag{9-45}$$

But (Equation 2-31)

$$\kappa = -\frac{dP}{d\Theta} \tag{9-46}$$

and

$$d\Theta = -\frac{d\rho}{\rho} \tag{9-47}$$

since $d\Theta$ is a dilatation and $d\rho$ an increase in density. Hence,

$$\frac{d\rho}{dP} = \frac{\rho}{\kappa} \tag{9-48}$$

But

$$\alpha^2 - \frac{4}{3}\beta^2 = \frac{\lambda + 2\mu}{\rho} - \frac{4}{3}\frac{\mu}{\rho} = \frac{\lambda + (2/3)\mu}{\rho} = \frac{\kappa}{\rho} = \phi \tag{9-49}$$

(See Equation 2-32.) Hence,

$$\frac{d\rho}{dP} = \frac{1}{\phi} = \frac{1}{\alpha^2 - (4/3)\beta^2} \tag{9-50}$$

$v_\phi = \sqrt{\phi}$ is sometimes called the bulk velocity, since κ is the bulk modulus. We thus can compute $d\rho/dP$ from the distribution of seismic velocities versus depth.

We can also compute dP/dr at any depth z; for hydrostatic conditions we have $P = \int_0^z \rho g\, dz$, hence $dP/dz = \rho g$. Since z and r point in opposite directions,

$$\frac{dP}{dr} = -\rho g \tag{9-51}$$

Note that this equation does not assume the constancy of ρ or g. But (see Chapter 8, Section B) at a point R inside the earth, the spherical shell above R exerts no gravitational attraction; g is caused only by the mass of the sphere below it. If m_b represents this mass, m_a the mass of the spherical shell, and M_e the mass of the earth, then

$$m_b = M_e - m_a = M_e - \int_r^{R_e} 4\pi r^2 \rho\, dr \tag{9-52}$$

where ρ may vary with r. Also,

$$g = G\frac{m_b}{r^2} = \frac{G(M_e - m_a)}{r^2} \tag{9-53}$$

Using Equations 9-53, 9-51, and 9-50, we obtain, from Equation 9-45,

$$\frac{d\rho}{dr} = -\frac{\rho g}{\alpha^2 - (4/3)\beta^2} \tag{9-54}$$

or

$$\frac{d\rho}{\rho} = -\frac{G}{\alpha^2 - (4/3)\beta^2} \times \frac{M_e - m_a}{r^2}\,dr \tag{9-55}$$

or

$$d(\log_e \rho) = \frac{G}{\alpha^2 - (4/3)\beta^2} \times \frac{M_e - m_a}{r^2}\,dz \tag{9-56}$$

This equation may be used to find ρ versus r as follows. Starting from a shallow depth where ρ_o is known, we assume that ρ remains constant over a depth δz (say, 100 km). The starting point might be the Moho, in which case m_a would be the mass of the "average" crust. We evaluate the right-hand side of Equation 9-56, replacing dz by δz and using the values of α, β, and r at the Moho. This yields $\delta(\log_e \rho)$: that is, $\log_e \rho_1 - \log_e \rho_o$. Since $\log_e \rho_o$ is known, we find $\log_e \rho_1$. By repeating the procedure, we determine ρ throughout the mantle. There are, of course, many possible refinements in this procedure, but they need not concern us. What is important here is to note the necessity for homogeneity and isentropy (adiabaticity).

The condition of homogeneity arises from the fact that only in a homogeneous material is the change in density caused solely by a change in pressure (Equation 9-48); clearly, the density may also vary as a result of phase changes or changes in composition. As a consequence, the Adams-Williamson equation cannot be applied to the whole mantle; independent data must be used to estimate the density changes across the 400- and 650-km discontinuities.

The need for isentropy (adiabaticity), pointed out by Birch (1952), results from the fact that seismic velocities are related to the isentropic (adiabatic) compressibility. Because the actual thermal gradient is steeper than the isentropic one (Chapter 14, Section B), Equation 9-54 becomes

$$\frac{d\rho}{dr} = -\left[\frac{\rho g}{\alpha^2 - (4/3)\beta^2}\right] + \alpha\rho s$$

where s is the superisentropic (superadiabatic) gradient—that is, the excess of the real temperature gradient over the isentropic one. This quantity is poorly known.

D. Gravity Measurements

Absolute Measurements

Because of advances in time and distance measurements, it is now possible to use free-fall devices to determine the absolute value of g in a few days. In principle, of course, it suffices to measure how far a body falls *in vacuo* in a given time interval, or how long it takes for a body to fall a certain distance. There are, however, a large number of practical difficulties. In the first such instrument built, for example, the body was a ruler with fine graduations engraved on it; when the ruler was released, it ceased to be "stretched" by gravity and started vibrating longitudinally, thus reducing the accuracy. If one uses small spheres the vibration problem is minimized, but it is very difficult to determine the centers of the spheres. Since 1 mgal $= 10^{-2}$ mm/s^2, the distance must be very accurately measured. Much ingenuity has been devoted to solving these and similar problems. After some experiments with "cat's eyes" (a lens in front of a mirror), experimenters have turned to corner reflectors (the corner of a cube), which always reflects the light parallel to the incoming beam. The distance is measured by counting interference fringes on a laser beam (Zumberge and Faller 1982).

The latest instrument is said to be portable because it weighs less than 1 ton. It determines g with an accuracy on the order of 0.01 mgal $= 10^{-7}$ m/s^2. Note that such accuracy involves additional difficulties, resulting from the variation of g along the path because of the free-air correction, the acceleration caused by microseisms, tidal effects (since measurements may take a few days), and other things.

Pendulum Measurements

The period of an ideal simple pendulum is given by

$$T = 2\pi\sqrt{\frac{l}{g}} \tag{9-57}$$

This equation supposes that the motion of the mass is small. The length l of the ideal simple pendulum is the distance between the axis of oscillation and the point mass. In a real pendulum the equation is still valid but l cannot be accurately determined. (Before the advent of free-fall apparatus, special pendulums were used that could oscillate on either of two knife edges, say, A and B, set near each end. The distance AB is equal to the length of the simple pendulum if the periods of oscillation around A and B are equal.) In order to determine l, it is thus necessary to take the pendulum to a reference station where g is known. Once this is done, the instrument can be used to measure g anywhere. Such an apparatus may thus be said to give a relative measurement—that is, relative to the reference station.

The advantage of a pendulum apparatus is that it is not subject to drift. The disadvantage is that the measurements take a long time (several hours). Consequently pendulum measurements are used only at secondary reference stations that are fairly widely spaced and are distributed throughout the world.

Gravimeters

As discussed in Chapter 7, a seismograph acts as an accelerometer for periods that greatly exceed its free period. Many present-day gravimeters are essentially based on the schematic diagram shown in Figure 7-5. In others, the axis of rotation consists of a torsion spring that functions as the main spring of the instrument.

All things being equal, the sensitivity s of such an instrument (displacement from zero for a given change in g) is proportional to the square of the period, for the following reason. The elongation e of a spring with spring constant c stretched by a mass m in a gravity field g is $e = mg/c$. The period of such a system when set in vibration is $T = 2(m/c)^{1/2}$. Hence,

$$e = \frac{gT^2}{4\pi^2}$$

And

$$s = \frac{de}{dg} = \frac{T^2}{4\pi^2} \tag{9-58}$$

If mechanical means of obtaining the displacement are used (as in the Worden gravimeter—see below), then it is advantageous for T to be relatively large (~ 6 s). If electronic amplification of the displacement is used, this is less important. Furthermore, as T increases, the instrument may become relatively more difficult to set up.

A superconducting gravimeter also has been built; its purpose is to measure small variations of g with time at a given place. In it, the position of a small sphere suspended in the magnetic field of two superconducting coils is sensed electronically. In the United States the Worden gravimeter is a commonly used instrument. The mechanism itself, including the springs, support, and other components, is built of fused quartz so as to be dimensionally stable with temperature changes. Some adjustments must still be made, though, to compensate for the variation with temperature of the elastic constant of the spring. A dial on the instrument can be turned (and read) to move the attachment point of a fine spring that nulls the instrument.

A gravimeter needs two reference stations. At the first one, a certain value of g is determined to correspond to a certain reading of the dial. At the second one, the change in g is determined to correspond to a certain number of dial divisions. For this reason, gravimeters may be said to be doubly relative.

The main difficulty with gravimeters is that they are subject to drift caused by the slow creep of the spring and/or by thermal effects. If this drift were exactly linear with time, it could easily be taken into account. Such is not always the case, however, and some instruments are affected by unpredictable jumps, or "tares," which can cause serious difficulties.

A gravimeter must be enclosed in a rigid, airtight case; otherwise, the variation of air density with atmospheric pressure will cause large errors resulting from the variable buoyancy of the mass. Alternatively, a hollow compensating volume may be installed on the opposite side of the pivot point from the mass.

All field measurements are now made with gravimeters and take only a few minutes.

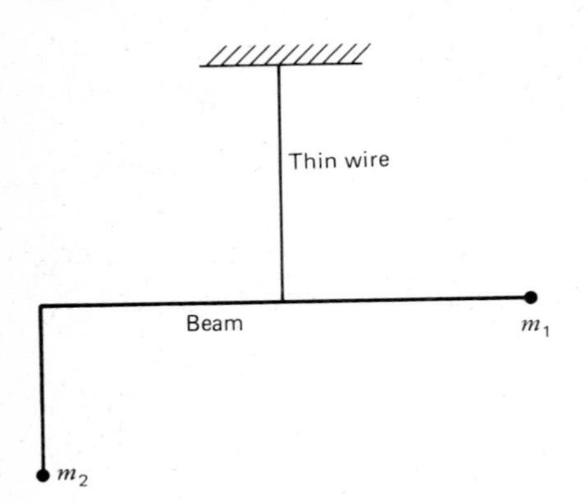

FIG. 9-27 *Schematic diagram of a torsion balance.*

Torsion Balances

The earliest instrument used in prospecting was the torsion balance, invented by the Hungarian physicist Roland von Eötvös (see Equation 9-20) in 1880. (Eötvös was primarily interested in the earth's gravitational and magnetic fields. He also wanted to find out whether the inertial and the gravitational masses of an object are the same. Using a torsion balance, he showed that the difference between them is less than 10^{-7}. This difference is now thought not to exist.) The instrument now has only a historical interest, but it is too clever to pass over completely. It consists of two equal masses, m_1 and m_2 (Figure 9-27), suspended at unequal heights by means of a thin wire and a beam.

The exact theory behind the instrument is complicated and no longer relevant, but some of its implications are still of academic interest. If the direction of gravity—that is, the vertical—were exactly the same at m_1 and m_2, then, by definition, $g_x = g_y = 0$ if the z-axis is vertical. This would imply that there is no horizontal force, and thus no force rotating the beam. In fact, though, the beam does rotate. Hence the vertical is not exactly the same at the two masses, which is the same as saying that the equipotential surfaces are warped and/or do not remain equidistant. But $g_x = \partial U/\partial x$ and $g_y = \partial U/\partial y$. The rotation of the beam thus permits one to compute such quantities as $\partial g_x/\partial x = \partial^2 U/\partial x^2$, $\partial g_z/\partial x = \partial g_x/\partial z = \partial^2 U/\partial x\, \partial z$, and so forth. (The system is compensated to allow the beam to remain horizontal.)

Although the torsion balance was very laborious to use by present standards (five hours per station), and the interpretation of the results equally difficult, up to 125 instruments were in use annually in the mid-1930s, and they greatly contributed to the discovery of 79 oil fields and salt domes.

A final note: The military uses of gravity and geodesy have been clear since about 1960. In order to shoot a ballistic missile at a target, it is necessary to know with precision both the position of the target and the gravity field along the trajectory. These uses have provided a great impetus for the creation of large data banks in these two branches of science.

Brief Review of the Contributions of Gravimetry to Plate Tectonics

Gravimetry has not yet had a major impact on plate tectonics. It appears possible that satellite altimetry may alter this situation for short wavelengths. Discovery of the significance of the general pattern of elevation and depression of the geoid could make a major contribution to the understanding of the dynamics of the earth's interior, hence also of plate motions.

PART THREE

Geomagnetism

10

Magnetostatics

A. Introduction

The fact that magnetite attracts iron was known to the ancient Greeks, and the invention of the compass, generally credited to the Chinese, occurred at least as early as the eleventh century. Observation of the magnetic field at the surface of the earth can have a variety of purposes: for example, to detect temporal electromagnetic disturbances originating outside the earth, especially from the sun; to record the large-scale spatial variations connected with plate tectonics; to detect the depth of the basement; or to attempt to locate magnetic mineral deposits.

There are both similarities and differences between gravity and magnetism. In gravity, all masses are positive and two masses (both positive) attract each other. In magnetism, instead of a mass one speaks of a pole: a positive pole and a negative one of equal strength always occur together; like poles repel each other and opposite poles attract each other. It is sometimes thought that the possibility of magnetic shielding is linked with this duality of sign, whereas the nonexistence of gravity shielding is linked with the absence of negative mass. Finally, both the magnetic field and the gravity field are the gradient of a potential.

The branch of magnetism that deals with the field of motionless magnetic bodies is called magnetostatics. This is the subject that we will be concerned with here, except when we consider the origin of the earth's field.

B. Magnetic Field

The Dipole

Consider a positive magnetic pole, $+p$, and negative one, $-p$, separated by a distance $\mathbf{d}$ (from $-p$ to $+p$). The product $\mathbf{M} = p\mathbf{d}$ is called the magnetic moment. When $\mathbf{d} \to 0$, the result is a dipole. (Sometimes this term is applied to the case in which $\mathbf{d}$ is not infinitesimal; if so, there is an infinitesimal dipole when $\mathbf{d} \to 0$). Any magnetic body may be thought of as consisting of an infinity of dipoles, and the magnetic moment per unit volume is the intensity of magnetization, $\mathbf{J}$ (also $\mathbf{I}$ or $\mathbf{M}$), often called simply the magnetization. The direction of magnetization is $\hat{\mathbf{j}}$.

The Magnetic Field

Just as the gravitational field at point P is the force exerted at P on a unit mass, so the magnetic field $\mathbf{H}$ at P is the force exerted at P on a unit positive pole, even though such an isolated pole does not exist. The magnetic field $\mathbf{H}$ is the gradient of the magnetic potential W.

Relation between the Gravitational and the Magnetic Potentials

Consider a point P and a finite dipole (Figure 10-1). The distance from the positive pole $+p$ to P is $\mathbf{r}_+$, and that from the negative pole is $\mathbf{r}_-$. The magnetic potential W at P is the difference of the potentials caused by the positive and negative poles.

$$W = A\left(\frac{1}{r_+} - \frac{1}{r_-}\right) = \frac{Ad\left(\frac{1}{r_+} - \frac{1}{r_-}\right)}{d}$$

where A is a constant to be determined. The expression in parentheses is the change in $1/r$ evaluated from $-p$ (where $r = r_-$) to $+p$ (where $r = r_+$). (A change is always the final value minus the initial one.)

But $\mathbf{d}$ is the vector from $-p$ to $+p$. Let $\mathbf{d} \to 0$. Then,

$$\frac{1}{r_+} - \frac{1}{r_-} \to 0$$

but

$$\frac{(1/r_+) - (1/r_-)}{d} \to \frac{\partial(1/r)}{\partial d}$$

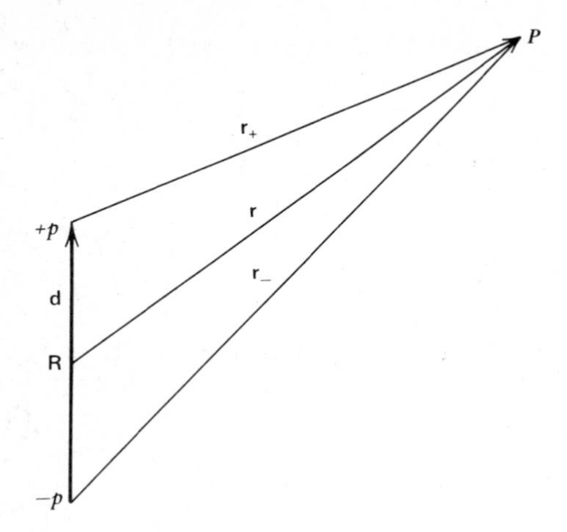

FIG. 10-1 *A finite dipole, an outside point P and the vectors* $\mathbf{d}, \mathbf{r}, \mathbf{r}_+, \mathbf{r}_-$.

This last expression is the directional derivative of $1/r$ in the direction of the dipole (see Appendix E). Hence, as $d \to 0$, we obtain

$$W = Ad\frac{\partial(1/r)}{\partial d}$$

However, $1/r$ is proportional to the gravity potential U at P caused by a point mass at R, and $\mathbf{r} = \mathbf{RP}$. Since $\mathbf{d} \to 0$, $\mathbf{r} = \mathbf{r}_+ = \mathbf{r}_-$. Hence,

$$W = C\frac{\partial U}{\partial d}$$

where the constant C remains to be determined. This equation states that the magnetic potential of a dipole is the directional derivative of the gravity potential of a point mass taken in the direction of the dipole, except for a constant factor C that takes into account the change from gravity to magnetism. But the directional derivative is equal to the projection of the gradient on the direction considered. Hence (see Figure 10-2),

$$W = C\ \nabla U \cos\Theta \tag{10-1}$$

where ∇U is the magnitude of $\nabla \mathbf{U}$, C remains to be determined, and Θ is the

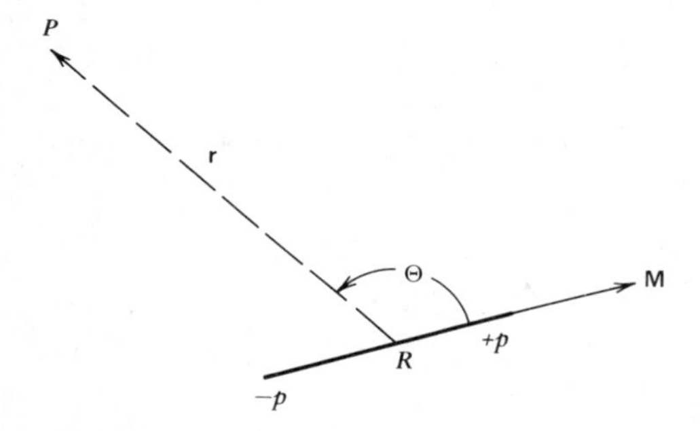

FIG. 10-2 *A dipole, the direction of its moment* **M**, *a point P, and the angle* Θ.

angle from **M** to **RP**. (**M** and **d** are in the same direction.) But for a point mass (in gravity, see Equation 8-11),

$$F = \nabla U = -\frac{Gm}{r^2}$$

Substituting in Equation 10-1, we obtain

$$W = -CGm\frac{\cos\Theta}{r^2} \tag{10-2}$$

where C must have the proper units: that is, it should be proportional to M and not include G or m. Hence,

$$W = -\frac{M\cos\Theta}{4\pi r^2} \tag{10-3}$$

The factor $1/4\pi$ is attributable only to the unit system in which we operate (SI; see Section 10C); in the emu (electromagnetic unit) system this factor does not appear.

The same reasoning may be applied to a uniformly magnetized body (constant **J**) consisting of an infinity of dipoles rather than a single dipole: M is replaced by $J\,dV$, and we obtain, from Equation 10-3,

$$W = -\int_V \frac{J\cos\Theta}{4\pi r^2}\,dV \tag{10-4}$$

But for an extended mass the gravitational attraction is (Equation 8-16)

$$\nabla U = F = -G\int_V \frac{\rho\,dV}{r^2} \tag{10-5}$$

Substituting Equations 10-4 and 10-5 into equation 10-1, we obtain

$$-\int \frac{J\cos\Theta}{4\pi r^2}\,dV = -CG\int \frac{\rho\cos\Theta}{r^2}\,dV$$

Equating the integrands gives

$$C = +\frac{J}{4\pi G\rho}$$

Equation 10-1 thus becomes

$$W = \frac{J}{4\pi G\rho}\,\nabla U\cos\Theta = \frac{J}{4\pi G\rho}\frac{\partial U}{\partial j} \tag{10-6}$$

It is easy to show that a similar relation holds for the magnetic and gravitational fields. For example,

$$H_x = \frac{\partial W}{\partial x} = +\frac{J}{4\pi G\rho}\frac{\partial}{\partial x}\frac{\partial U}{\partial j} = \frac{J}{4\pi G\rho}\frac{\partial}{\partial j}\frac{\partial U}{\partial x}$$
$$= \frac{J}{4\pi G\rho}\frac{\partial}{\partial j}(F_x)$$

where F_x is the x-component of the gravitational field. A similar relation holds for the y- and z-components. These three relations can be conveniently written together as

$$\mathbf{H} = \frac{J}{4\pi G\rho} \frac{\partial \mathbf{F}}{\partial j} \tag{10-7}$$

That is, the magnetic field is proportional to the directional derivative of the gravitational field in the direction of magnetization.

The relation between the gravitational and magnetic potentials (Equation 10-6), or between the gravitational and magnetic fields (Equation 10-7), is known as Poisson's theorem (or Poisson's relationship).

If J or $\hat{\mathbf{j}}$ varies in the magnetic body, Equations 10-6 and 10-7 are still valid but represent only the infinitesimal potential or field due to an element of volume.

Mathematical Note The sign of the potentials are a subject of "puzzling confusion" (Kellogg 1953, p. 53) in gravity and magnetism. The same is true for the sign in the relation between the field and the potential and for the direction of the vector **r**. In this text, both fields are taken to be the positive gradient of a potential and to be in the direction of the force acting on a positive point mass or pole. Furthermore, the potential is taken to be at a point of coordinates (x, y, z), which is the end point of **r**. These conventions are by no means universal, but they seem to be the simplest.

Field of a Dipole

Because the magnetic field **H** is the gradient of the magnetic potential W (i.e., $\mathbf{H} = \nabla \mathbf{W}$), Equation 10-3 can be used to find the field **H** of a dipole (Figure 10-3).

$$H_r = \frac{\partial W}{\partial r} = \frac{2M\cos\Theta}{4\pi r^3} \tag{10-8}$$

$$H_\Theta = \frac{\partial W}{r\,\partial\Theta} = \frac{M\sin\Theta}{4\pi r^3} \tag{10-9}$$

where H_r is the component of the magnetic field in the radial direction (>0 outward), H_Θ in the tangential direction (>0 for increasing Θ), and Θ is counted from the $-p$ to the $+p$ direction.

We can easily justify Equation 10-8 by orienting the x-axis along r. Then,

$$H_r = H_x = \frac{\partial W}{\partial x} = \frac{\partial W}{\partial r} = \frac{\partial}{\partial r}\left(-\frac{M\cos\Theta}{4\pi r^2}\right) = \frac{2M\cos\Theta}{4\pi r^3}$$

With this orientation of the x-axis, we orient the y-axis along the direction of

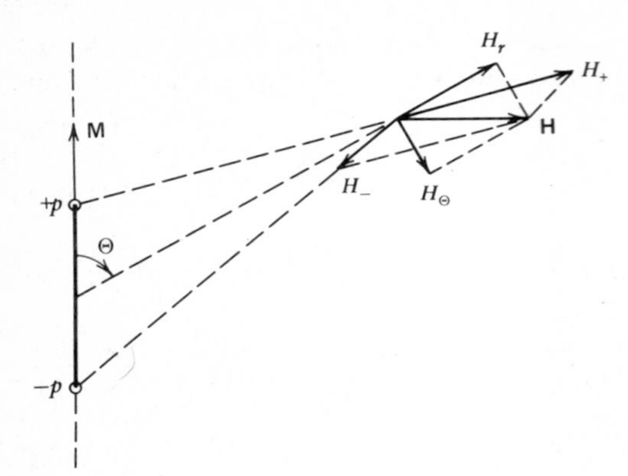

FIG. 10-3 *The field* **H** *at a point due to a dipole can be approximately computed from the attraction H_-, due to the negative pole $-p$, and the repulsion H_+, due to $+p$. H_r is the radial component of* **H** *and H_Θ its tangential component.*

increasing Θ, and obtain $dy = r\,d\Theta$. Hence,

$$H_\Theta = H_y = \frac{\partial W}{\partial y} = \frac{\partial W}{r\,\partial\Theta} = \frac{1}{r}\frac{\partial}{\partial\Theta}\left(-\frac{M\cos\Theta}{4\pi r^2}\right) = \frac{M\sin\Theta}{4\pi r^3}$$

thus proving Equation 10-9.

As shown in Figure 10-3, the total field **H** may also be viewed as the resultant of the repulsive field H_+ due to the positive pole and of H_- due to the negative pole. (Of course, Equation 10-8 and 10-9 are strictly valid only for an infinitesimal dipole.)

Equations 10-8 and 10-9 show that the magnetic field of a dipole decreases according to the inverse of the cube of the distance, whereas the gravitational field decreases with the inverse of the square of the distance. They also show that the average value of H_r is twice that of H_Θ.

Finally, note that two dipoles of equal moment that are located at the same distance from a point yield different magnetic fields at that point, depending on their orientations (Θ). The same is true for extended magnetic bodies of constant magnetization. This phenomenon has no counterpart in gravity. In reality, as we shall see, the magnetization is often in the direction of the earth's field, whose direction (inclination) varies with latitude (Chapter 11, Section A); hence, the field resulting from a given magnetic mass differs greatly at different latitudes.

Field of a Sphere

To show that the magnetic field has a much more complex expression than the gravitational field, we will compute the magnetic field arising from a sphere of constant magnetization. The algebra is tedious but elementary.

Given that the center of the sphere (see Figure 10-4) is at $O\ (a, b, c)$ and its volume is V, we wish to find the magnetic field at $P\ (x, y, z)$. Poisson's theorem shows that the magnetic field is the directional derivative of the gravity field times a constant factor. But the gravitational field of a sphere is the same as that of point mass located at the sphere's center. It follows that the magnetic field of a

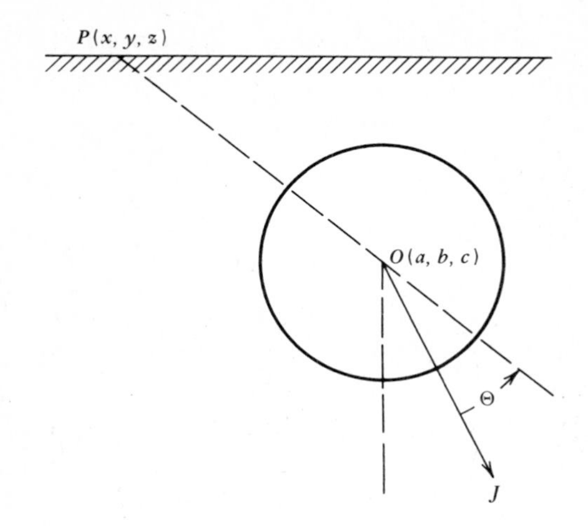

FIG. 10-4 *A magnetized sphere of center O and magnetization* **J**, *a neighboring point P, and the angle Θ.*

uniformly magnetized sphere is the directional derivative of that of a point mass located at its center.

Our first step is to use Equation 10-6 to compute W. The gravitational potential of a sphere at an external point is

$$U = \frac{GM}{r} = \frac{G\rho V}{r}$$

Hence, Equation 10-6 yields

$$W_{\bullet} = \frac{JV}{4\pi}\frac{\partial}{\partial j}\left(\frac{1}{r}\right) \tag{10-10}$$

Now, **J** can have any direction. Assuming first that it is vertical and of magnitude J_z, we write the corresponding potential W_z. Hence,

$$W_z = \frac{J_z V}{4\pi}\frac{\partial}{\partial z}\left(\frac{1}{r}\right) \tag{10-11}$$

where $r^2 = (x-a)^2 + (y-b)^2 + (z-c)^2$. This yields, as in the case of gravity (see Equations 8-8 and 8-9)

$$W_z = \frac{J_z V}{4\pi}(c-z)r^{-3} \tag{10-12}$$

Writing the horizontal field for this case as $(H_z)_x = (\partial/\partial x)(W_z)$ (with similar notations for the other components of the field), we obtain

$$4\pi\frac{\partial}{\partial x}(W_z) = 4\pi(H_z)_x = J_z V(c-z)\frac{\partial(r^{-3})}{\partial x} \tag{10-13}$$

$$4\pi\frac{\partial}{\partial y}(W_z) = 4\pi(H_z)_y = J_z V(c-z)\frac{\partial(r^{-3})}{\partial y} \tag{10-14}$$

$$4\pi\frac{\partial}{\partial z}(W_z) = 4\pi(H_z)_z = J_z V\frac{\partial}{\partial z}\left[(c-z)r^{-3}\right] \tag{10-15}$$

which can easily be computed (see Equations 10-22 and 10-23).

Similarly, if **J** is along the x-axis and of magnitude J_x, Equation 10-11 becomes

$$W_x = \frac{J_x V}{4\pi} \frac{\partial}{\partial x}\left(\frac{1}{r}\right) \tag{10-16}$$

$$= \frac{J_x V}{4\pi}\left[(a-x)r^{-3}\right]$$

and Equations 10-13 to 10-15 become

$$4\pi \frac{\partial}{\partial x}(W_x) = 4\pi (H_x)_x = J_x V \frac{\partial}{\partial x}\left[(a-x)r^{-3}\right] \tag{10-17}$$

$$4\pi \frac{\partial}{\partial y}(W_x) = 4\pi (H_x)_y = J_x V (a-x)\frac{\partial (r^{-3})}{\partial y} \tag{10-18}$$

$$4\pi \frac{\partial}{\partial z}(W_x) = 4\pi (H_x)_z = J_x V (a-x)\frac{\partial (r^{-3})}{\partial z} \tag{10-19}$$

Note that from Equations 10-16 and 10-18,

$$4\pi (H_x)_y = \frac{\partial}{\partial y}\left[J_x V \frac{\partial}{\partial x}\left(\frac{1}{r}\right)\right] = J_x V \frac{\partial^2}{\partial x\,\partial y}\left(\frac{1}{r}\right)$$

and

$$4\pi (H_y)_x = \frac{\partial}{\partial x}\left[J_y V \frac{\partial}{\partial y}\left(\frac{1}{r}\right)\right] = J_y V \frac{\partial^2}{\partial x\,\partial y}\left(\frac{1}{r}\right)$$

hence,

$$\frac{(H_x)_y}{J_x} = \frac{(H_y)_x}{J_y}, \qquad \frac{(H_z)_y}{J_z} = \frac{(H_y)_z}{J_y}, \qquad \text{etc.}$$

The case in which **J** is along the y-axis is now evident. The general case, in which **J** has an arbitrary orientation,

$$\mathbf{J} = \hat{\mathbf{i}} J_x + \hat{\mathbf{j}} J_y + \hat{\mathbf{k}} J_z \tag{10-20}$$

can be obtained by addition: for example,

$$H_y = (H_x)_y + (H_y)_y + (H_z)_y \tag{10-21}$$

where $(H_x)_y$ is given by Equation 10-18, $(H_z)_y$ is given by Equation 10-14, and $(H_y)_y$ can easily be written. It may be useful to write out two typical equations: for example,

$$4\pi (H_z)_x = +3 J_z V (c-z)(a-x) r^{-5} \tag{10-22}$$

$$4\pi (H_z)_z = -J_z V\left[r^2 - 3(c-z)^2\right] r^{-5} \tag{10-23}$$

All other equations can be obtained through circular permutations of these two.

The case of a uniformly magnetized sphere is the simplest one imaginable, but the resulting formulae are vastly more involved than are the comparable ones for gravity, because of the fact that a spherical mass may be represented by a scalar, whereas a magnetized sphere may be represented by a vector. As we shall see (Chapter 11, Section F), the scalar magnetic anomaly caused by a magnetized body is the projection of its field on the earth's field—which further complicates the picture.

PROBLEM

10-1 The center of a magnetized sphere of 0.5 km radius is at a depth of 2 km. Suppose that J is unity and exactly in the plane of the meridian. It dips 60°N. Write a program to find H_z right above the center of the sphere (point A) and at every 500 m to the north and south for distances up to 4 km from A. Plot the results.

C. Magnetic Quantities and Units

Magnetic Induction, B; Permeability, μ; and Susceptibility, χ

When a body is placed in an outside field **H**, the number of lines of force inside the body depends both on **H** and on the resulting magnetization in the body. The field inside the body is the magnetic induction, or magnetic flux density, **B**. *In vacuo*,

$$\mathbf{B} = \mu_o \mathbf{H} \tag{10-24}$$

where μ_o is the magnetic permeability of free space; whereas inside a body,

$$\mathbf{B} = \mu_o(\mathbf{H} + \mathbf{J}) \tag{10-25}$$

Alternatively, we may say that the induction **B** is proportional to **H**: that is,

$$\mathbf{B} = \mu \mathbf{H} \tag{10-26}$$

where μ is the magnetic permeability of the medium, expressing the "ease" with which the magnetic field passes through the medium. In free space,

$$\mu = \mu_o = 4\pi \times 10^{-7} \mathrm{N/A^2} = 4\pi \times 10^{-7}\ \mathrm{H/m} = 1\ \mathrm{G/Oe} \quad (\mathrm{H} = \text{henry})$$

The first two sets of units are both SI, the last one is in the emu system (see below and Appendix B). The permeability of some materials varies with **H**, in which case **B** ceases to be proportional to **H**.

From equations 10-25 and 10-26,

$$\mu \mathbf{H} = \mu_o(\mathbf{H} + \mathbf{J})$$

Hence,

$$\mu = \mu_o(1 + \chi) \tag{10-27}$$

with

$$\chi = \frac{\mathbf{J}}{\mathbf{H}} \tag{10-28}$$

which is the magnetic susceptibility, a scalar. Finally, we also define

$$\mu_r = \frac{\mu}{\mu_o} = 1 + \chi \tag{10-29}$$

as the relative permeability, which is nondimensional and invariable in all systems of units. (This definition of χ is valid when the dipole is defined in terms of current in a loop. See Appendix B.)

SI and emu

There are at least five systems of electromagnetic units, and as Jackson (1962, p. 611) notes, "The question of units and dimensions in electricity and magnetism has exercised a great number of [scientists] over the years." The resolutions adopted by the International Union of Pure and Applied Physics (IUPAP) and its Commission on Symbols, Units and Nomenclature are reproduced in current editions of the *Handbook of Chemistry and Physics*; Markowitz (1973) is also a valuable reference.

The two most important systems are the electromagnetic unit (emu) system and the international system (Système International, or SI). The basic difference between them lies in the number of "basic quantities" that each system uses. The emu system uses only *three* basic quantities: mass, length, and time. The magnetic permeability of free space, μ_o, is taken as dimensionless and assigned a value of unity. Thus the relation *in vacuo* between the magnetic induction **B** and the magnetic field **H**, normally $\mathbf{B} = \mu_o \mathbf{H}$, becomes $\mathbf{B} = \mathbf{H}$. The units for these basic quantities and their internationally accepted abbreviations are the centimeter (cm), the gram (g), and the second (s).

In the international system, or SI, there are *four* basic quantities: mass (m), length (l), time (t), and electric current (I). As a result, some quantities, such as the magnetic field, have different dimensions and values in the two systems (see Appendix B). The units for the basic quantities and their abbreviations are the meter (m), the kilogram (kg), the second (s), and the ampere (A). The ampere is defined in terms of the other three units as the current that creates a force of 2×10^{-7} newtons (N) per meter of length when passing in two infinitely thin parallel wires *in vacuo* separated by 1 m. This definition could be interpreted to give the (ampere)^2, or A^2, the dimensions of the newton ($1 \text{ N} = 1 \text{ kg m s}^{-2}$). This is not done, however. Instead, the ampere is taken as a "basic quantity" with its own independent dimension, whereas μ_o is given the dimension of N/A^2. (It *is* confusing, but that cannot be helped!) Since a small loop of wire carrying a current produces the same field as a magnetic dipole, it is possible to define the magnetic field in terms of the four basic units.

The changeover from emu to SI units is far from universally accepted or practiced in geophysics. Appendix B can be used for conversions from SI to emu; it also gives internationally adopted symbols, units, abbreviations, and dimensions. As yet, there is no complete agreement on the units of dipole moment and magnetization; the alternatives listed in Appendix B are both correct (Smythe 1950, p. 587), although the first one is more commonly used. As will be seen, you may convert one to the other by giving A^2 the dimensions of N, as just described.

D. Diamagnetism, Paramagnetism, and Ferromagnetism

Most materials tend to repel the magnetic lines of force ($\chi < 0$): that is, the magnetic induction in the body is less than that in free space (in the air) around it. Such an effect is to be expected; in a way, it is comparable to inertia in gravity; In electromagnetism, Lentz's law states that the field produced by a conductor moving in a magnetic field tends to oppose the external field. The moving conductor here consists of the electrons orbiting around the nucleus. Such materials are said to be diamagnetic.

There is, however, a small class of materials that tends to concentrate the lines of force ($\chi > 0$). The most important of these materials are the elements in the chemical table from Ca to Ni and from Nb to Rh, along with some compounds formed from them. Such materials are said to be paramagnetic. Paramagnetism arises from the spin of orbital electrons on their axes. These electrons are normally arranged in pairs, of which one has a positive spin (or net magnetic moment) and one a negative spin; thus, there is no net effect. However, if the number of orbital electrons is odd, the result is a net magnetic moment, which tends to align itself with the external field and reinforce it.

Neither diamagnetic nor paramagnetic materials retain a permanent magnetization when the outside field vanishes. Out of these two classes the only geological material that can produce any magnetic anomaly is salt, which is strongly diamagnetic.

Most observed magnetic anomalies are caused by the presence of magnetite, a typical ferromagnetic material. In such materials, atoms in neighboring molecules tend to be systematically aligned (the so-called cooperative phenomenon) within a magnetic "domain," with dimensions on the order of 10 μm, that is sometimes dendritic and sometimes compact. The domains are separated by thin regions (Bloch walls) in which the direction of magnetization varies rapidly; this implies that the walls store a relatively large amount of magnetic energy.

Ferromagnetic materials retain a permanent magnetization after the external field is removed; in addition, they very strongly concentrate the lines of force ($\chi \gg 0$). Only a few naturally occurring minerals are ferromagnetic. By far the most important one is magnetite, but ilmenite, hematite, pyrrhotite, and some minerals of intermediate composition (collectively called titano-magnetites) are also ferromagnetic.

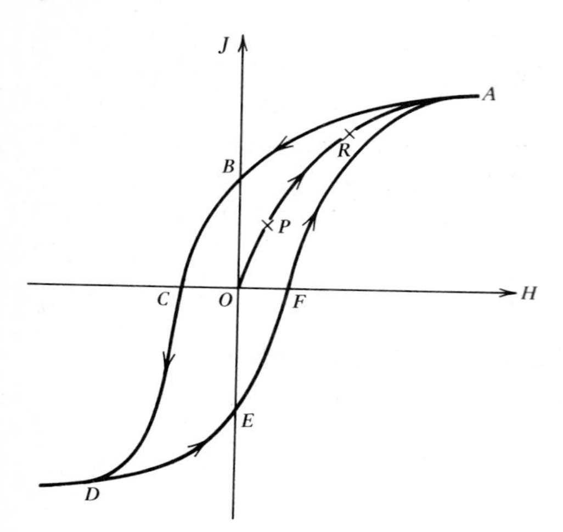

FIG. 10-5 *A typical hysteresis curve.*

It is of some interest to note that most of these minerals actually belong to a subclass called the ferrimagnetic. In this class there are two different sites in the lattice (corresponding to iron in, respectively, fourfold and sixfold coordination with oxygen); because the moments on these sites are opposite and unequal, a net effect results. In pure iron, of course, only one type of site can exist.

When a ferromagnetic material with zero magnetization is placed in an external magnetic field (Figure 10-5), its magnetization is at first approximately proportional to the applied field (segment OP). This corresponds to a reversible expansion of the favorably oriented domains. As the field increases, this expansion continues but becomes largely irreversible (segment PR). Near saturation (see below), the direction of magnetization of the domains rotates reversibly toward the direction of the field. As this process nears completion, it encounters increasing resistance; hence, the curve of J versus H becomes increasingly flat. When J no longer increases with increasing H, the material is said to be saturated. When the field is decreased and then reversed, the magnetization does not follow the same path, but instead follows curve $ABCD$. The magnetization remaining for $H=0$—that is, OB—is the remanent magnetization J_r; the field necessary to bring J back to zero is the coercive field, or coercivity, OC. Increasing the field starting at D results in the curve $DEFA$, which is symmetric of $ABCD$ around O. The complete loop is known as the hysteresis curve (or loop); the area that it encloses is proportional to the energy dissipated in one cycle of field reversal. The material used in electrical transformers in which the field reverses 60 times per second is thus designed to have a very small hysteresis loop. (A transformer changes the voltage of an electric current. It consists of two coils of wire with different numbers of turns around a common ferromagnetic core that is closed upon itself to minimize energy losses.) Conversely, permanent magnets have a very large J_r. The fact that $J_r \neq 0$ makes magnetic tapes and magnetic disks possible.

11

The Earth's Magnetic Field

In the emu system the permeability of air and that of free space are the same, and both are unity and dimensionless (see Appendix B). It is thus customary to speak of the earth's *field* rather than the earth's *induction*. In the SI, however, $\mu_o \neq 1$ and has the dimensions of N/A^2. To be precise, then, one should speak not of the earth's magnetic field but of the earth's induction, and measure it in teslas (T). It seems clear, however, that the name is not going to be changed, and scientists will continue to speak of the "field" measured in teslas (T), which is inconsistent (Appendix B). To indicate that it is the induction and not the field that is measured, we will use the symbol **B**, rather than **H**, following the example of Stacey (1977). A commonly used unit of magnetic "field" is the gamma, now often called the nanotesla.

$$1\gamma = 10^{-5}\text{G} = 1\text{ nT}$$

A. Components

The earth's field at a point may be defined by its three components, north, east, and vertical—often X, Y, and Z, and here B_x, B_y, B_z. (Note that B_x points north and B_y points east.) More frequently, it is given in terms of the magnitude of the total field, B; the declination, d (the angle between the horizontal field and the meridian); and the inclination, i (the angle between **B** and the horizontal). On marine charts, d is called the magnetic variation (see Figure 11-2b).

Although like poles repel each other, in the case of the earth the north-seeking pole of a compass is conventionally called the North Pole. We will call this the North Pole convention.

As early as 1600, the British physician and scientist William Gilbert (1544–1603) had established that the earth's magnetic field is similar to that of a magnetized sphere, by taking a sphere made of magnetite (lodestone) and measuring the field at its surface. He published his results in 1600 in the famous treatise *De magnete*.... Since the magnetic field of a uniformly magnetized sphere is the same as that of a centered dipole, Equations 10-8 and 10-9 become

$$B_r = \frac{2\mu_o M \cos\Theta}{4\pi r^3}$$

$$B_\Theta = \frac{\mu_o M \sin\Theta}{4\pi r^3}$$

(The values of μ_o and M are given in Appendix C.) We thus find (see Figure 11-1a) that

$$\frac{B_r}{B_\Theta} = \tan i = 2\cot\Theta \tag{11-1}$$

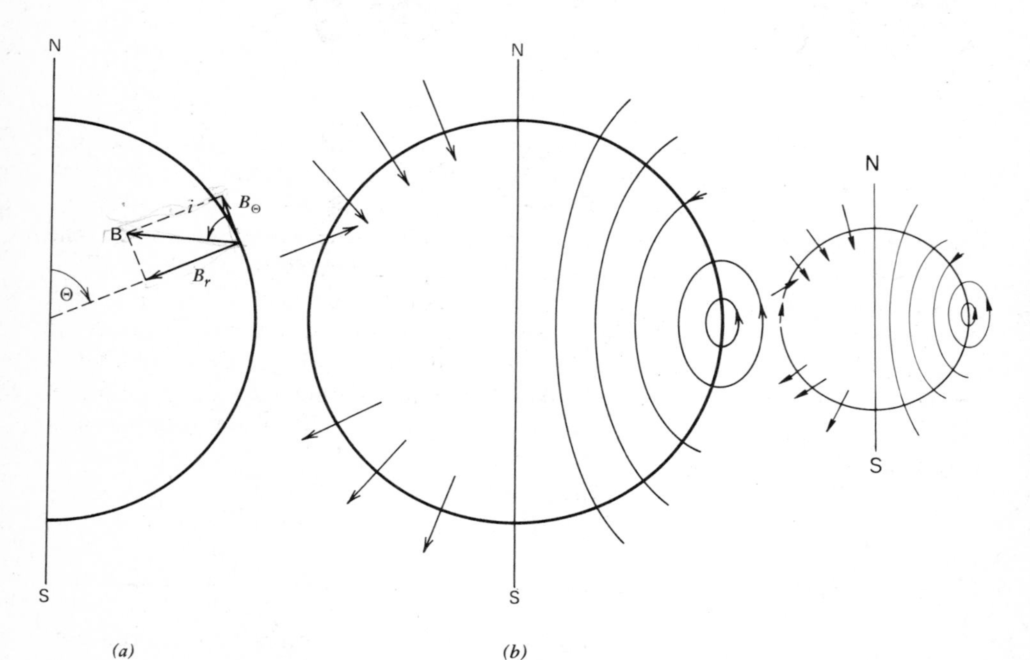

FIG. 11-1 *(a) The horizontal (B_Θ) and vertical (B_r) components of the earth's field, the colatitude Θ, and the inclination i. (b) Left: the value and direction of the earth's field. Right: magnetic lines of force (From Nettleton 1971.)*

where Θ is the colatitude of the station—that is, the angle measured from the north pole. Comparing Figures 10-3 and 11-1*a*, we see that in the former, H_r points outward and H_Θ southward, whereas in the latter, B_r points inward and B_Θ northward. This is the result of the north pole convention. Figure 11-1*b* shows the magnetic lines of force (right half) and the value and orientation of **B** over the surface of the homogeneous magnetized sphere (left half).

Although the largest part of the earth's field at the surface is that of a centered dipole aligned along the axis of rotation of the earth, such a dipole has no physical existence (see Section 11C). As the number of observations of the field over the surface of the earth increased, the complexity of the field became apparent. Figure 11-2 shows the total intensity, the declination (variation), and the inclination as observed over most of the world in 1975, along with the rates of change of these quantities.

Internal Origin

In the early nineteenth century, Gauss (see Chapter 9, Section C) became interested in determining whether the Earth's field is *entirely* of internal origin or whether there is an external component. The exact method by which he settled the question is mathematically involved, but its basic principle is similar to that of Fourier analysis. We will assume, for simplicity, that $B_y = 0$. B_x may be decomposed in the product of two sets of functions, one that varies with longitude only (the familiar sines and cosines) and another that varies with the colatitude, Θ, only (much like sines and cosines, but modified to take into account the fact that the earth is a sphere rather than a plane). We can do the same with B_z. In each case, we may compute the amplitude of the fundamental mode of variation with Θ—that is, the mode that varies with Θ itself rather than with 2Θ, 3Θ, and so on. If the field is of purely internal origin, these amplitudes should be, respectively, $\mu_o M/(4\pi r^3)$ and $2\mu_o M/(4\pi r^3)$. Thus, each determines the strength M of the main dipole. If the field is of partly internal and partly external origin, these two equations together determine M and the external field. The process can be repeated for higher harmonics, in longitude and latitude.

By this method, Gauss was able to show (a) that the earth's field is of purely internal origin, and (b) that the field is more closely approximated by the field of a tilted centered dipole than by one along the earth's axis. The angle of tilt is approximately 11.5°. This is the field referred to as the "dipole field." (Subsequent investigation has shown that a still better approximation can be obtained through the use of an eccentric tilted dipole.)

The points at which the axis of the tilted centered dipole pierces the surface of the earth are the geomagnetic poles. In contrast, the magnetic poles (or, preferably, dip poles) are the points at which the actual inclination is 90°. The average magnetic field is to some extent conventional: it does not attempt, for instance, to take local magnetic anomalies into account. Because it varies slowly with time (see Section 11B), a new reference field, (the International Geomagnetic Reference Field), is adopted approximately every 10 years (Peddie 1983). The present one (IGRF75) contains 80 terms. This is the field shown in Figure 11-2.

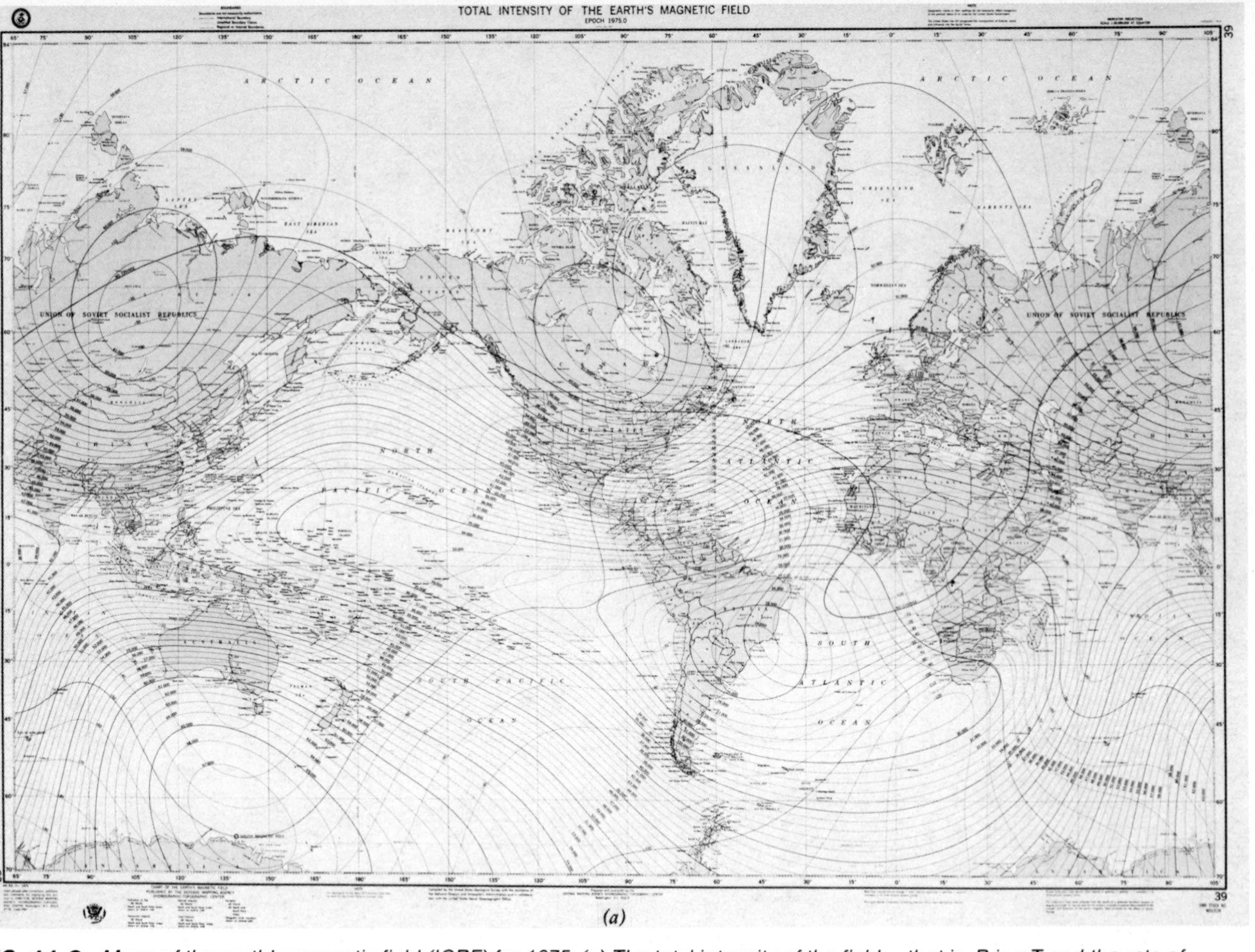

FIG. 11-2 *Maps of the earth's magnetic field (IGRF) for 1975. (a) The total intensity of the field— that is, B in nT and the rate of change of B in nT/a. (The lines of constant B are roughly parallel to the equator; the values range from ~30,000 to 50,000 nT.) (b) Magnetic declination (variation) in degrees, and its change in degrees/a. (The line of zero declination goes ~ NS across the Americas.) (c) Magnetic inclination in degrees (2° interval), and its changes in minutes/a. (The line of 70° N inclination passes close to New York City.) (From Defense Mapping Agency Hydrographic Office 1975.)*

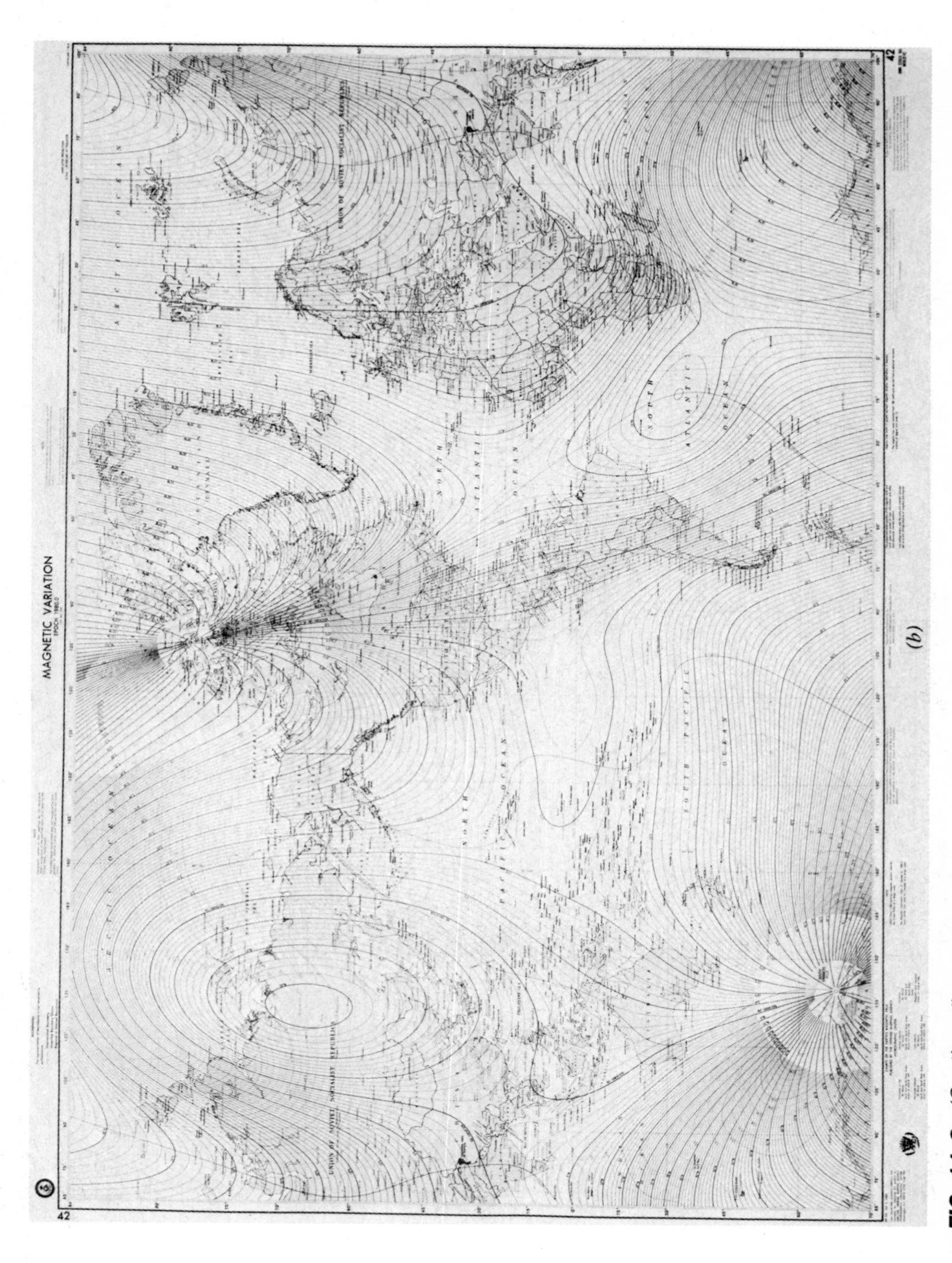

FIG. 11-2 *(Continued)*

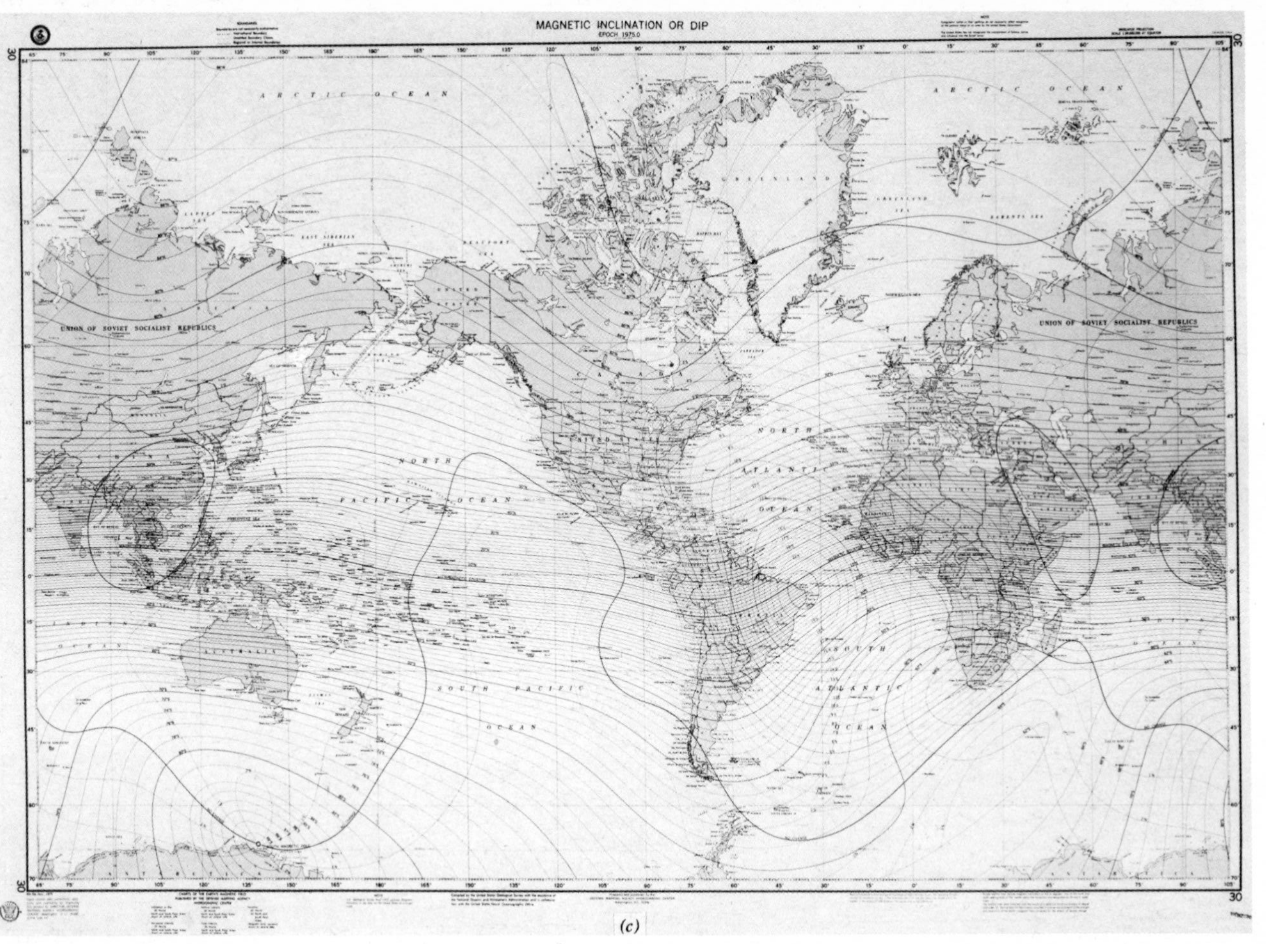

FIG. 11-2 *(Continued)*

B. Fluctuations of the Geomagnetic Field

Daily Variations

An important difference between the earth's magnetic and gravity fields lies in the nature of the temporal fluctuations. While the temporal fluctuations of the gravity field are due to the highly predictable luni-solar attraction, those of the magnetic field are neither highly predictable nor strictly periodic. There are minor daily variations (amplitude on the order 10 nT to 50 nT—or γ) associated with changes in conductivity of the upper atmosphere caused by sunlight. An even smaller variation is associated with atmospheric tides caused by the moon. The daily variation fluctuates rather unpredictably over distances of a few hundred kilometers. It may be necessary to record the variation at a base station when attempting a high-precision magnetic survey.

Storms

By far the most important variations, however, are irregular changes known as magnetic storms, which last a few hours to a few days. They are caused by charged particles emitted by the sun, especially near sunspots. These particles may also greatly disturb long-distance radio communications. The amplitudes of the variations are on the order of hundreds of nanoteslas. Because of the erratic nature of these variations, no magnetic survey should be conducted during a magnetic storm.

Secular Variation

The direction of the magnetic field has been recorded in London for about 500 years. During that period the declination has changed from about 10° E to 20° W. Similar changes (over shorter time-spans) have been recorded elsewhere. More recently, it has become possible to compute the rate of change of the magnetic field everywhere through comparison of measurements made a few years apart. These changes (also shown in Figure 11-2) are best analyzed after removal of the dipole field (the remaining field is the nondipole field). Analysis of such changes has suggested that at present, there is an *average* westward drift of the nondipole field of approximately 6 minutes of longitude per year. This corresponds to an average motion of 6 km/a at the core-mantle boundary at the equator. (The reason for this indication is explained in Section 11C.) By geological standards, this is an extremely rapid motion. The westward drift appears to vary with time: from about 9 minutes/a in 1945, down to ~ 3 minutes/a in 1970, up to ~ 6 minutes/a in 1980 (Le Mouël et al. 1983). Changes over longer periods are discussed in Section 11E.

C. Origin of the Earth's Field

The Geomagnetic Dynamo

Although the main field of the earth is that of a magnetized sphere, or a centered dipole, it is known that no ferromagnetic material can retain a remanent magnetization at temperatures greater than ~ 600° C (see Section 11D)—that is, in most of the earth. On the other hand, the essential coincidence (over periods of a few centuries) between the earth's axis of rotation and the dipole direction suggests a coupling mechanism. (Blackett suggested in 1947 that magnetization is an inherent property of a mass in rotation, but this suggestion was rapidly disproven by its own author.)

It is now accepted that the field is created by rapid motions (on the order of km/a—i.e., at least as rapid as the westward drift—) in the liquid core. In a large body of highly conductive fluid the magnetic lines of force are "frozen" in the body. In other words, as the liquid moves, it carries the lines of force with it. This may be easier to understand if temperature is used as an analogy (see Chapter 13). In any material, temperatures tend to equalize by conduction (diffusion of heat); how fast this equalization takes place depends on the thermal conductivity of the material. Such diffusion also takes place in a moving liquid. However, if a liquid particle moves over a certain distance much faster than its temperature can equalize by conduction, the liquid essentially carries its temperature with it: the isotherms are "frozen" in the liquid. In the case of the magnetic field, a similar phenomenon occurs. This analogy has a mathematical basis grounded in Maxwell's equations of electromagnetism. However, these equations assume a knowledge of vector analysis, so we will confine ourselves to two limiting cases.

For our first case, we will assume that the material is at rest and that the components of the magnetic field **H** are H_x, H_y, and H_z. We also will designate any one component by H_i, where $i=1$ stands for H_x, $i=2$ for H_y, and so forth —just as was done in the case of elasticity (see Equation 2-15). Starting from Maxwell's equations, we can show that any component of the magnetic field obeys the equation

$$\frac{\partial H_i}{\partial t} = \kappa_m \nabla^2 H_i \tag{11-2}$$

Alternatively, all three equations may be written together as the vector equation

$$\frac{\partial \mathbf{H}}{\partial t} = \kappa_m \nabla^2 \mathbf{H} \tag{11-3}$$

where $\kappa_m = 1/\sigma\mu$ is the magnetic diffusivity, σ is the electrical conductivity, and μ is the magnetic permeability.

The symbol ν_m usually is used in place of κ_m, but the latter has the advantage of reminding us that the diffusion of the magnetic field and that of the temperature obey the same equation (see Equation 13-7). Sometimes $\kappa_m = \nu_m$ is called the magnetic viscosity.

As we will see when discussing heat conduction (see Chapter 13, Section B), the characteristic time for heat conduction is $\tau_c = L^2/\kappa$, where L (called D in that chapter) is a characteristic linear dimension of the body and κ is the thermal diffusivity. Similarly, in a body at rest the magnetic field decays (is "conducted away") in a time on the order of

$$\tau_c = L^2/\kappa_m = L^2\sigma\mu \tag{11-4}$$

The decay time thus increases as the *square* of the linear dimensions. This explains why it is difficult to maintain such a field in the laboratory (except where $\sigma = \infty$; that is, in superconductors), whereas in the earth's core such a field takes approximately 30,000 years to decay (Parker 1983).

In an alternative limiting case, we assume that the fluid is in motion and has infinite electrical conductivity. The equation governing the magnetic field then becomes identical to the equation of vorticity transport governing the motion of the vortex field in an inviscid fluid. This equation shows that in such a fluid the vortices ("whirls") move at the same velocity as the liquid. In the same way, the magnetic field moves at the same velocity as the fluid—that is, the field is "frozen" in the liquid.

The change from the transport of the magnetic field by the fluid to the leakage of the field through the fluid is very similar to the one we will encounter when considering the relative importance of the heat transported by the motion of the fluid (advected heat) versus that of the heat conducted through the fluid (conducted heat—see Chapter 13, Section B). This relative importance is measured by the Peclet number (see Equation 13-21).

$$\mathrm{Pe} = \frac{\text{Advected heat}}{\text{Conducted heat}} = \frac{LV}{\kappa}$$

where L is a characteristic linear dimension, V is the velocity of the fluid, and κ is the thermal diffusivity. In the present case we have a *magnetic* Peclet number.

$$\mathrm{Pe}_m = \frac{LV}{\kappa_m}\left(= \mathrm{Re}_m = \frac{LV}{\nu_m}\right)$$

It has become common to refer to this quantity as the magnetic Reynolds number, Re_m (see Chapter 12, Section C). However, the analogy between the ratio of inertial to viscous forces and the ratio of transport to diffusion of the magnetic field is not obvious. Because the analogy with heat transport is much clearer, the term "magnetic Peclet number" appears preferable.

What we have, then, is a large volume of highly conductive liquid—the earth's core—in which motions are more or less random. We assume that initially there was a small external magnetic field, caused perhaps by the sun, perhaps by the galaxy. In this field the motion of one part of the conductive liquid induces an electrical current in it, and the magnetic field induced by this current is carried by the moving fluid. In favorable circumstances this moving magnetic field may produce a stronger electric current elsewhere in the fluid. The process may thus be self-reinforcing (positive feedback). The magnetic field cannot grow indefinitely via this process, because it is not perfectly "frozen" in the moving liquid, any more than the isotherms are; it tends to diffuse away. In the absence of an energy source, the motions would, of course, rapidly cease.

Such a generator is not purely fanciful, although it is difficult to build one in the laboratory because the relatively small dimensions imply a rapid decay of the field (see Equation 11-4). Michael Faraday's disk generator is an example of such a device. Figure 11-3 shows how a copper disk rotated through the magnetic lines of force of a bar magnet can generate a small electric current. This occurs because each radius of the disk may be considered a conductor moving through a magnetic field. If the current thus produced is itself used to create the magnetic field, the result is a very simple (and inefficient!) electric generator. Although it is true that such a simple system is far from representative of what goes on in the earth's core, the fact that it can be built suggests that such a mechanism is not impossible. The idea that such a self-reinforcing dynamo exists in the earth's core was first suggested by Larmor in 1919. Several errors were eventually found in his reasoning, and it was only in 1946 that the idea was revived by Elsasser. An excellent semiqualitative description is given by Inglis (1955), and a more quantitative one by Jacobs (1975, pp. 115–183—especially pp. 134–136).

This "self-sustaining geomagnetic dynamo" is generally accepted as the source of the earth's field. The branch of science that deals with the interaction between a moving conductive medium and a magnetic field is called hydromagnetics, or magnetohydrodynamics (MHD). In space the most common conductive medium is ionized gas or plasma; some engineers believe that electrical generators based on this idea (MHD generators) may have great advantages over conventional ones.

In the case of the earth, the field is caused by a large number of eddies in the core. The field at the core-mantle boundary is thus extremely "rough"; that is, its spectrum is rich in short wavelengths. At the surface of the earth, on the other hand, the field is essentially that of a dipole. This difference can be explained by reference to the fact that a gravity fluctuation at depth decreases exponentially with distance (Equation 9-36), with the exponent being proportional to the wave number. In the present case the essentially dipolar character of the surface magnetic field is caused, similarly, by the exponential decrease suffered by the short-wavelength components of the magnetic field.

This mode of generation also explains why the central dipole approximately coincides with the axis of rotation. On the one hand, the earth's rotation has a statistical influence on the motion and may tend to align those of the lowest wave number along the axis. On the other hand, there is no reason why in such a system, the North Pole could not be located near where the South Pole now is (see Section 11-E). Indeed, some simple models of self-exciting dynamos have shown random self-reversals of their fields.

The equations governing the process are so complex, however, that the above theory has not been conclusively proven. It is by far the most reasonable hypothesis, however, and it is supported by some experimental evidence.

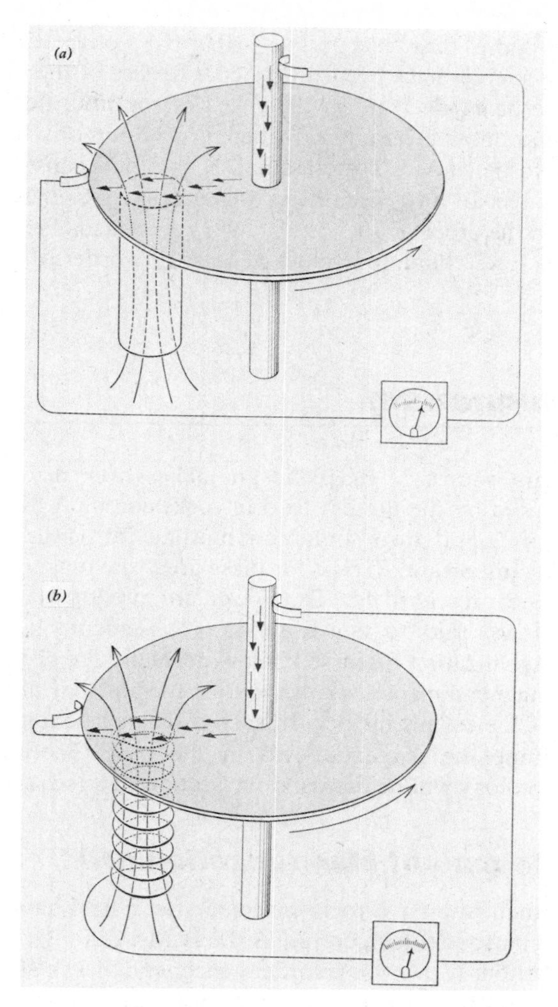

FIG. 11-3 *Faraday's disk generator. (a) A bar magnet produces the required magnetic field. The direction of the electric current is shown by the short arrows. (b) The electric current produced by the generator itself is used to produce the magnetic field. (From Elsasser 1958. Copyright © 1958 by Scientific American Inc. All rights reserved, reprinted by permission.)*

Energy Sources for the Earth's Field

The energy sources that produce the motions in the core remain hypothetical and debatable. The most likely source is the gradual solidification of the earth's inner core, the implication being that the core is gradually cooling and solidifying (Carrigan and Gubbins 1979). Most of this energy (about 80%) is probably derived from the sinking of the denser material of slightly different composition, which goes to form the inner core. The rest comes from the release of the heat of solidification. Thus, the earth's dynamo may be driven by chemical or compositional convection more than by ordinary thermal convection (Frank 1982). Various other heat sources have been suggested—release of the heat of formation (so-called primitive heat), decrease in the speed of rotation, tidal effects, among others—but these are now generally considered to be minor. Only radioactivity, generally thought to be caused by potassium, K^{40}, is still considered by some to be the primary source. In any case, the power necessary to maintain the earth's field is probably on the order of 10^{10} to 10^{11} W (Verhoogen 1980, p. 74), although some geophysicists have computed values at least one order of magnitude larger (Olson 1981).

D. Rock Magnetism

The most important source of magnetic anomalies over the continents is the magnetization induced by the present field in rocks containing magnetite; exceptions are volcanic areas and areas containing plutonic intrusives, where TRM (see below) is often very important. Except for these areas the magnetization is thus in the direction of the present field. The rocks are predominantly igneous and metamorphic, and are said to constitute the "basement." Consequently, the aspects of rock magnetization discussed below normally are not important in the interpretation of magnetic maps over the continents, but they are fundamental to a reconstruction of the earth's history. In such a reconstruction, rock magnetism is used to determine the magnetic field in the past. The relevance of this reconstruction to geology will be discussed in Section 11E (see also Tarling 1971).

Isothermal Remanent Magnetization (IRM)

The remanent magnetization in a rock specimen after a field has been applied and then nulled at constant room temperature is the IRM. Although very important in commercial applications (e.g., tape recorders, magnetic disks), IRM is of no value in determining the past field, since it is "unstable"—that is, it can be modified by any subsequent field of the same or greater strength.

Thermo-Remanent Magnetization (TRM)

By far the most interesting and important magnetization is acquired when a rock containing magnetite or another ferromagnetic mineral cools in a magnetic field. At a sufficiently high temperature, these minerals are paramagnetic, because the thermal agitation is then so strong that neighboring atoms can no longer interact (see Chapter 10, Section D). As the rock cools, it passes through a temperature known as the Curie point or Curie temperature (named after Pierre Curie, the French scientist who subsequently codiscovered radioactivity with his wife, Marie).

Cooperative phenomena become possible, and the mineral becomes ferromagnetic. The Curie point of magnetite is approximately 580° C. As the temperature further decreases, domains that are smaller or less favorably oriented gradually become magnetized. This means that over a range of about 200° C a succession of different partial TRMs (PTRMs) may be acquired; in exceptional cases, these may even possess different directions. We can better understand the mechanism by which this type of magnetization—along with two others—is acquired by introducing the concept of relaxation time, τ. This is the time necessary for the magnetization of a magnetic particle to return to its thermal equilibrium value and direction. It may be shown that

$$\tau = Ae^{BV/T}$$

where A and B are constants of the material, V is the volume of the particle, and T is the absolute temperature. For a given particle volume, there is thus a certain temperature at which the relaxation time becomes $\gtrsim 1$ Ga. The magnetization is then "permanent." This, of course, is the TRM process. Contrastingly, the relaxation time of very small particles is so short that they cannot remain magnetized even at normal temperature. If such a particle grows, however it may acquire a permanent magnetization, if its volume becomes large enough. This mechanism is called chemical remanent magnetization (CRM). Finally, it is possible to induce relaxation through an alternating magnetic field rather than through thermal agitation. A constant magnetic field over the specimen can then result in a magnetization that becomes permanent when the alternating field and the constant field are removed. This is anhysteretic remanent magnetization (ARM).

The equation for the relaxation time is valid only for an isolated single domain particle. For this reason it is generally possible to separate TRM, CRM, and ARM by careful laboratory techniques.

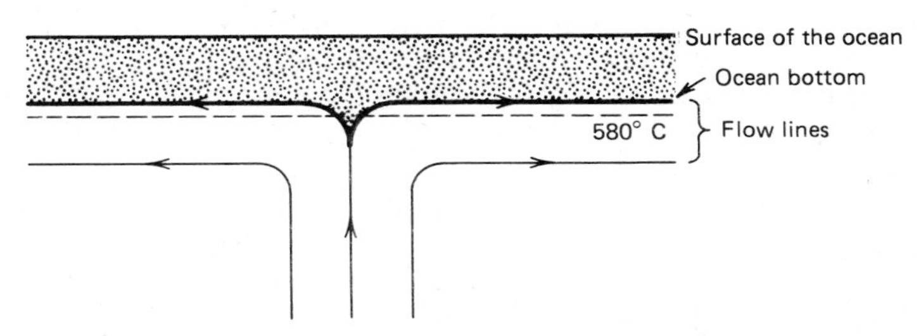

FIG. 11-4 *Simple concept of the "freezing" of the earth's field near a midocean ridge.*

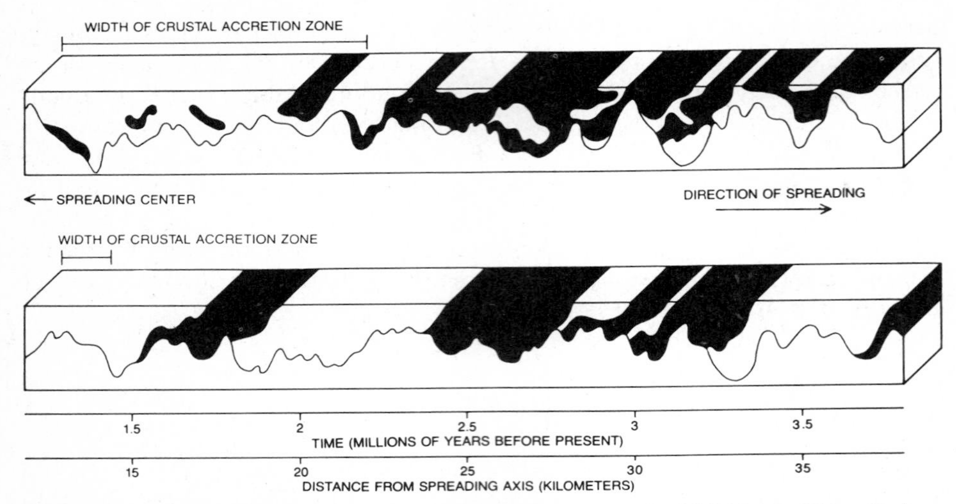

FIG. 11-5 *Presumed distribution of magnetic polarities near the Mid-Atlantic Ridge (top) and near the East Pacific Rise (bottom). (From Macdonald and Luyendyk 1981. Copyright © 1981 by Scientific American Inc. All rights reserved, reprinted by permission.)*

TRM is basically responsible for the magnetic stripes in the ocean floor. For many years the processes at midoceanic ridges were viewed rather simplistically. The material was seen (Figure 11-4) as rising vertically (above its Curie point), turning at right angles, passing through its Curie point, and thus freezing the direction of the field at the time. This picture is gratifyingly straightforward, and it explains the magnetic observations very satisfactorily (as will be seen below); however, drilling has shown that the situation is more complex. In the Atlantic, where spreading is slow, the direction of magnetization varies quite chaotically with depth; in the Pacific, where spreading is much faster, the direction of magnetization is less variable (Macdonald and Luyendyk 1981; see also Figure 11-5). Presumably, the chaos stems partly from motions that take place after the material cools below the Curie point, partly from injections of dikes at different times and places, and partly from a succession of lava flows over a few thousand years. Spreading does not take place as pictured in Figure 11-4, but by a combination of all the phenomena just described. Why fast spreading produces a more orderly arrangement remains speculative. In any case, it is remarkable that such disorderly phenomena produce such clear magnetic stripes—at least when the stripes are recorded at the ocean's surface (see Problem 11-3).

PROBLEM

11-1 It has often been said that the ocean floor records the earth's field like a tape recorder. To what extent are the magnetization processes similar?

Chemical Remanent Magnetization (CRM)

Some ferromagnetic minerals—in particular, hematite—may be deposited as cement long after the deposition of the rock itself. This phenomenon is not uncommon in sandstone. As explained above, CRM and TRM may both endure through geologic periods. The existence of CRM may, on the one hand, greatly alter the total magnetization of the specimen and thus create serious difficulties. On the other hand, if the time of formation of the "magnetic cement" is known, the stability of CRM can be of great advantage (for example, because of the wide distribution of sandstone).

Detrital Remanent Magnetization (DRM) and "Bioremanent" Magnetization (BRM)

Sediments often contain enough small grains of magnetite to enable sensitive instruments to detect a magnetization. Except in fast-flowing streams, such sediments tend to become oriented in the magnetic meridian as they are deposited (Tarling 1971, p. 35). Whether the magnetic inclination is affected by subsequent compaction is not clear. It has recently been found that some bacteria contain a string of minute grains, each consisting of a single domain of magnetite and all oriented in the same direction (Blakemore and Frankel 1981). The organisms are thereby able to orient themselves; they are said to be magnetotactic. (Homing pigeons orient themselves by a combination of magnetic and visual clues.) It appears possible that some of the magnetite grains in sediments are attributable to such organisms. This phenomenon might thus be called "bioremanent" magnetization (BRM).

E. The Earth's Field in the Past

Determination of the Pole

In order to determine the past position of the earth's pole, it is necessary to assume that the main field in the past was dipolar. This assumption appears reasonable. It is also generally assumed that over periods of a few thousand years, the average dipole is in the direction of the earth's axis of rotation, and this also seems highly plausible. Thus, if the direction of magnetization of rocks is known, the position of the earth's pole at the time of formation can be found. Using Equation 11-1 and knowing i, we can find Θ. It is, of course, necessary that the magnetization be stable. Moreover, the resultant pole is relative to the position of the plate on which the specimens have been collected. The branch of science dealing with the collection and analysis of rocks for this purpose is known as paleomagnetism. (Archeomagnetism deals with the same subject as applied to human artifacts.)

The results of paleomagnetic research are fundamental to a reconstruction of plate positions in the past, and the difficulties encountered in this field should be very briefly mentioned. All such difficulties stem from the fact that in general only TRM is of interest. To eliminate the other causes of magnetization, three procedures are commonly used. In one, known as alternating field demagnetization or magnetic cleaning, the specimen is rotated (tumbled) around two, three, or four axes simultaneously in a slowly decreasing alternating field. As explained above, it is important that the earth's field be canceled in an appropriate way while this process is going on; if not, a net magnetization (ARM) might be induced. In the second method, "thermal cleaning," the material is heated to a fairly low temperature (approximately 100 to 200° C) and allowed to cool in zero field. Finally, leaching or chemical cleaning may sometimes be used in order to gradually dissolve the magnetic cement and thus to determine the direction of magnetization of the detrital grains.

In all these methods, it is possible to operate with progressively stronger "cleaning agents" and to ascertain the constancy or inconstancy of the direction of the magnetization after each cleaning. When (and if) this direction remains essentially constant, it may generally be presumed to date back to the formation of the specimen. The magnetic poles computed through paleomagnetism are termed virtual geomagnetic poles (VGP), or paleopoles.

Reversals

Between approximately 1950 and 1960, there was much debate over whether the magnetic reversals observed in some rock specimens were caused by reversals of the earth's field or to some self-reversal mechanism. Theoretical reasoning showed that self-reversal mechanisms were possible, and eventually a dacite from Mt. Haruna, Japan, became famous when it was shown to acquire a reverse magnetization on cooling in a magnetic field. By that time, however, it had become clear that this was a rare phenomenon.

Over many years, meanwhile, Cox and Doell had been carefully investigating the pattern of magnetic reversals in lavas dated by the K-Ar method. They had come to the conclusion that this pattern was characteristic and worldwide (see Figure 11-6*c*).

In 1963, Vine and Matthews realized that the magnetic anomalies on either side of several midocean ridges were close to being mirror images of each other (Figures 11-6*a* and 11-6*b*). They thus proposed, as already noted, that the magnetic anomalies record the direction of the earth's field (normal—i.e., as at present—or reversed) at the time that the rocks cooled through the Curie point. (A layer of rock magnetized in the direction of the present field will produce a positive anomaly and vice versa; see Section 11 F. The magnetized layer is taken to be about 2 km thick.) As the figure shows, the model that they proposed to explain the magnetic anomalies coincides exactly with the magnetic chronology of Cox and Doell.

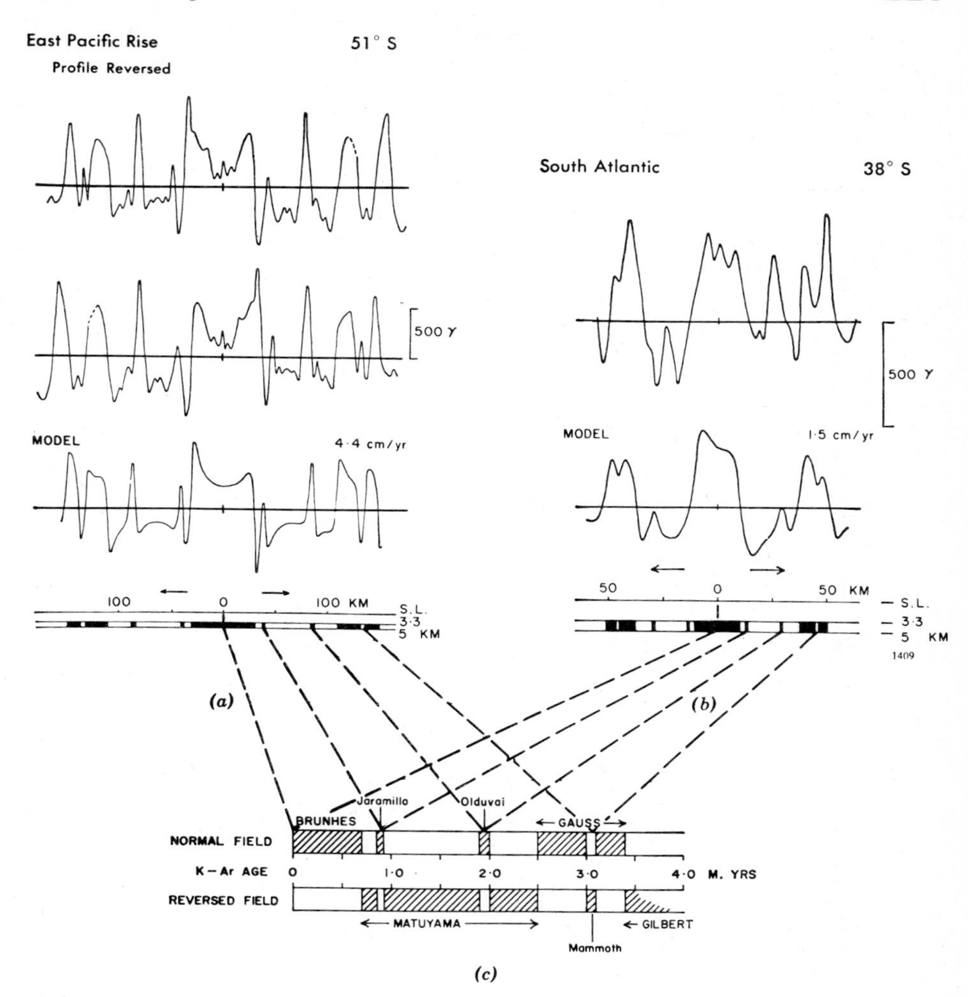

FIG. 11-6 *(a) Magnetic anomalies across the East Pacific rise at 51° S; spreading velocity, 4.4 cm/a. From top to bottom: profile reversed east-to-west, observed profile, computed profile based on the magnetized layer shown at the bottom. (b) Magnetic anomalies across the Mid-Atlantic Ridge at 38° S; spreading velocity, 1.5 cm/a. The same sequence as in Figure 11-6a, except that there is no reversed profile. Note the effect of the change in spreading rate. (c) Geomagnetic-polarity epochs deduced from paleomagnetic results and potassium-argon dating. (After Vine 1966. Copyright © 1966 by the American Association for the Advancement of Science, reprinted by permission.)*

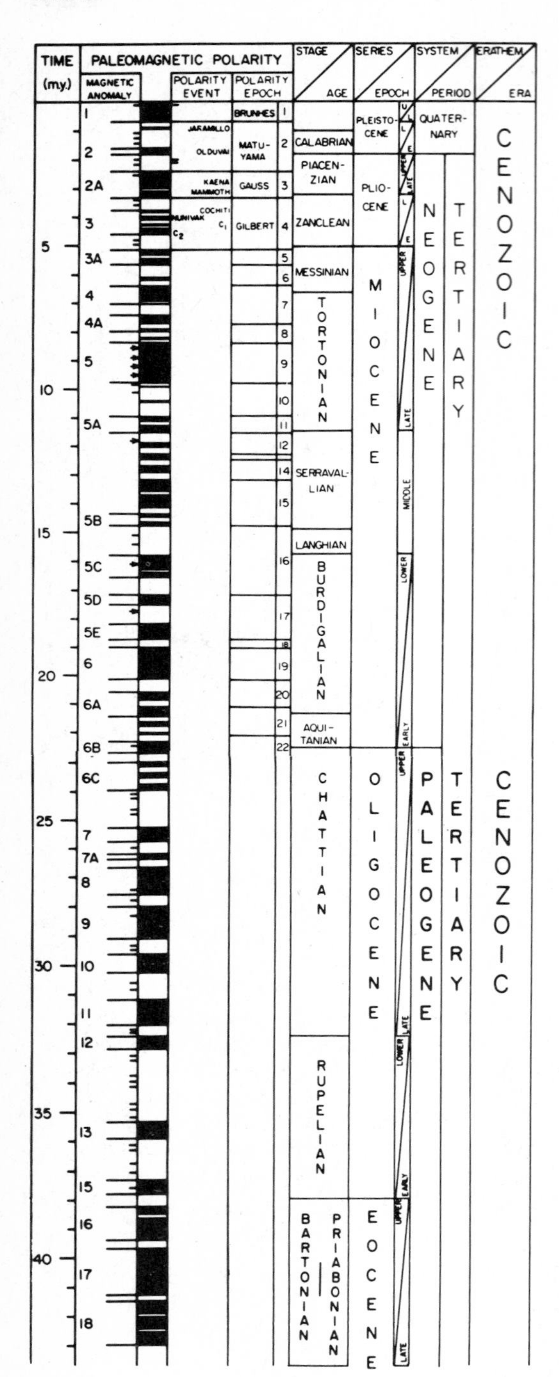

FIG. 11-7 *Paleomagnetic chronology for the last 83* Ma. *The anomalies are numbered; letters following the digits indicate further subdivisions; arrows in this column indicate short (less than 40,000 a) events (not otherwise shown). Dark blocks: normal polarity. (From La Brecque, Kent, and Cande 1977. Reprinted by permission.)*

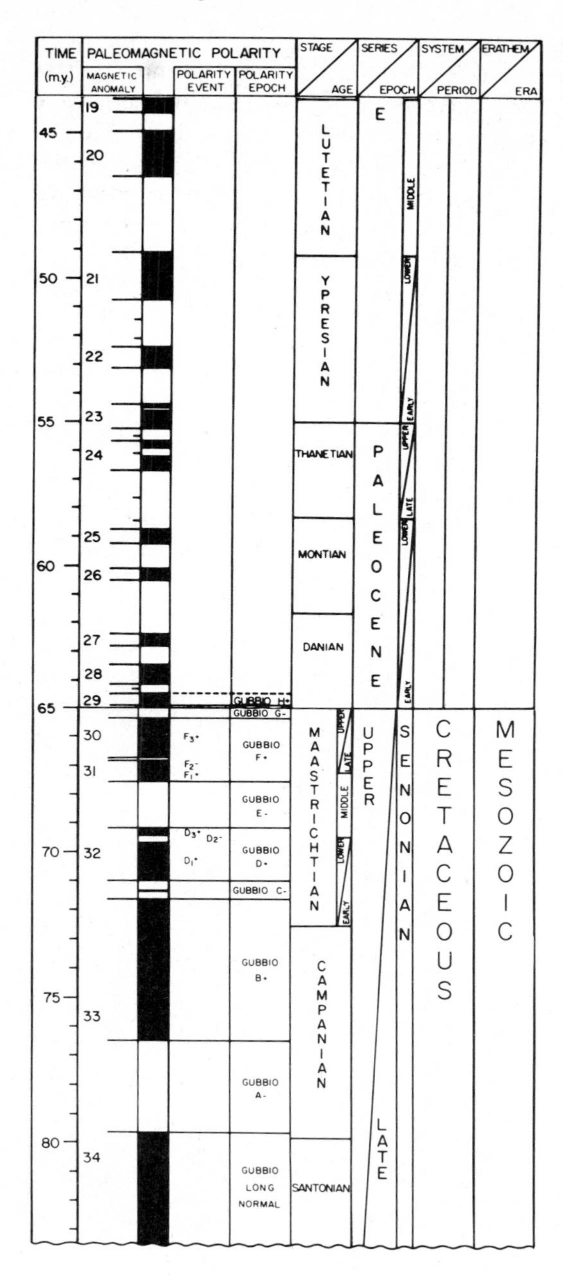

FIG. 11-7 *(Continued)*

Through the assumption (essentially correct, as it happened) that the South Atlantic Ocean opened at a constant rate, the magnetic geochronology was subsequently extended to roughly 80 Ma and then to 160 Ma. These results were confirmed through dating of the cores obtained by deep-ocean drilling. Figures 11-7 and 11-8 show recent revisions of the magnetic chronology; the anomalies are indicated by numbers ranging from 1 for the most recent to 34 for ~ 80 Ma B.P., and from M0 at ~ 110 Ma to M25 at ~ 155 Ma. In 1966 the same characteristic pattern was found in deep-ocean sediments (Figure 11-9).

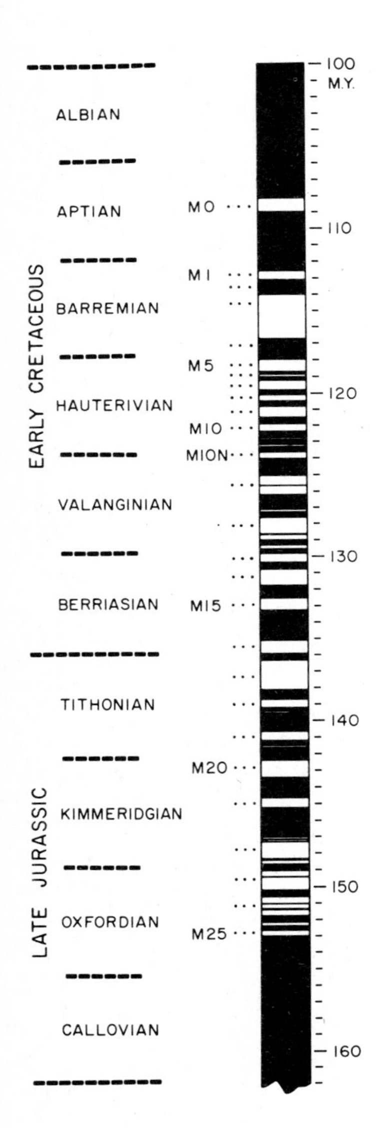

FIG. 11-8 *Paleomagnetic chronology from 100 to 160 Ma B.P. These anomalies are numbered Mi (i = 1 . . . 25). Dark blocks: normal polarity. (From Larson and Hilde 1975. Copyright © by the American Geophysical Union.)*

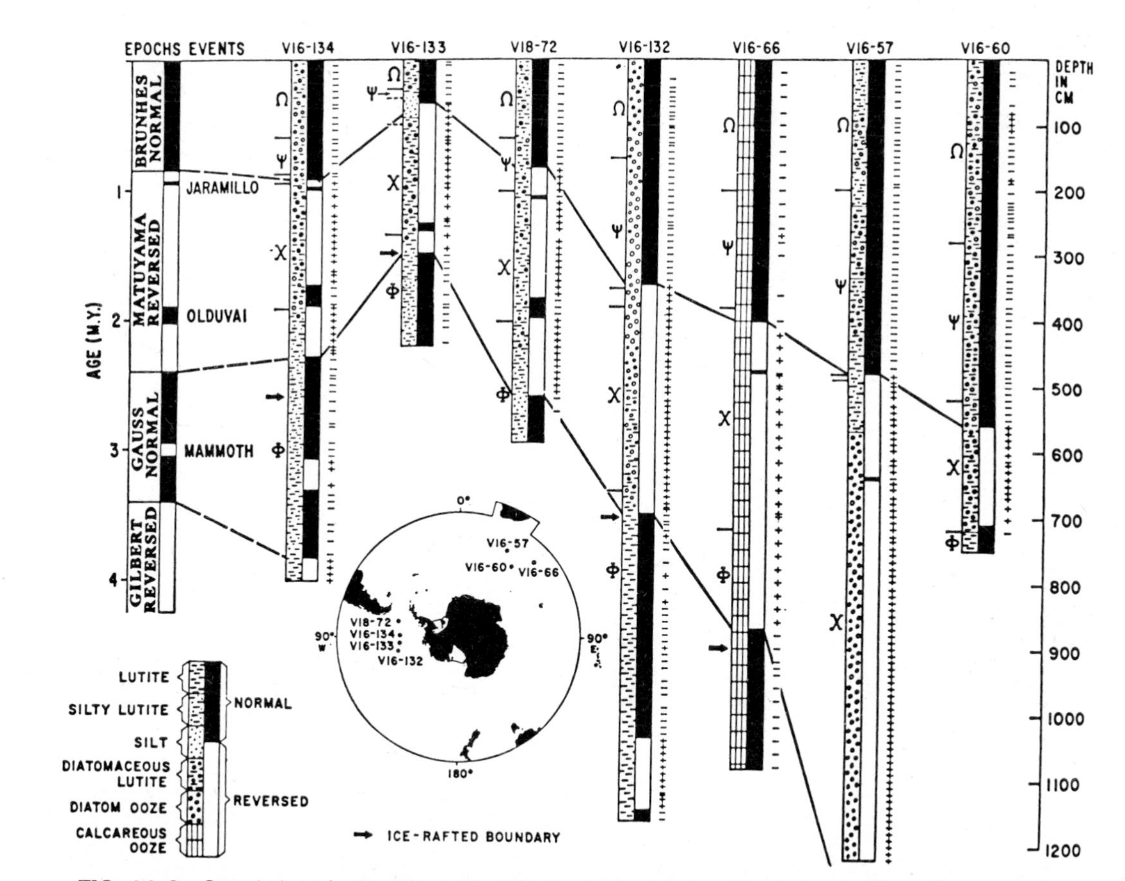

FIG. 11-9 *Correlation of magnetic stratigraphy in seven cores from the Antarctic. (From Opdyke et al. 1966. Copyright © 1966 by the American Association for the Advancement of Science, reprinted by permission.)*

This triple coincidence provides a convincing proof of the plate tectonic model. Furthermore, by "removing" the oceanic crust between two anomalies of the same age, we can determine the history of oceanic areas. As will be explained in Chapter 14, Section D, there is a strong correlation between the depth of the ocean and the age of the anomalies, which means that the age of the anomalies can also be used for paleobathymetric reconstructions of oceanic areas. Such reconstructions also rely on transform faults (Chapter 3, Section C) for determination of the pole of rotation of the plates. A recent reconstruction of the opening of the Atlantic is shown in Figure 11-10.

As stated previously, magnetic anomalies look very different at different latitudes. The reason for this phenomenon in the case of a dipole was discussed in Chapter 10, Section B, and it is discussed further in Section 11F. This effect greatly complicates the task of identifying magnetic anomalies.

Polar Wandering

Most magnetic stripes in the ocean can be adequately explained by the assumption that the magnetization of the rocks is either in the direction of or opposite to the present field. This is due to the fact that the age of the ocean bottom does not exceed roughly 200 Ma and that most plates have not greatly changed in latitude and orientation during that period.

When we examine the direction of magnetization for older rocks, the situation becomes appreciably different. We must first make sure that only the magnetization imparted at the time of the formation of the rock is being measured; therefore, the specimens must be magnetically, thermally, and/or chemically cleaned. The effects of subsequent motions, if any (e.g., folding) must also be taken into account.

Finally, as has been mentioned, only the virtual geomagnetic pole (VGP) of the plate can be determined; the paleolongitude cannot be. In Figure 11-11*a*, then, if the pole of rotation P of plate B with respect to plate A coincides with one VGP, the change in distance of these plates will not be reflected in any change in the VGP. Although such a case is unlikely, it illustrates one extreme possibility; indeed, the change in position of Africa and South America between 165 and 80 Ma B.P. (Figures 11-10*a* and 11-10*b*) is fairly close to it.

Another extreme case occurs when the relative pole of rotation P of B with respect to A is on the equator. This case can be most easily visualized in the plane, as shown in Figure 11-11*d*, although it is unaltered when applied to a sphere (Figure 11-11*b*). Taking a point R on plate B, and setting the north direction as RN, let us suppose that RN becomes fixed with respect to plate B, after which plate B rotates by an angle Θ around P on plate A. R moves to R' and RN to $R'N'$. The true north direction at R' is $R'N$, which makes an angle Θ with $R'N'$. In this case, therefore, the rotation of the VGP is the same as the rotation of plate B with respect to plate A. This is essentially what happened during the Tertiary period, when Australia separated from Antarctica.

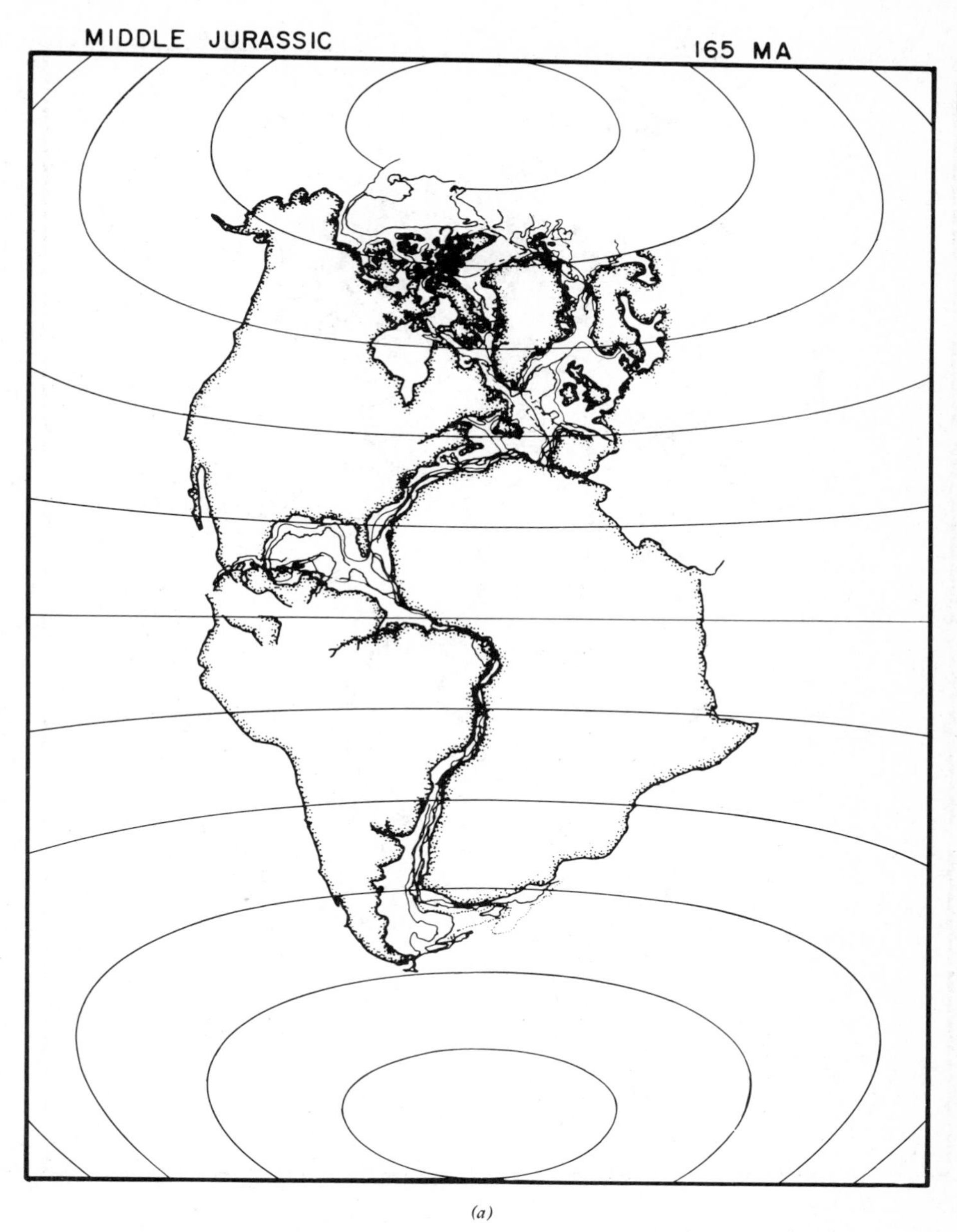

(a)

FIG. 11-10 *Reconstruction of the opening of the Atlantic Ocean, based on magnetic anomalies and transform faults. Continental edges are stippled; 200-m depth, thin line; 2000-m depth, heavy line; 3000-m depth, dotted line; 4000-m depth, dashed line; 5000-m depth, hatched line. The black triangles indicate JOIDES drilling sites. (a) Middle Jurassic, 165* Ma *B.P. (b) (On next page) Late Cretaceous, 80* Ma *B.P., anomaly 34 (see Figure 11-7). (c) Late Miocene, 10* Ma *B.P., anomaly 5. (From Sclater et al. 1977. Copyright © by the University of Chicago, reprinted by permission.)*

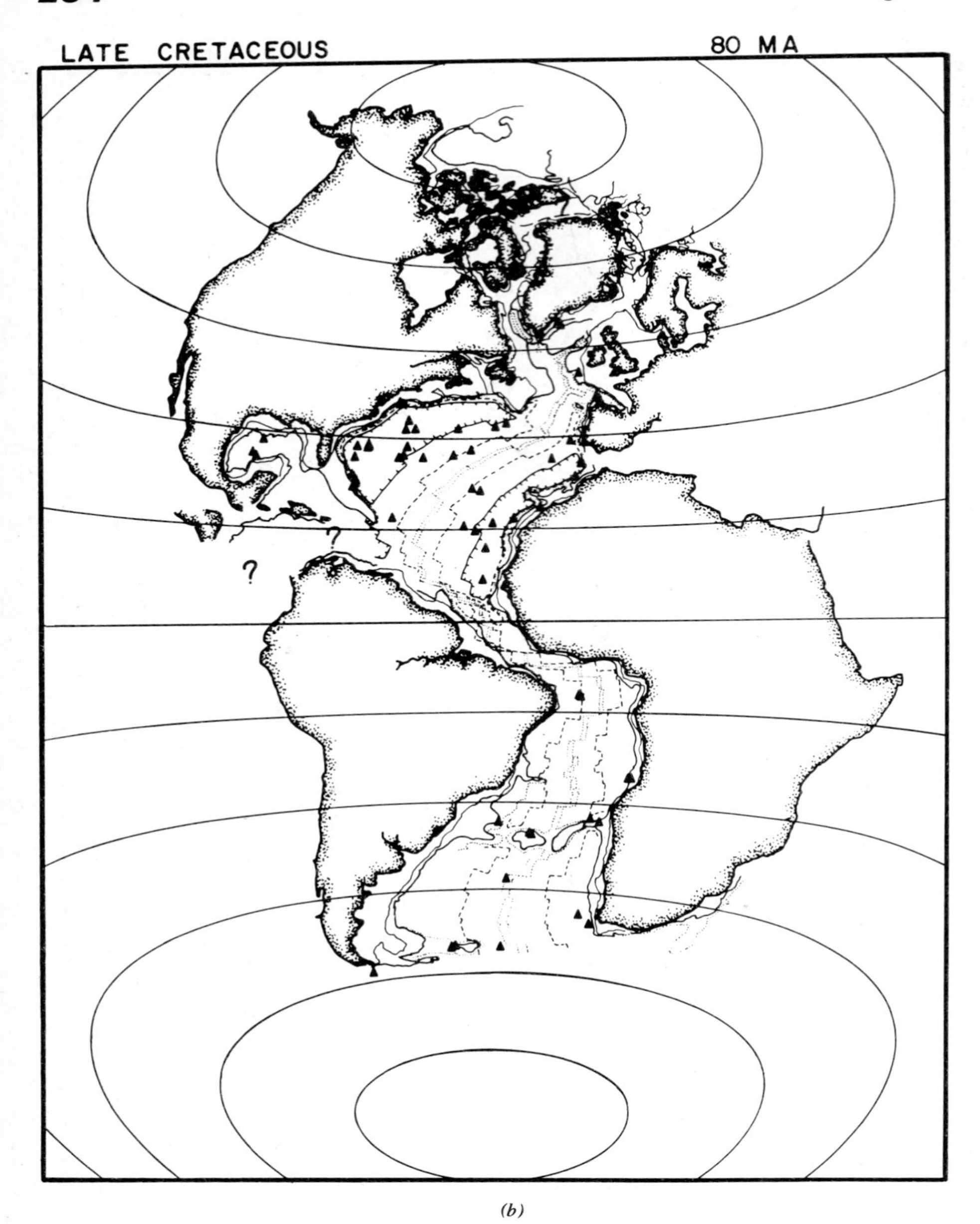

(b)

FIG. 11-10 *(Continued)*

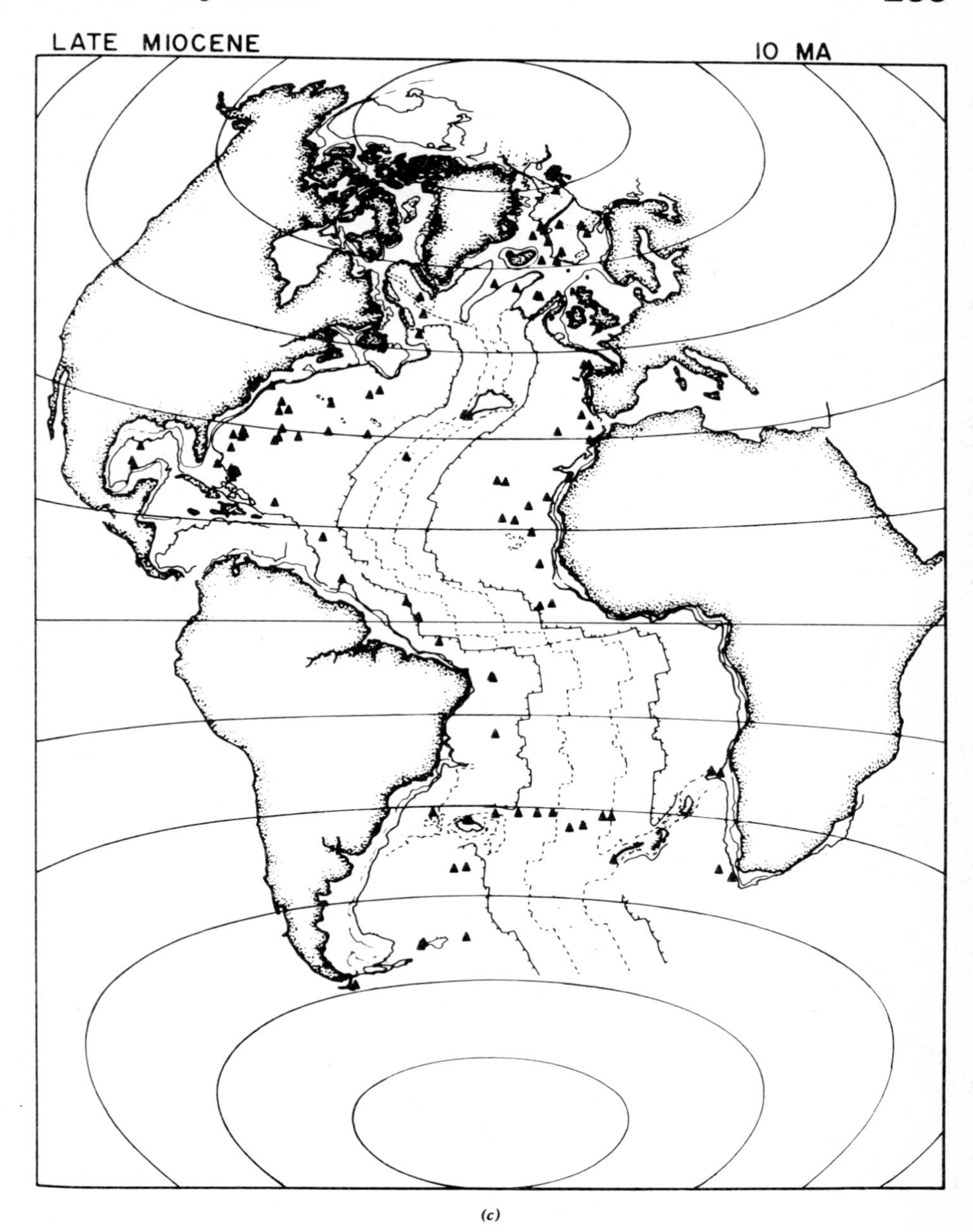

(c)

FIG. 11-10 *(Continued)*

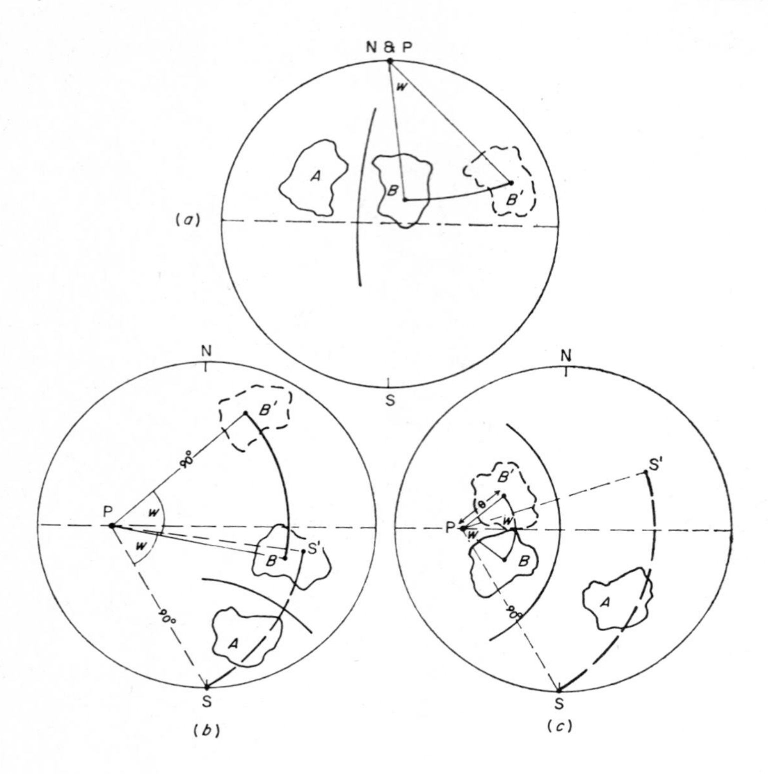

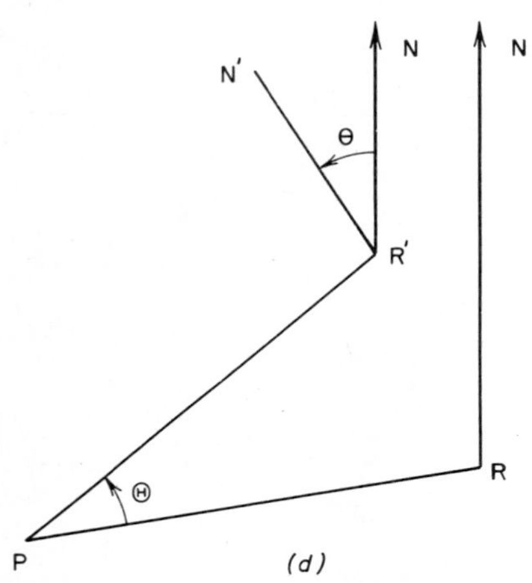

FIG. 11-11 *Some extreme cases of polar wandering. Plate A is supposed to be fixed, P is the pole of rotation of plate B with respect to plate A. (a) Plate B rotates around the earth's axis with respect to plate A. (b) The pole of rotation of plate B is on the equator (see also d). (c) The pole of rotation of B is on the equator, and plate B is close to its pole. (d) Same as b, but in a plane. (Figures a, b, and c from McElhinny 1973. Copyright © by Cambridge University*

Finally, if the pole of rotation P is on the equator, and plate B is close to it (Figure 11-11c), a small displacement of the plate will correspond to a large rotation of its VGP. This was closely approximated when the Iberian Peninsula (present-day Spain and Portugal) rotated in relation to most of the rest of Europe during the Permian and Triassic periods.

A detailed reconstruction of the relative positions of the continents thus cannot be based exclusively on paleomagnetic data, but these data provide important constraints. The best example of such constraints involves the polar paths of Europe and North America. When the paths of the geomagnetic poles

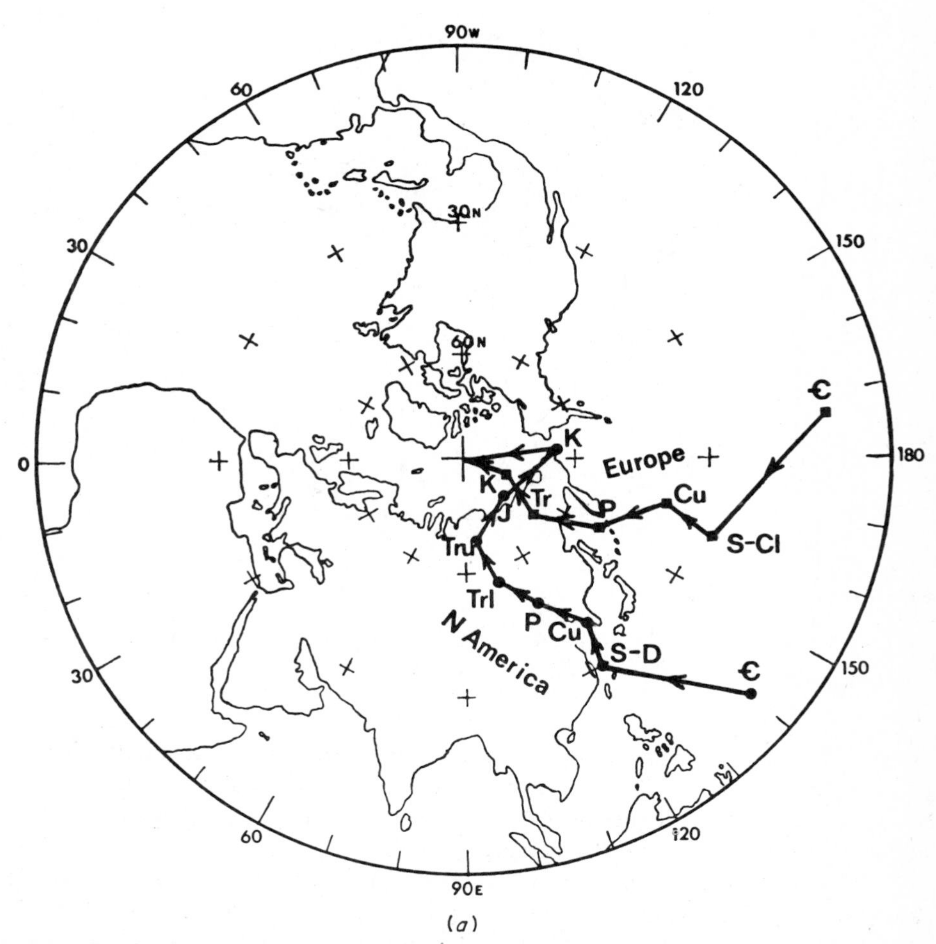

FIG. 11-12 *Comparison of the polar wander paths for Europe and North America. (a) With Europe and North America in their present positions. (b) (see next page) Taking into account the gradual opening of the Atlantic. (From McElhinny 1973. Copyright © by Cambridge University Press, reprinted by permission.)*

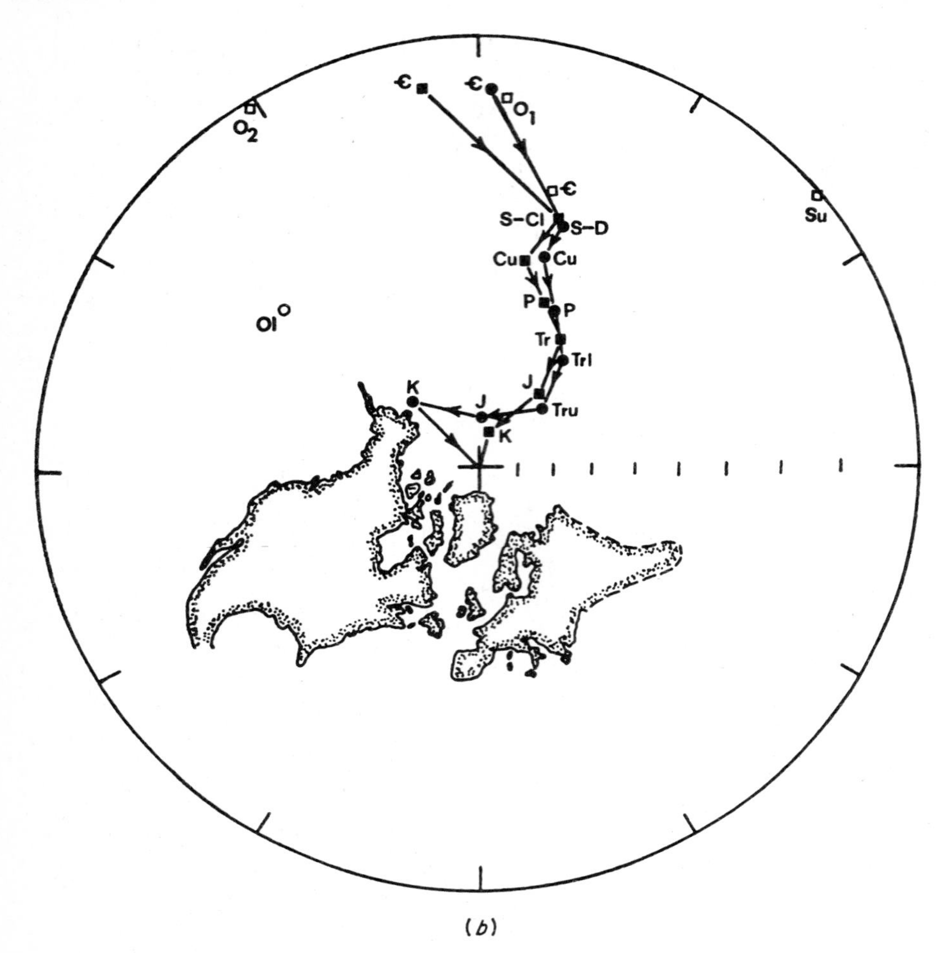

FIG. 11-12 (*Continued*)

are plotted for these continents in their present positions, they gradually diverge (Figure 11-12*a*); when the plot takes the opening of the Atlantic into account, however, the paths are brought essentially into coincidence (Figure 11-12*b*). McElhinny (1973) discusses this subject in detail.

It may be objected that it is not the pole that wanders, since presumably the geomagnetic pole remains close to the geographic pole, and the latter is essentially fixed with respect to the "fixed stars" because of the flattening of the earth. "Polar wandering," then, is merely a reflection of the motion of the plates on the earth's surface. This objection is correct, but the expression "polar wandering" is now as firmly embedded in geophysics as that of the "earth's field."

F. Anomalies Due to Some Simple Shapes

As already noted (Chapter 10, Section B), magnetic anomalies are much more difficult to compute and interpret than are gravity anomalies, especially on the continents. This situation arises principally from the following facts.

- On the continents, magnetic anomalies are generally caused by the present field acting on magnetite-bearing igneous and metamorphic rocks. These may vary both in depth and in susceptibility. On the one hand, the relation $\mathbf{J} = \chi \mathbf{H}$ (Equation 10-37) shows that a change in χ causes a proportional change in J. On the other hand, the susceptibility of rocks is generally proportional to their magnetite content, although a few other ferromagnetic minerals are sometimes important. This is true because in a magnetic field as weak as the earth's, the magnetic susceptibility of magnetite is essentially constant at $\sim 3 \ldots 10$ (in SI units). It follows that the magnetization is approximately given by

$$\mathbf{J} = (3 \ldots 10) 10^{-2} \, P\mathbf{H}$$

  where P is the percent magnetite content in the rock. P varies greatly in the same rock type, and may even vary laterally in a given rock unit over distances of a few kilometers. Thus, it may sometimes be difficult or impossible to distinguish the effect of "basement relief" from that of magnetite content. The table of susceptibilities in Appendix C should thus be used with considerable prudence. As would be expected from their mineralogy, basalts normally have the highest susceptibility.
- Because the inclination of the earth's field varies with the latitude (or the colatitude—i.e., angle with the North Pole—see equation 11-1), the same distribution of magnetic matter will give rise to different anomalies at different latitudes.
- When the components of the earth's field are measured, three components of the magnetic anomaly are obtained. This is now very rarely done. As will be explained in Section 11-G, most modern instruments measure the magnitude of the total field. The direction of the latter is always close to that of **B**, because the field caused by any anomalous body is much smaller than the earth's field. The "total" or "scalar" anomaly, B_{an}, is thus the projection of the field due to the body on the earth's field whose direction varies with latitude.

Qualitative Methods

Before discussing an exact method for computing some magnetic anomalies, we must examine two methods that yield qualitative results.

Poisson's Imaginary Magnetic Coating Poisson showed that the magnetic field created by any volume of matter possessing a constant magnetization is identical (except for a scale factor) to the gravitational attraction created by an imaginary coating whose surface density, σ_{im}, is equal to the projection of the outer normal to the surface upon the direction of magnetization. Figure 11-13 should help clarify this concept. Let the magnetization be **J**, presumed to be in the direction of the earth's present field. At point A the outer normal AB to the surface is perpendicular to **J**, and σ_{im} is thus null. At point C the outer normal to the surface is approximately opposite to **J**, and σ_{im} is thus negative and large. At E the outer normal is at an angle to **J**, hence σ_{im} is positive but not as large as at C (in absolute value). The density of the imaginary coating is thus roughly proportional to the distance between the body (solid line) and the dashed line in the figure. The approximate field at any point P can be estimated from the relative attraction (for $\sigma_{im} < 0$) and repulsion (for $\sigma_{im} > 0$) at P caused by the coating. For example, B_- for $\sigma_{im} < 0$, B_+ for $\sigma_{im} > 0$, keeping in mind the inverse square (not cube!) law. This yields **B** due to the magnetized body. As explained above, it is further necessary to project **B** on **J** if B_{an} is desired. For a complex shape the result is not very good, but for a simple one it can be. In some cases this method yields a simple, analytic way of computing the anomaly.

Figure 11-14a shows the scalar anomaly for a sphere magnetized in different directions of **J**, and Figure 11-14b shows $\mathbf{B}_+$, $\mathbf{B}_-$, **B**, and B_{an} at two points in the

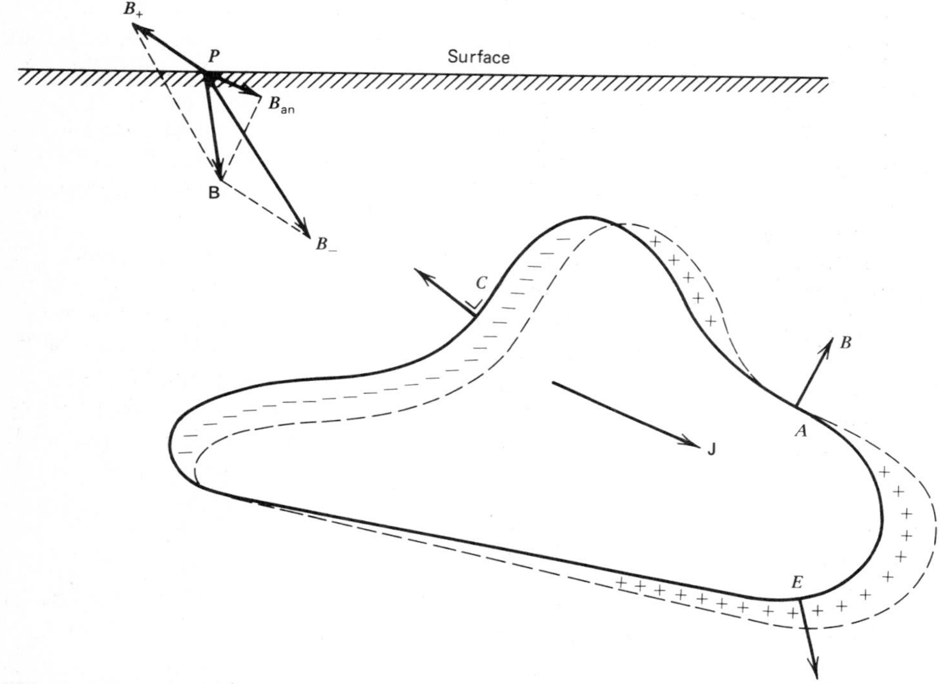

FIG. 11-13 *The surface density of Poisson's imaginary magnetic coating is proportional to the distance between the surface of the body (solid line) and the dashed line. It may be positive or negative. The "total," or "scalar," anomaly B_{an} is obtained by projecting* **B** *caused by the body on the direction of the earth's field (i.e., on* **J***).*

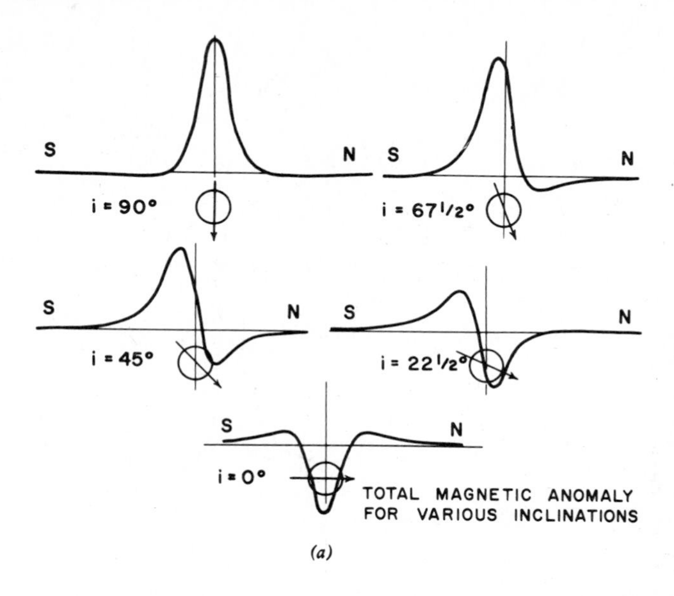

(a)

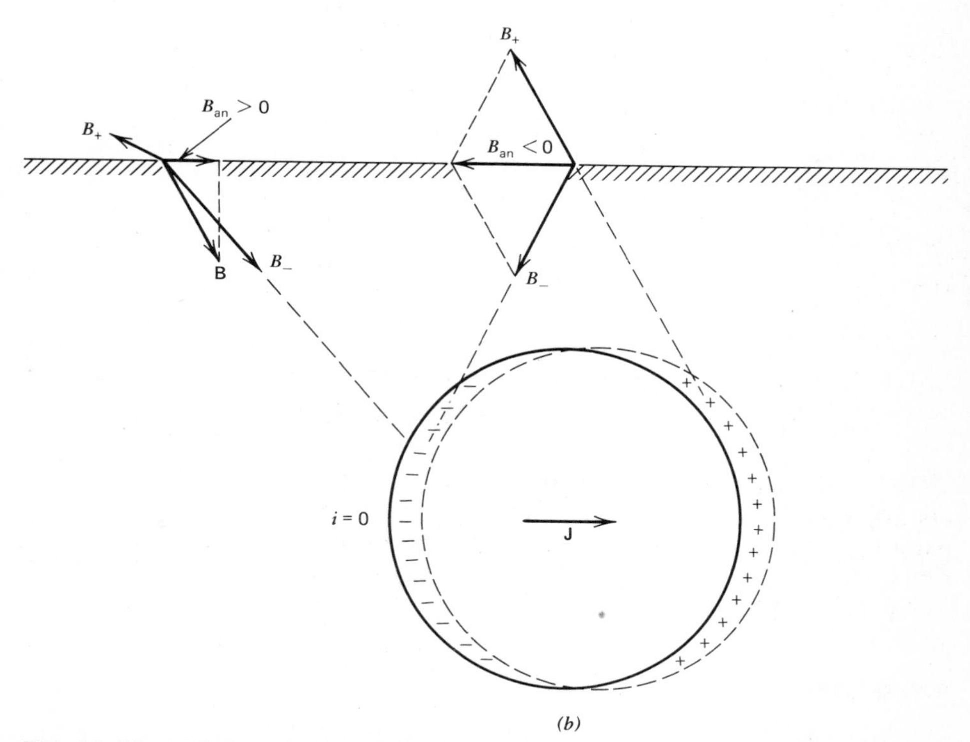

(b)

FIG. 11-14 *(a) Total anomaly caused by a magnetized sphere for various inclinations. (From Nettleton 1962. Reprinted by permission.) (b) Approximate computation of the total anomaly at two points when $i = 0$.*

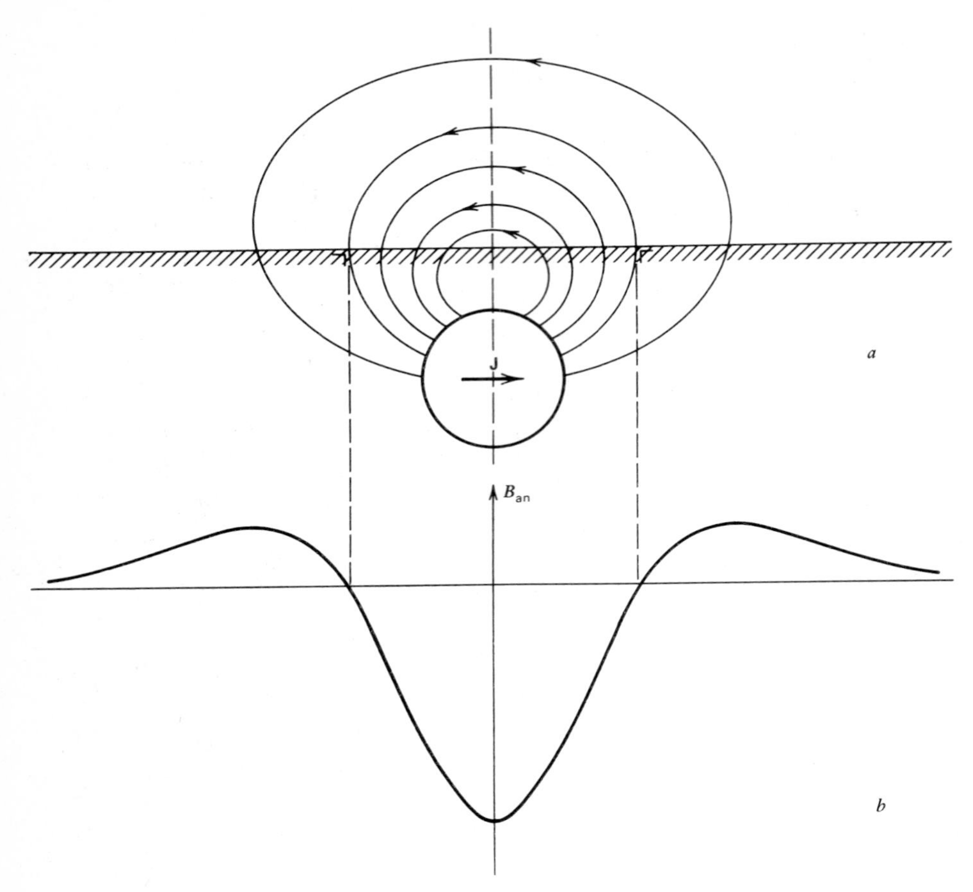

FIG. 11-15 *(a) "Lines of force." (b) The resulting anomaly. Compare with Figure 11-14(a).*

$i = 0$ case. The case of the vertical magnetization is straightforward; that of horizontal magnetization is perhaps unexpected.

The Lines-of-Force Method In this very rough method, approximate lines of force (from the + to the − poles) are drawn. Where the lines of force are parallel to the direction of the earth's field, the anomaly is maximum (positive); where they are perpendicular to it, the anomaly is zero (see Figure 11-15); and where they are opposite, it is negative.

PROBLEMS

11-2 Why are the magnetic anomalies near the equator in the Atlantic Ocean very difficult to identify? Is this true for other oceanic ridges near the equator? Why or why not?

11-3 **(a)** Magnetic records in oceanic areas are normally made by a proton magnetometer near the surface, towed behind a ship. Would these records have the same character if the magnetometer was near the ocean bottom? Why? (Give the relevant equation if possible.)

(b) What does it mean, exactly, to say that the character is different?

(c) How could you transform an ocean-bottom magnetic record into a surface record? Be specific. (This part of the problem supposes that you have read Chapter 6.)

Quantitative Method

The magnetic field of two-dimensional bodies of arbitrary shape can be computed (Talwani and Heirtzler 1964) by a method entirely similar to the one given in Chapter 9 (Equation 9-31), and a program to perform these computations has been published. If the body is parallel to the y-axis and the two components of magnetization in the x-z plane are J_x and J_z (see Equation 10-20), the x and z components of the resulting field will be (Figure 11-16)

$$B_z = 2(J_x Q - J_z P) \tag{11-5}$$
$$B_x = 2(J_x P + J_z Q)$$

where

$$P = \sum_{i=1}^{n} (A_i B_i + C_i D_i)$$

$$Q = \sum_{i=1}^{n} (C_i B_i - A_i D_i)$$

with (see Figure 9-21)

$$A_i = \frac{(z_{i+1} - z_i)^2}{(z_{i+1} - z_i)^2 + (x_i - x_{i+1})^2}, \qquad B_i = \gamma_i - \gamma_{i+1}$$

$$C_i = \frac{(z_{i+1} - z_i)(x_i - x_{i+1})}{(z_{i+1} - z_i)^2 + (x_i - x_{i+1})^2}$$

$$D_i = \frac{1}{2}\log_e\left(\frac{x_{i+1}{}^2 + z_{i+1}{}^2}{x_i^2 + z_i^2}\right)$$

and $\gamma_i = \tan^{-1}(z_i/x_i)$.

For a determination of the scalar magnetic anomaly B_{an}, B_x and B_z must be projected on **J** (Figure 11-16): that is,

$$B_{\text{an}} = B_x \cos i + B_z \sin i - B_{\text{ref}} \tag{11-6}$$

where (see Equation 11-1)

$$i = \tan^{-1}(J_x/J_z) = \tan^{-1}(2\cot\Theta) \tag{11-7}$$

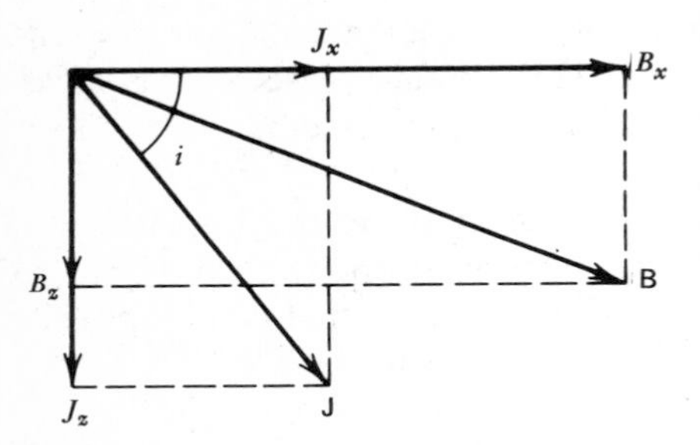

FIG. 11-16 *Computation of the magnetic anomaly caused by a two-dimensional body:* **J**, J_x, J_z *represent the magnetization and its components along the x- and z-axes;* **B**, B_x, B_z *represent the resulting "field" and its components along the x- and z-axes.*

Here, B_{ref} is the normal field of the Earth (IGRF), and Θ is the colatitude of the station.

Plouff (1976) has given a method of computing the magnetic field of three-dimensional bodies composed of right polygonal prisms (see also Chapter 9, Section C).

G. Magnetic Measurements

Prior to the miniaturization of electronic components, the components of the earth's field were measured from the equilibrium position of a magnetic needle subjected, on the one hand, to a component of the earth's magnetic field and, on the other hand, either to the action of gravity or to the torque of a suspension wire. Such instruments are rarely used now.

Flux-Gate Magnetometer

Victor Vacquier developed the first operational flux-gate magnetometer during World War II to detect submerged submarines, although the principle behind the device had been known earlier. This principle is extremely simple. Assume, first, that we are operating in zero ambient field. Now consider two identical coils wound in opposite directions around identical strips of high-permeability metal (Figure 11-17), through which the same alternating current is passed. The magnetic fields being equal and opposite in both strips, they follows their hysteresis loops (Figure 10-5) in opposite directions and reach saturation at the same time.

Next, a winding is wound around both coils (sensor winding). When the magnetic field changes in either coil, it tends to induce a current in the sensor coil; in zero ambient field, the currents in the two strips constantly cancel each other. In the presence of an ambient field, one strip reaches saturation first, and thus ceases to induce a current in the sensor coil. The output, then, is only due to the other strip; it is proportional to the component of the ambient field in the direction of the strip. The device can thus be used to measure the field in any

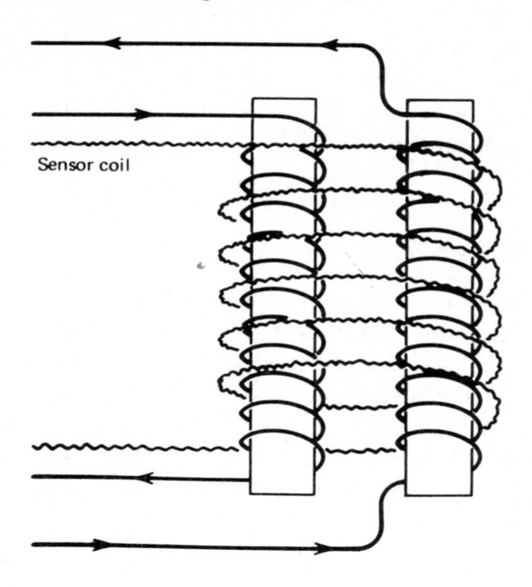

FIG. 11-17 *Two identical coils wound in opposite directions around strips of high-permeability metal, along with one sensor coil wound around both, form the heart of the flux-gate magnetometer.*

particular direction. If high sensitivity is desired, the greatest part of the ambient field can be canceled out, leaving only its variations. For historical reasons, this instrument is known as a flux-gate magnetometer.

This instrument is still the device most commonly used in aeromagnetic surveys, for which it is enclosed in a "bird" towed behind an aircraft. Because a strictly horizontal or vertical orientation cannot then be maintained, three flux-gate magnetometers are mounted at right angles to each other, with axes and electric motors arranged so that any orientation relative to the "bird" can be obtained. If the main flux gate is aligned with the earth's field, the outputs of both auxiliary flux gates are zero. The system senses the outputs of the auxiliary flux gates, constantly reorients them to maintain zero output, and records the output of the main flux gate. In fieldwork, though, a single flux gate may be used in any desired orientation to measure a particular component of the earth's field directly.

Finally, it should be noted that this is a (doubly) relative instrument. Neither the zero nor the scale factor are immediately known. (See gravimeters, Section 9-D.)

Helmholtz Coil(s)

We have repeatedly alluded to the fact that it is necessary to be able to cancel the earth's field. What one wishes to do, in fact, is to produce an essentially uniform field of a given orientation in a volume sufficiently large to conduct whatever experiment is planned. The instrument used for this purpose is the Helmholtz coil (or coils), which consists of two coaxial coils separated by a distance equal to their radius, r. Although analysis of the magnetic field inside such coils yields functions that cannot be evaluated exactly, it can be shown that the field near the center of the system (at less than $\sim 0.1r$) remains remarkably constant and is oriented along the axis of the pair of coils. The field can be computed in terms of the current, the radius, and the number of turns, but today it is usually measured directly. Three such coils mounted at right angles to each other can thus create any desired field.

Proton Precession Magnetometer

This instrument measures the total magnetic field, B, in magnitude but not in direction. The principle is simple. A container filled with some fluid (water) containing protons is surrounded by a coil. When a sufficiently strong current is passed through the coil, the induced magnetic field causes most of the protons to align themselves parallel to the imposed field. This occurs in a fraction of a second. When the current is interrupted, the protons precess much like gyroscopes, around whatever ambient field is present. The frequency of precession, called the Larmor frequency, is given by the formula

$$B = 23.4874\nu \tag{11-8}$$

for B in nT (or γ) and ν in Hz. After a short time (~ 1 s) the thermal agitation destroys the coherency of motion of the protons. The only restriction is that the induced field not be parallel to the earth's field. For an average value of the earth's field—say, 50,000 nT—ν is ~ 2 kHz. In order to obtain a precision of 1 nT (1/50,000), one measures the time interval for a fixed number of cycles, inverts the resulting period, and obtains B with sufficient accuracy. In contrast to the flux-gate magnetometer, this instrument gives the absolute value of the earth's field, but not its direction.

Optically Pumped Magnetometers

The phenomena of "optical pumping" and "lasing" (or laser action) were first investigated in the late 1950s. The term *pumping* reflects the fact that atoms are raised from lower to higher states of internal energy. This process is accomplished by means of light waves, hence the word *optical*. The change of orbit of one electron causes an atom to absorb or emit a photon of light; the absorption of a photon of radiowaves (which is much less energetic, because its frequency is much lower) causes the spin of the electron on its orbit to be reversed, but the orbit is left unchanged.

We now turn to the idealized situation shown in Figure 11-18. An atom has only three energy levels, A, B, and C. The energy change from C to A or C to B is large (orbit change) and corresponds to the emission of light (absorption is from A to C or B to C). The energy difference between A and B is small (spin reversal) and corresponds to emission or absorption of radio-frequency (RF) waves. Taking a gas consisting of an assemblage of atoms in their ground states—that is, some in A, some in B (Figure 11-18*a*)—we irradiate it with light that does *not* contain the frequency, or spectral line, BC. Atoms in state B thus cannot absorb energy. Those in A can, and they rise to C (b). After a short time (about 10 μs), they spontaneously drop down to a lower state, some to A and some to B (c). Those in A are pumped back up, and so on. Eventually (e), all the atoms are in state B; that is, the material (gas) is completely pumped and is thus transparent to the light beam. (All of this may seem to have nothing to do with the magnetic field, but it will!) To induce the electrons to jump from B to A, we must now irradiate them with an RF wave of the appropriate frequency. When this is done, the electrons will jump back down to the A state and again be able to absorb the light beam.

With sodium, cesium, and rubidium, there is a single unpaired electron in the outermost shell. This electron can have its magnetic field, and thus its spin axis, in two positions only: either parallel or antiparallel to the ambient magnetic field. These two states correspond to our A and B states. Thus, the energy difference between A and B, as well as the corresponding RF frequency, depend on the outside field. Furthermore, these two states are also distinguished by the

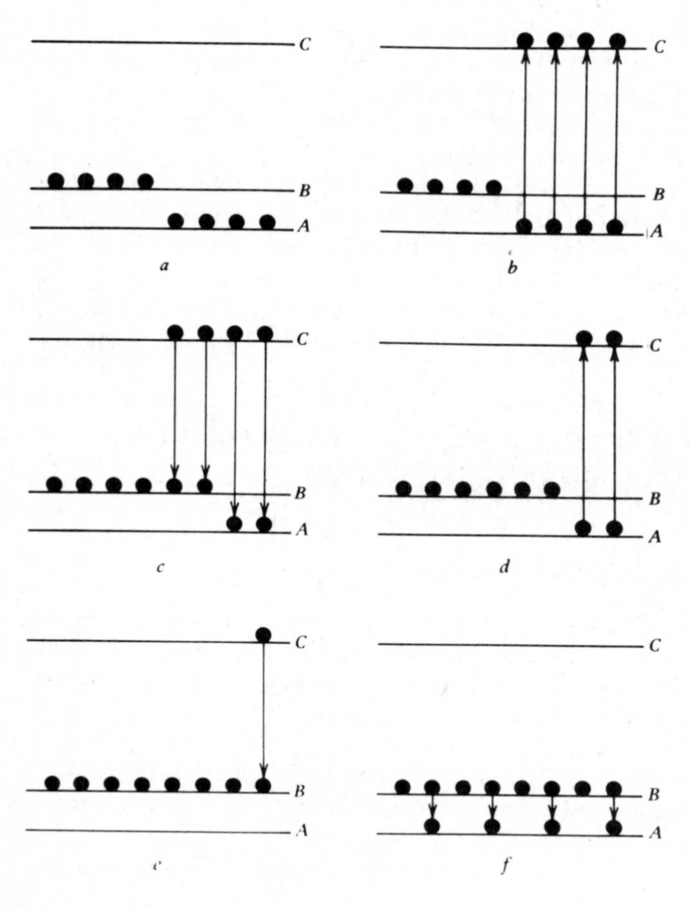

FIG. 11-18 *The optical pumping principle. The atoms in their ground state may be in either A or B (a). Those in A are optically pumped to C (b), and eventually (e), all end up in B. They can then detect (by precessing at a certain frequency) the magnetic field corresponding to the AB transition. (After Bloom 1960. Copyright © 1960 by Scientific American, Inc. All rights reserved, reprinted by permission.)*

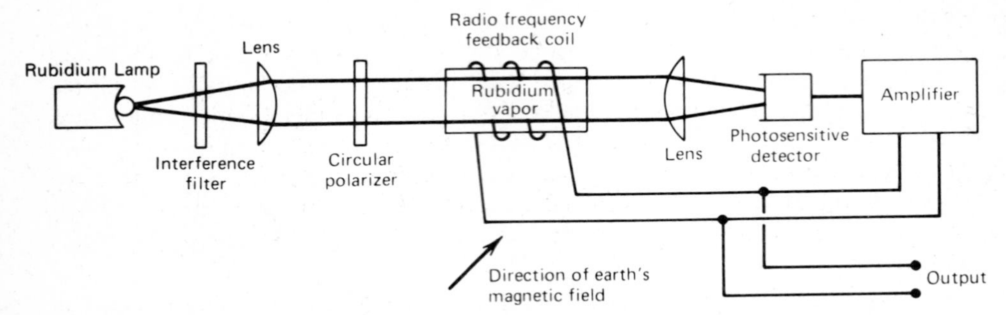

FIG. 11-19 *The radio frequency of a rubidium (or cesium) vapor magnetometer is proportional to the earth's field. (From Bloom 1960. Copyright © 1960 by Scientific American, Inc. All rights reserved, reprinted by permission.)*

fact that one of them—say, B—has one more unit of angular momentum than the A state, whereas the C state has the same angular momentum as the B state.

Now, light photons also have one spin unit of angular momentum, and if the light is circularly polarized, all the spin axes will point in the same direction. Thus, if we shine a beam of circularly polarized light on atoms in a mixture of A and B states, only those in A will be able to absorb the light, because those in B already have their maximum angular momentum.

To complete the picture, we should note that instead of dropping from B to A, the electrons can be considered to "wobble" between these states, thus causing the light to flicker at the same frequency as the RF waves. We thus need only detect this flicker on a photocell, amplify it, and reinject it as RF waves into the container in order to obtain an atomic oscillator whose frequency is governed by the external magnetic field (Figure 11-19). It is thus an absolute, nondirectional instrument.

In addition to Na, Cs, and Rb, metastable He can also be used in such magnetometers. Such an instrument was manufactured in the United States in the early 1960s.

Superconducting Magnetometer

It is not possible to describe this apparatus here, and the reader should consult articles by Goree and Fuller (1976) and Zimmerman and Campbell (1975) on the subject. We should note, though, that these instruments depend on the phenomenon of superconductivity, which occurs at or near liquid helium temperature. They come in a variety of forms, all of which are known as Squids (superconducting quantum interference devices). They are directional instruments—that is, they are capable of measuring any component of the earth's field—and have by far the highest sensitivity of any magnetometer: 10^{-14} T $= 10$ fT $= 10$ $\mu\gamma$ (f $= 10^{-15}$ = "femto") in a period as short as 1 ms. Such sensitivity makes them particularly suitable for applications such as magnetotelluric prospecting. However, they are not absolute instruments.

Measuring the Magnetism of Rocks

By far the most frequently made measurement in rock magnetism is that of the direction of **J**. For this measurement it is necessary that the samples taken in the field (e.g., cores) be precisely oriented. Measurement of the value of J is done far less often, usually in order to determine the value of the earth's field in the past. For this purpose, the rock first has to be reheated past its Curie point and then cooled in a known field without any chemical alteration. This is extremely difficult, and the scatter in the measurements is large. Nevertheless, these measurements appear to show that the dipole moment of the earth has on the average been steadily increasing in the last 500 Ma (Figure 11-20). Yet over periods on the order of 1000 to 10,000 years, greater fluctuations have taken place (Figure 11-21). These measurements, including some made on pottery, bricks, and other artifacts (archeomagnetism), have suggested that the main dipole field has been decreasing for the past 1000 years; according to some authors, a field reversal may occur in another 1200 years.

Measurements of the direction of **J** form a fairly specialized field of geophysics. Only two of the most commonly used instruments will be briefly mentioned here.

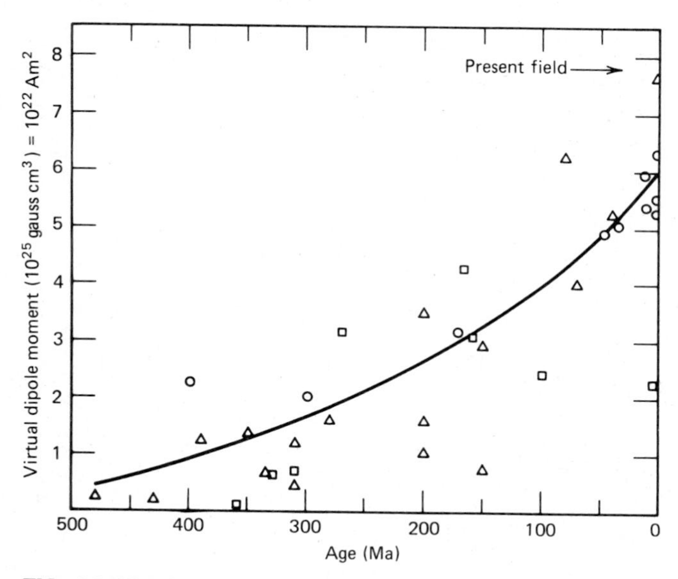

FIG. 11-20 *On a 10-Ma time scale the earth's magnetic field seems to have increased by a factor of ~10 in the last 500* Ma, *but the scatter is large. The scale on the left is the equivalent dipole moment in units of 10^{22}* A m^2 *(see Appendix B). (From McElhinny 1973. Copyright © by Cambridge University Press, reprinted by permission.)*

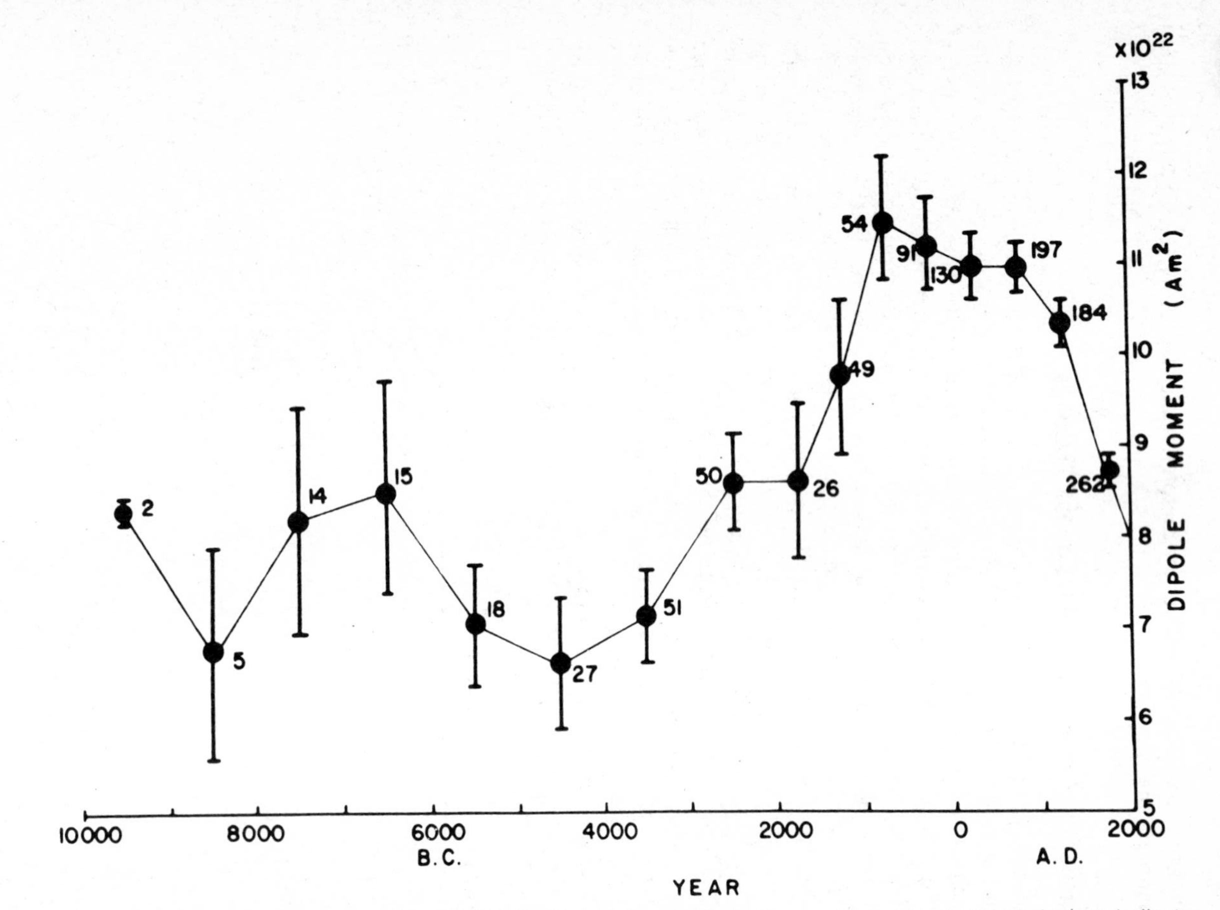

FIG. 11-21 *Global mean dipole moments from the present to 10,000* B.C. *The numbers indicate the number of samples used; the bars indicate the 95% confidence limits. (From McElhinny and Senanayake 1982. Reprinted by permission.)*

Spinner Magnetometer In this instrument a small oriented sample—for example, a cube 2 cm on a side—is spun successively around three orthogonal axes. In each case a small sensor (for example, a flux gate) senses the angular position at which the maximum field occurs (relative to a reference mark) and the relative strength of the field. This is recorded by, for example, a dedicated microcomputer. The direction of **J** can thus be found, the determination of paleopole (VGP) follows. In order to minimize the noise, some generally evident precautions must be taken. The spinning sample must be far from the electric motor (because of its magnetic field) and linked to it by a nonmagnetic rod; the apparatus should be enclosed in Helmholtz coils designed to cancel the earth's field; and the measurements should (if possible) be carried out in a "magnetically quiet" building.

Squid Magnetometers The Squid magnetometer mentioned above has also been used (with appropriate modifications) to measure rock magnetism. This instrument may contain one, two, or three sensors. With three sensors, it is capable, after one insertion of the sample, of yielding **J** in both direction and magnitude. The instrument is extremely sensitive, convenient, and fast, but it is expensive to purchase and operate. It can detect magnetic moments as small as 2×10^{-10} A/m^2 (or 2×10^{-7} G/cm^3)—that is, in the usual sample, J as small as 2×10^{-17} A/m (or 2×10^{-8} G). Because spinning of the sample is not required, this instrument is well suited to the analysis of unconsolidated sediments.

Brief Review of the Contributions of Geomagnetism to Plate Tectonics

- Geomagnetism gave birth to the plate tectonic theory.
- The recognition of magnetic anomalies in the oceans and the determination of the paleopoles have made possible the reconstruction of the paleogeography of several large areas.
- Detailed study of the magnetic field and of rock specimens near oceanic ridges is clarifying the process of sea-floor spreading.

PART FOUR

Flow and Temperatures in the Earth

12

Viscous Flow

A. Introduction

It is obvious that folds in rocks are permanent: they do not vanish when the stresses vanish. Thus, for large deformations and times on the order of 1 Ma or greater, elasticity is not a good model. Experience with liquids such as oil shows that in such bodies the stresses are proportional to the strain rates, rather than to the strains. Instead of (Equation 2-20)

$$p_{xy} = 2\mu e_{xy}$$

the constitutive relation is thus of the general form

$$p_{xy} = 2\eta \frac{\partial}{\partial t}(e_{xy}) = 2\eta \dot{e}_{xy} \tag{12-1}$$

(In this chapter a dot over a letter designates a partial derivative with respect to time.) Both pure elasticity and pure viscosity are idealized states, but they closely approximate important real situations. In the case of an elastic body, *all* of the mechanical energy used in deformation is recoverable—*none* is transformed into heat and lost. In the case of a viscous body, conversely, *none* of the deformational energy is recoverable—*all* is transformed into heat and lost. It appears logical to surmise that reality lies between these two extremes (see Section 12E). Nevertheless, the pure-viscous case provides a very useful approximation.

In light of Equations 2-20 and 12-1, then, we see that just as μ measures the resistance to shear strain, so η measures the resistance to shear strain rate. This quality is called the viscosity or, sometimes, the dynamic viscosity. It is measured in Pa s in SI units, and in poises (P) in the cgs system (1 Pa s = 10 P—see Appendix A).

The question arises whether the relations between normal stress and normal strain that exist in the elastic case (Equation 2-21) can be transposed to the viscous case simply by replacement of the strain by the rate of strain, as is done for shear stress and strain. One difficulty is immediately apparent: in the elastic case,

$$p_{xx} = \lambda\theta + 2\mu e_{xx} = \left(\kappa - \tfrac{2}{3}\mu\right)\theta + 2\mu e_{xx}$$

But many liquids are essentially incompressible, which implies both $\kappa = \lambda = \infty$ and $\theta = 0$ (Chapter 2, Section C); this renders the above equation meaningless. Another formulation of the problem is thus needed. In the classical formulation, the constitutive equation is written in terms of the deviators.

B. Deviators

Constitutive Equation

We define the stress deviators as

$$P_{ij} = p_{ij} + P\delta_{ij} \tag{12-2}$$

(See Equation 2-22.) The shear stress deviators are thus identical to the shear stresses, but the normal stress deviators are equal to the normal stresses plus the pressure. In geology the normal stresses are often counted as positive along the *inner* normal, so Equation 12-2 becomes

$$P_{ij} = p_{ij} - P\delta_{ij} \tag{12-3}$$

PROBLEM

12-1 In a laboratory, a cylindrical specimen 10 cm long and 6 cm in diameter is set on its base on a horizontal surface. A mass of 10 kg then is set on its top. What are the normal stress and the normal stress deviator along the axis of the cylinder? (Assume ideal conditions, and do not neglect air pressure.) Why do these differ, if they do?

The strain rate deviator can be defined similarly as

$$\dot{E}_{ij} = \dot{e}_{ij} - \tfrac{1}{3}\dot{\theta}\delta_{ij}$$

If the fluid is incompressible, $\theta = 0$; hence $\dot{\theta} = 0$, and we have

$$\dot{E}_{ij} = \dot{e}_{ij} \tag{12-4}$$

That is, the strain rate deviators are equal to the strain rates. The constitutive equation then becomes

$$P_{ij} = 2\eta \dot{E}_{ij} = \eta\left(\frac{\partial \dot{u}_i}{\partial x_j} + \frac{\partial \dot{u}_j}{\partial x_i}\right) \tag{12-5}$$

to which must be added the condition that the rate of dilatation is zero:

$$\dot{\theta} = \frac{\partial \dot{u}}{\partial x} + \frac{\partial \dot{v}}{\partial y} + \frac{\partial \dot{w}}{\partial z} = 0 \tag{12-6}$$

For normal stresses, Equation 12-5 yields, for example,

$$P_{xx} = p_{xx} + P = 2\eta \dot{e}_{xx}$$

which shows that the previous difficulty arising from the incompressibility no longer exists.

Validity

The approximate validity of the constitutive equation of elasticity is relatively easy to establish in the laboratory. Since the phenomenon is time-independent, measurements can be carried out as rapidly as is convenient. In principle this can also be done to check the validity of Equation 12-5. No difficulty arises when relatively ordinary liquids are examined; the strain rates are then easily measurable. Geological strain rates, however, are very small; a strain rate of 10% in 100,000 years would be quite large in most geological cases. But such a rate is only on the order of 3×10^{-9}/day. At this rate, a specimen 5 cm long subjected to a deviatoric normal stress along its axis would shorten by only 1.5×10^{-8} cm $= 0.15 \times 10^{-3}$ μm in one day. This distance is much too small to be measured. Laboratory experiments must thus be conducted at strain rates a thousand to a million times greater than geological rates, and it is very difficult to establish with certainty that the results thus obtained are entirely valid in the earth. This means that in the earth, Equation 12-5 may not be exactly correct; though the stresses may still be functions of the strain rates rather than of the strains, this relationship may not be exactly linear. In such cases, we speak of generalized viscosity.

Viscosity further implies that the total strain may be obtained as

$$E_{ij} = \int_o^t \dot{E}_{ij}\, dt = \int_o^t \frac{\partial E_{ij}}{\partial t}\, dt \tag{12-7}$$

If $\dot{E}_{ij}$ does not vary with time, the total strain is thus proportional to the strain rate and to the time interval.

In seismology, where the stresses are small (except near the focus) and of short duration, elasticity is a good approximation. In tectonics, where the stresses may be large and last for hundreds of thousands of years, generalized viscosity is normally the best model.

Despite the difficulties involved in making accurate measurements of the viscosity in rocks and minerals, we can be reasonably certain of the following facts.

- The viscosity decreases exponentially with a rise in temperature.
- Rocks above ~ 500° C are closer to being viscous than to being elastic. Below that temperature, the opposite is true.
- The viscosity increases exponentially with pressure, but such an increase is slight.
- The viscosity is inversely proportional to a power of the stress deviator.

Power Law Viscosity

The relation between the viscosity and the stress deviator given above is generally referred to as power law viscosity. (Note that it is a negative power!) This concept must be clarified, for the stress deviator is a tensor and the viscosity is a scalar, and such different quantities cannot be powers of each other.

The exact meaning of the concept of power law viscosity may be stated as follows. If an "average stress deviator" (there is no standard term) is defined as

$$\tau = \frac{1}{\sqrt{2}}\left(P_{xx}^{\ 2} + P_{yy}^{\ 2} + P_{zz}^{\ 2} + 2P_{xy}^{\ 2} + 2P_{yz}^{\ 2} + 2P_{xz}^{\ 2}\right)^{1/2} \tag{12-8}$$

it may be shown that τ is a scalar and that it is invariant for a rotation of the coordinate axes, just as the pressure is. It may thus be put in relation with the viscosity.

To understand the physical meaning of the concept, we will examine a very simple viscous flow shown in Figure 12-1—that of a viscous liquid slowly flowing in a two-dimensional conduit (laminar flow). The liquid adheres to the sides,

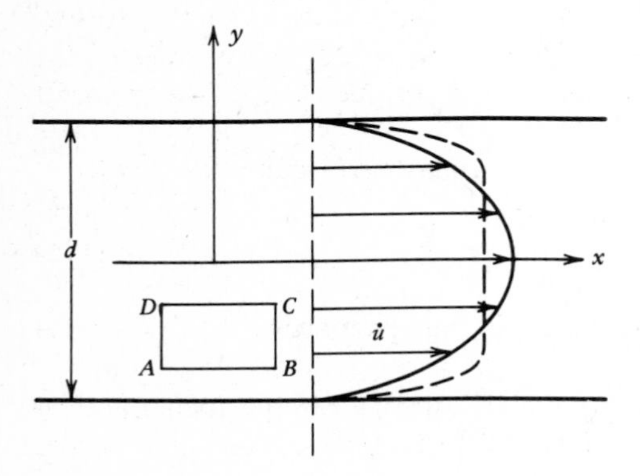

FIG. 12-1 Velocity profiles for a laminar viscous flow in a two-dimensional duct. The velocity is zero at the sides and increases toward the center. The solid line shows the profile for a constant viscosity; the dashed line for a power law viscosity. The element of volume ABCD is used in the computation of the Reynolds number.

hence its velocity vanishes there. By symmetry, the velocity is maximum at the center. Because the phenomenon is independent of x and z, all $\partial/\partial x$ and all $\partial/\partial z$ equal zero. Since $w = 0$ (two-dimensional flow), it follows that only $e_{xy} \neq 0$. Hence only $\dot{E}_{xy} \neq 0$ (which we write $\dot{\varepsilon}$) and only $P_{xy} \neq 0$. Thus, in this case only,

$$P_{xy} = \tau = 2\eta\dot{\varepsilon} = 2\eta\dot{E}_{xy} \tag{12-9}$$

Generalizing from this equation, we conclude that τ is a measure of the stress deviator, and that $\dot{\varepsilon}$ is a measure of the strain rate (deviator).

Returning to power law viscosity, we define it by the equation

$$\eta = k\tau^{-(n-1)} \tag{12-10}$$

k is a constant and n varies with pressure, temperature, and the nature of the material. For olivine under upper-mantle conditions, n is between 3 and 5.

As shown below, the effect of this law is to cause the strain rate to become concentrated in narrower zones than would otherwise be the case. This property is known as strain (rate) softening, or thixotropy (a term used particularly in the paint industry, where thixotropy is a much-soughtafter quality). Another way to look at this process is to say that the viscosity decreases where the strain rate increases.

PROBLEM

12-2 What is the advantage of a highly thixotropic paint?

To show why power law viscosity causes the material to be thixotropic, we need to express η as a function of $\dot{E}_{ij}$ rather than of τ. Starting thus with Equation 12-8 and using Equation 12-5, we find

$$\tau^2 = \tfrac{1}{2}\left(P_{xx}{}^2 + \cdots + 2P_{xy}{}^2 + \cdots\right) \tag{12-11}$$
$$= \tfrac{1}{2}(4\eta^2)\left(\dot{E}_{xx}{}^2 + \cdots + 2\dot{E}_{xy}{}^2 + \cdots\right)$$

Let

$$\dot{\varepsilon} = \frac{1}{\sqrt{2}}\left(\dot{E}_{xx}{}^2 + \cdots + 2\dot{E}_{xy}{}^2 + \cdots\right)^{1/2}$$

Then, from the last equation,

$$\tau = 2\eta\dot{\varepsilon} \tag{12-12}$$

from which, using Equation 12-10,

$$\eta = k\tau^{-(n-1)} = k\ 2^{1-n}\ \eta^1\ \eta^{-n}\ \dot{\varepsilon}^{1-n}$$

and thus,

$$\eta = 2^{(1/n)-1}\ k^{1/n}\ \dot{\varepsilon}^{(1/n)-1} \tag{12-13}$$

For simplicity, examine the case of $n \gg 1$. Then $\eta = C(1/\dot{\varepsilon})$, where C is a constant. Thus, when the strain rate is large, η is small, which shows that the material is thixotropic.

By introducing Equation 12-13 into Equation 12-12, we obtain

$$\dot{\varepsilon} = C\tau^n \tag{12-14}$$

where C is a constant. An equation of this type is frequently taken as the constitutive equation for power law viscosity, but it is important that the meaning of $\dot{\varepsilon}$ and τ be recognized. A more explicit form is

$$\dot{E}_{ij} = C\tau^{n-1}P_{ij} \tag{12-15}$$

This equation can easily be obtained from Equation 12-5 by replacing η by its value from Equation 12-10; C is an appropriate constant.

In a thixotropic material an increase in the strain rate leads to a decrease in η. This may lead to a further increase in $\dot{\varepsilon}$, a further decrease in η, and so forth. In two-dimensional laminar flow, this results in a pluglike motion (Figure 12-1). In the earth, it may result in narrow zones of high strain rates separating wide zones of low strain rates. Faulting is sometimes thought of as an extreme case in this process, but the mechanism of fault propagation differs greatly from that of viscous flow.

By including the temperature dependence in Equation 12-15, we obtain, for near-surface conditions (upper ~ 100 km),

$$\dot{E}_{ij} = A\exp\left(-\frac{Q}{\mathrm{RT}}\right)\tau^{n-1}P_{ij} \tag{12-16}$$

where, for dry olivine, $A \cong 1.2 \times 10^{10}$, $Q = 5.02 \times 10^5$ J/mole, R (Boltzmann's constant) = 8.343 J/°K mole, and $n \cong 4.8$. At high pressures, it is necessary to let Q increase with the pressure.

C. Reynolds Number

The difference between waves breaking on a beach and the flow of molasses is obvious. The physical reason for this difference is that in the former case, inertial forces are much larger than viscous forces, whereas the opposite is true in the latter case. The Reynolds number characterizes flow conditions. It is defined as

$$\mathrm{Re} = \frac{\text{Inertial forces}}{\text{Viscous forces}} = \frac{F_i}{F_v} \tag{12-17}$$

Hence, for $\mathrm{Re} \gg 1$, the inertial forces are dominant, and for $\mathrm{Re} \ll 1$, the viscous forces are dominant.

To compute Re, we must turn again to the situation shown in Figure 12-1, taking an element of volume $ABCD$—say, near the bottom of the conduit. As a first approximation, we can say that the only viscous force F_v acting on this volume is the difference in the shearing stresses on DC and on AB: that is,

$$F_v \times \delta V = \left[\left(p_{xy} + \frac{\partial p_{xy}}{\partial y} \delta y \right) - p_{xy} \right] \delta S = \frac{\partial p_{xy}}{\partial y} \delta V$$

But

$$p_{xy} = P_{xy} = 2\eta \dot{E}_{xy} = \eta \left(\frac{\partial \dot{u}}{\partial y} + \frac{\partial \dot{v}}{\partial x} \right)$$

and since $\partial v/\partial x = 0$, $\partial \dot{v}/\partial x = 0$. Hence,

$$F_v = \frac{\partial}{\partial y} \left(\eta \frac{\partial \dot{u}}{\partial y} \right) = \eta \frac{\partial^2 \dot{u}}{\partial y^2}$$

The inertial forces per unit volume are

$$F_i = \rho a$$

where a is the acceleration. We then make the following order-of-magnitude estimations.

$$F_v = \eta \frac{\partial^2 \dot{u}}{\partial y^2} = \eta \frac{V}{L^2} = \frac{\eta L}{t L^2} = \frac{\eta}{L t} \tag{12-18}$$

where V is an average velocity, L is a characteristic length, and t is a characteristic time.

Similarly,

$$F_i = \rho a = \frac{\rho V}{t} \tag{12-19}$$

From which, using Equations 12-17, 12-18, and 12-19,

$$\text{Re} = \frac{\rho L V}{\eta} = \frac{L V}{\nu} \tag{12-20}$$

where $\nu = \eta/\rho$ is the kinematic viscosity.

Since Re is a ratio of forces per unit volume, it is nondimensional, hence independent of the system used (the units must, of course, be consistent).

For plate tectonics,

$$\rho \cong 3000 \text{ kg/m}^3$$
$$L \cong 1000 \text{ km} = 10^6 \text{ m}$$
$$V \cong 10 \text{ cm/a} \cong 3 \times 10^{-9} \text{ m/s}$$
$$\eta \lesssim 10^{19} \text{ P} = 10^{18} \text{ Pa s}$$

Hence,

$$\text{Re} = \frac{3 \times 10^3 \times 10^6 \times 3 \times 10^{-9}}{10^{18}} \approx 10^{-17}$$

The inertial forces are thus on the order of 10^{17} smaller than the viscous forces, hence entirely negligible. This point must be emphasized: *in tectonics there is no inertia.*

D. Stokes Equation

In the "solid" earth, where motions are very slow, the Reynolds number is vanishingly small. Under these conditions the law governing fluid motions is greatly simplified. We need only write that every element of volume is in equilibrium under the influence of the stresses acting on its surface and of the gravity acting on its volume—all other terms are negligible. This is not the case in general fluid flow.

PROBLEM

12-3 Why does it make sense to state that "an element of volume is in equilibrium" and at the same time assume that it is in motion?

Consider the x-components of the forces acting on the parallelepipedon shown in Figure 12-2. (The following reasoning resembles somewhat that used in Chapter 2, Section A, to show the symmetry of the stress tensor.)

On the $\hat{\mathbf{x}}$-faces (or y-z faces) the only stress component in the x-direction is p_{xx}. The difference in the value of this component between the two faces is

$$-p_{xx} + \left(p_{xx} + \frac{\partial p_{xx}}{\partial x} \delta x \right) = \frac{\partial p_{xx}}{\partial x} \delta x$$

This component acts on the area of the face $\delta y \, \delta z$. The force in the x-direction due to this component is thus

$$\frac{\partial p_{xx}}{\partial x} \delta x \, \delta y \, \delta z = \frac{\partial p_{xx}}{\partial x} \delta V \tag{12-21}$$

where δV is the element of volume. Similarly, the only stress component in the x-direction acting on the $\hat{\mathbf{z}}$-faces is p_{xz} and the difference is

$$-p_{xz} + \left(p_{xz} + \frac{\partial p_{xz}}{\partial z} \delta z \right)$$

The area of the face is $\delta x \, \delta y$, and thus the force in the x-direction is

$$\frac{\partial p_{xz}}{\partial z} \delta z \, \delta x \, \delta y = \frac{\partial p_{xz}}{\partial z} \delta V \tag{12-22}$$

The result for the $\hat{\mathbf{y}}$-face is obviously

$$\frac{\partial p_{xy}}{\partial y} \delta V \tag{12-23}$$

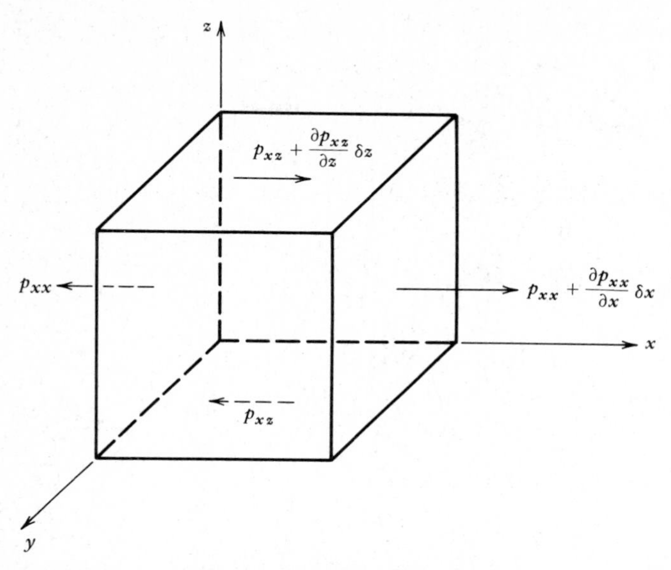

FIG. 12-2 *A parallelepipedon of sides δx, δy, and δz, and the stress components in the x-direction. To simplify the drawing, the stress component p_{xy} and $p_{xy} + (\partial p_{xy}/\partial y)\delta y$ has not been drawn.*

The force of gravity on δV along the x-axis (which need not be horizontal) is

$$g_x \rho \, \delta V \tag{12-24}$$

Since the element of volume is in equilibrium, all the forces along the x-axis balance out, and we find, from Equations 12-21 through 12-24,

$$\frac{\partial p_{xx}}{\partial x} + \frac{\partial p_{xy}}{\partial y} + \frac{\partial p_{xz}}{\partial z} + \rho g_x = 0 \tag{12-25}$$

and along the y-axis, similarly,

$$\frac{\partial p_{yx}}{\partial x} + \frac{\partial p_{yy}}{\partial y} + \frac{\partial p_{yz}}{\partial z} + \rho g_y = 0 \tag{12-26}$$

The equation along the z-axis is obvious. To show why the pressure varies in a liquid in motion—that is why the pressure in the earth is not exactly hydrostatic—we need to take one more step. Using Equation 12-2 in the form

$$p_{xx} = P_{xx} - P, \quad p_{xy} = P_{xy}, \quad p_{xz} = P_{xz}$$

we find, from Equation 12-25,

$$\frac{\partial P_{xx}}{\partial x} + \frac{\partial P_{xy}}{\partial y} + \frac{\partial P_{xz}}{\partial z} - \frac{\partial P}{\partial x} + \rho g_x = 0 \tag{12-27}$$

and finally, from Equation 12-5,

$$2\frac{\partial}{\partial x}\left[\eta\left(\frac{\partial \dot{u}}{\partial x}\right)\right]+\frac{\partial}{\partial y}\left[\eta\left(\frac{\partial \dot{u}}{\partial y}+\frac{\partial \dot{v}}{\partial x}\right)\right]+\frac{\partial}{\partial z}\left[\eta\left(\frac{\partial \dot{u}}{\partial z}+\frac{\partial \dot{w}}{\partial x}\right)\right]-\frac{\partial P}{\partial x}+\rho g_x=0 \tag{12-28}$$

If we restrict ourselves to two dimensions and to a constant viscosity, Equation 12-28 will yield

$$\eta\left(2\frac{\partial^2 \dot{u}}{\partial x^2}+\frac{\partial^2 \dot{u}}{\partial y^2}+\frac{\partial^2 \dot{v}}{\partial x\,\partial y}\right)-\frac{\partial P}{\partial x}+\rho g_x=0$$

or

$$\eta\left[2\frac{\partial^2 \dot{u}}{\partial x^2}+\frac{\partial^2 \dot{u}}{\partial y^2}+\left(\frac{\partial^2 \dot{u}}{\partial x^2}+\frac{\partial^2 \dot{v}}{\partial x\,\partial y}\right)\right]-\frac{\partial P}{\partial x}+\rho g_x=0 \tag{12-29}$$

But for an incompressible fluid, $\theta=\dot{\theta}=0$; hence

$$\frac{\partial \dot{\theta}}{\partial x}=0=\frac{\partial}{\partial x}\left(\frac{\partial \dot{u}}{\partial x}+\frac{\partial \dot{v}}{\partial y}\right)=\frac{\partial^2 \dot{u}}{\partial x^2}+\frac{\partial^2 \dot{v}}{\partial x\,\partial y}=0$$

Thus the term in parentheses in Equation 12-29 vanishes, and we obtain

$$\eta\nabla^2\dot{u}-\frac{\partial P}{\partial x}+\rho g_x=0 \tag{12-30}$$

An entirely similar treatment for the y-axis yields

$$\eta\nabla^2\dot{v}-\frac{\partial P}{\partial y}+\rho g_y=0 \tag{12-31}$$

where

$$\nabla^2=\frac{\partial^2}{\partial x^2}+\frac{\partial^2}{\partial y^2}$$

since the case is two-dimensional. Note that these two equations contain *three* variables: the velocities $\dot{u}$ and $\dot{v}$ and the pressure P. This appears abnormal, and indeed, the equation for zero rate of incompressibility,

$$\dot{\theta}=\frac{\partial \dot{u}}{\partial x}+\frac{\partial \dot{v}}{\partial y}=0 \tag{12-32}$$

must be added to complete the problem.

PROBLEM

12-4 Suppose that all stresses in Figure 12-2 are positive. Why have the arrows been drawn as shown, rather than in the opposite directions? (*Hint*: see Chapter 2, Section A.)

Dynamic Pressure

The fact that the pressure plays a role in the first two equations and not in the third is remarkable and contributes to the difficulty of fluid dynamics. A simple example will show that the pressure term may not be negligible in some geological situations. For a very viscous rock mass moving under the influence of pressure differences, consider the case shown in Figure 12-1, taking the x-axis to be horizontal and the y-axis to be vertical. (The use of the constant viscosity approximation is justified here only by the rough nature of the computations.)

Since the flow is purely in the x-direction, Equation 12-30 yields

$$\eta\left[\frac{\partial^2 \dot{u}}{\partial x^2}+\frac{\partial^2 \dot{u}}{\partial y^2}\right]=\frac{\partial P}{\partial x} \tag{12-33}$$

since $g_x = 0$. If the horizontal velocity, $\dot{u}$, does not change along the x-axis, $\partial \dot{u}/\partial x = \partial^2 \dot{u}/\partial x^2 = 0$. Hence,

$$\eta\frac{\partial^2 \dot{u}}{\partial y^2}=\frac{\partial P}{\partial x} \tag{12-34}$$

It is easy to show that the velocity along the x-axis follows the law

$$\dot{u} = ay^2 + b \tag{12-35}$$

If the velocity in the middle of the conduit is v_M, then $b = v_M$. And if the width of the conduit is d, then at the sides, $y = \pm d/2$, $\dot{u} = 0$, hence $a = -4v_M/d^2$. Thus,

$$\dot{u} = -\frac{4v_M}{d^2}y^2 + v_M$$

and from Equation 12-34,

$$\frac{\partial P}{\partial x} = -\frac{8v_M\eta}{d^2} \tag{12-36}$$

This is the rate of decrease of the pressure in the conduit. The pressure decreases from left to right, as expected.

Using reasonable geological values, for example,

$$v_M = 10 \text{ cm/a}, \quad \eta = 10^{20} \text{ P} = 10^{19} \text{ Pa s}, \quad d = 500 \text{ m}$$

we find

$$\frac{\partial P}{\partial x} = 10^6 \text{ Pa/m} = 10 \text{ bar/m} = 10 \text{ kbar/km} \tag{12-37}$$

It is especially important to note that this pressure change is completely independent of the lithostatic pressure; it arises because a highly "viscous" rock mass has been forced to move slowly through a conduit 500 m wide over a distance of 1 km. If the conduit were 100 m wide, the pressure change would be 25 times larger. Such a change might arise in the earth from a "funnel" situation, such as the one schematized in Figure 12-3. If the motion ceases, the pressure equalizes. The

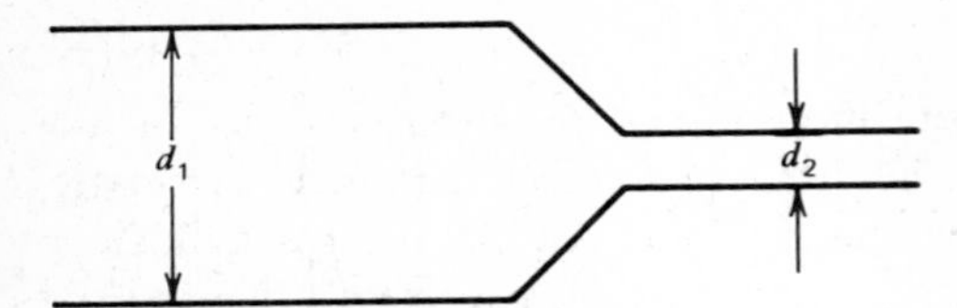

FIG. 12-3 A situation in which a rock mass moving through a large conduit is forced to move through a much narrower one. This leads to a higher velocity in the narrow conduit and necessitates a high dynamic pressure.

total pressure is equal to the static pressure plus the pressure linked to the motion, called the dynamic pressure or—in geology—the tectonic overpressure. (A static pressure of 1GPa = 10 kbar occurs at a depth of approximately 30 km.) The existence of such dynamic pressure in the earth is thus a direct consequence of the fact that the material is in motion, but its magnitude remains uncertain. Note that this dynamic pressure is a scalar but that it varies (here, along the x-axis; the lithostatic pressure increases in the vertical direction).

PROBLEMS

12-5 Lava in a volcano is assumed to be expelled at a maximum velocity of 10 m/s. The conduit is assumed to be two-dimensional (for simplicity) and 10 m wide. The reservoir is at a depth of 5 km. Assume $\rho = 2.5 \times 10^3$ kg/m^3, $g_y = 10$ m/s^2, and $\eta = 10^4$ Pa s. What is the pressure in the reservoir? What is the "overpressure" (i.e., the dynamic pressure)? At what depth would this reservoir pressure arise if only the lithostatic pressure were operating? (The viscosity of lava above 1100° C ranges from $\sim 10^3$ to $\sim 10^7$ Pa s; Clark 1966, p. 299.)

12-6 Repeat the derivation of Equation 12-36, but assume that the flow is vertical and upward. Is there a term corresponding to the increase in lithostatic pressure?

E. Combination of Elasticity and Viscosity

There are several ways to combine the elastic and viscous constitutive relations in order better to approximate the behavior of real bodies. Symbolizing an elastic body (or Hooke body) by a spring and a viscous body (or Newton body) by a dashpot, we can put the two in series (Maxwell body) or in parallel (Kelvin body). (A dashpot consists of a piston perforated with small holes that moves in a closed cylinder filled with a viscous fluid. It is actually a simple shock absorber.) Figure 12-4 shows that in the Maxwell body the stresses on both elements are the same, and that the strains are additive; in the Kelvin body the strains on both elements are the same, and the stresses are additive. Given these facts, it is not difficult to write the corresponding constitutive equations (Reiner 1960, p. 81–89).

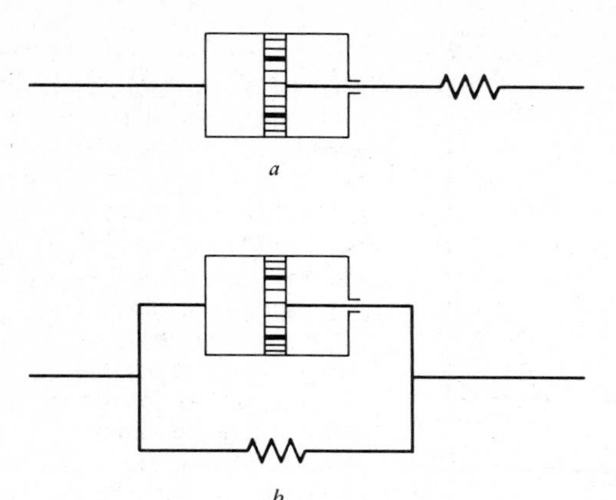

FIG. 12-4 *(a) The Maxwell body. (b) The Kelvin body.*

The Maxwell Body or Viscoelastic Body

Over long periods of time the behavior of a Maxwell body approximates that of the crust and mantle. Bucher (1956) appears to have been the first American geologist to realize this, and used stitching wax in some interesting experiments. We now examine the Maxwell body in more detail.

Constitutive Equation, Stress Relaxation If, for simplicity's sake, both the elastic and viscous constituents of the Maxwell body (Figure 12-4*a*) are incompressible, then, for the Hooke constituent of the body,

$$P_{ij} = 2\mu E_{ij} \tag{12-38}$$

and for the Newton constituent of the body,

$$P_{ij} = 2\eta \dot{E}_{ij} \tag{12-39}$$

But the total strain is the sum of the elastic and viscous strains. Taking the time derivative of Equation 12-38 and adding it to Equation 12-39, we obtain

$$\dot{E}_{ij} = \frac{P_{ij}}{2\eta} + \frac{\dot{P}_{ij}}{2\mu} \tag{12-40}$$

whose solution is

$$P_{ij} = e^{-\mu t/\eta}\left[P_{ij}^{o} + 2\mu \int_{o}^{t} \dot{E}_{ij} e^{\mu t/\eta}\, dt\right] \tag{12-41}$$

where P_{ij}^{o} is the initial stress. This equation may be thought of as the constitutive equation of the Maxwell body, or of viscoelasticity.

For constant strain, $\dot{E}_{ij} = 0$ (see Figure 12-5), and we obtain

$$P_{ij} = P_{ij}^{o} e^{-\mu t/\eta}$$

which shows that in a Maxwell body subjected to a constant strain the stress

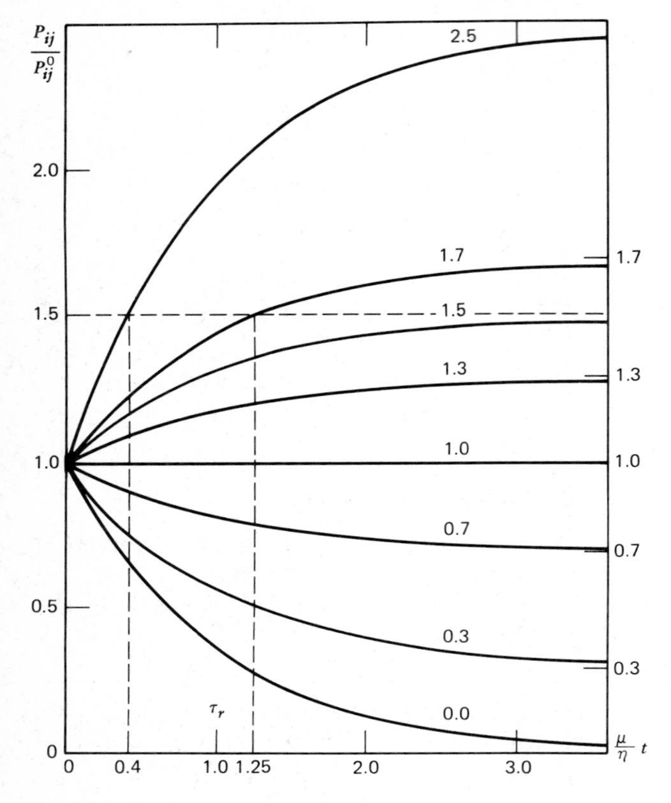

FIG. 12-5 *Variation of the stress deviator with time in a Maxwell body. At $t = 0$, the material is given a stress deviator P^o_{ij}, after which the strain rate $\dot{E}_{ij}$ remains constant. At low strain rates, P_{ij} decreases with time; it increases at high strain rates. The horizontal axis represents time in units of μ/η, the vertical axis represents the stress deviator in units of P_{ij}/P^o_{ij}; "0.7" on the curve means $\dot{E}_{ij} = 0.7\, P^o_{ij}/2\eta$. Curve 0.0 is the relaxation curve, τ_r the relaxation time. The horizontal dashed line represents P^f_{ij} for a value of $1.5\, P^o_{ij}$. Each curve is asymptotic to the corresponding value of P_{ij}/P^o_{ij}, except for curve 1.0, which is a straight line.*

gradually relaxes. For

$$t = \tau_r = \frac{\eta}{\mu}$$

the stress has decreased by a factor of $e = 2.718\ldots$; τ_r is the relaxation time of a Maxwell body. As expected, a high viscosity implies a slow relaxation or a large relaxation time. A low rigidity carries the same implication.

In many geological situations (see below), the strain rate is (approximately) constant. Equation 12-41 thus yields, by integration,

$$P_{ij} = 2\eta\dot{E}_{ij} + \left(P_{ij}^{\,o} - 2\eta\dot{E}_{ij}\right)e^{-\mu t/\eta} \tag{12-42}$$

The change of P_{ij} with time in some typical cases is plotted in Figure 12-5. For $\dot{E}_{ij} = P_{ij}^{\,o}/2\eta$, the stress deviator remains constant. For $\dot{E}_{ij} > P_{ij}^{\,o}/2\eta$, the stress deviator increases with time; for $\dot{E}_{ij} < P_{ij}^{\,o}/2\eta$, it decreases with time. Note that these curves suppose that the material is stressed by $P_{ij}^{\,o}$ at $t = 0$, after which $\dot{E}_{ij}$ remains constant.

Fracture The behavior of a Maxwell body helps us to quantify the dynamic theory of fracture ("leaky bucket theory"), which was previously described qualitatively (Chapter 3, Section C). Reiner's basic assumption is that fracture can result only from the conserved (i.e., elastic) energy per unit volume (Reiner 1960, p. 140). This is equivalent to saying that, in the model shown in Figure 12-4*a*, only the spring can break; the dashpot cannot.

To quantify this idea, we must first compute the total power (energy or work per unit time).

$$\dot{W}_t = P_{ij}\dot{E}_{ij} \tag{12-43}$$

in which the meaning of the RHS is (see Equation 3-11)

$$P_{xx}\dot{E}_{xx} + P_{xy}\dot{E}_{xy} + P_{xz}\dot{E}_{xz} + P_{yx}\dot{E}_{yx} + \cdots$$

Using Equation 12-40, we find

$$\dot{W}_t = P_{ij}\left(\frac{P_{ij}}{2\eta} + \frac{\dot{P}_{ij}}{2\mu}\right) \tag{12-44}$$

The viscous part of the deformation cannot cause fracture. The elastic part of the power is thus the total power $\dot{W}_t$ minus the viscous power $P_{ij}P_{ij}/2\eta$. It is this elastic power $\dot{W}_e$ that may cause fracture as it accumulates. Thus,

$$\dot{W}_e = \dot{W}_t - \left(\frac{P_{ij}P_{ij}}{2\eta}\right) \tag{12-45}$$

and

$$W_e = \int_o^t W_e\,dt = \int_o^t\left(\dot{W}_t - \frac{P_{ij}P_{ij}}{2\eta}\right)dt$$

Replacing $\dot{W}_t$ with its value from Equation 12-44, we obtain, if P_{ij} does not change with time,

$$W_e = \frac{1}{2\mu}\int_o^t P_{ij}\dot{P}_{ij}\,dt = \frac{P_{ij}P_{ij}}{4\mu} \tag{12-46}$$

But $(P_{ij}P_{ij}/2)^{1/2} = \tau$ (see Equation 12-11. Thus, if

$$W_e = \frac{P_{ij}P_{ij}}{4\mu} = \frac{\tau^2}{4\mu} \geq F \tag{12-47}$$

fracture occurs.

If μ and F are known, then fracture occurs if

$$\tau \geq 2\sqrt{\mu F} = \tau_f \tag{12-48}$$

In other words, a Maxwell body fractures when the average stress deviator reaches a definite limit, τ_f. Note particularly that there is no strain limit so long as the average stress deviator remains below τ_f. The reason is easy to understand from two points of view.

- The model in Figure 12-4 shows that the body breaks if and only if the spring is too highly stressed. So long as this is not the case, the dashpot can yield indefinitely, since the viscous energy is not conserved.
- The same conclusion may be reached through examination of Figure 12-5. It is evident that if all P_{ij} equal zero except for one shear stress—say, P_{xy}—then $\tau = P_{xy}$ (see Equation 12-8). Thus, if $P_{xy} > \tau_f = P_{ij}^f$, fracture will occur. In the case in which $P_{ij}^f = 1.5P_{ij}^o$, the figure shows that if $\dot{E}_{ij}$ is less than $1.5P_{ij}^o/2\eta$, fracture cannot occur, since P_{ij} (here P_{xy}) cannot reach P_{ij}^f. If $\dot{E}_{ij} = 1.7P_{ij}^o/2\eta$, fracture will occur at $(\mu/\eta)t \cong 1.25$: that is, $t = 1.25\eta/\mu$. It will do so for $(\mu/\eta)t = 0.4$ if $\dot{E}_{ij} = 2.5P_{ij}^o/2\eta$. Thus, for a given $\dot{E}_{ij}$, fracture cannot occur for a relatively small value of η. It will do so for a larger η, and fracture will occur increasingly rapidly as η increases. Vice versa, for a given η, fracture cannot occur for a small $\dot{E}_{ij}$ but will occur increasingly rapidly as $\dot{E}_{ij}$ increases.

The preceding theory provides a semiquantitative explanation for a number of observations. For example:

- Earthquakes occur along some segments of the San Andreas fault but not along others (Chapter 3, Section C). The aseismic parts may correspond to a low η because of the rock type.
- Earthquakes occur in the upper crust and upper mantle but rarely in the lower crust (Chen and Molnar 1983). The low η of the lower crust compared with that of the upper crust would stem from the rise in temperature. In the upper mantle the change in composition more than compensates for the higher temperature. It is perhaps not impossible that the absence of earthquakes below 700 km is similarly due to a slight decrease in viscosity, but this remains a matter of speculation.

Application 12-1

Following a large earthquake, a certain area is left with a stress of 0.5 kbar. The rock can support a stress of 0.8 kbar. The viscosity of the rock mass is 10^{22} P, the $S-$wave velocity is 3.1 km/s, and the density is 2.65 g/cm³. What is the maximum strain rate $\dot{E}_{ij}^M$ that the material can support without fracture? How long will it take for the material to break if the real strain rate is twice $\dot{E}_{ij}^M$?

Solution

$$P_{ij}^o = 50 \text{ MPa} \qquad P_{ij}^f = 80 \text{ MPa}$$

$$\eta = 10^{21} \text{ Pa s} \qquad \mu = \beta^2\rho = 2.55 \times 10^{10} \text{ Pa}$$

$\dot{E}_{ij}^M = P_{ij}^f/2\eta = 4 \times 10^{-14}$/s. At this strain rate, P_{ij} will never exceed P_{ij}^f.

Proof: If $t \to \infty$ in Equation 12-42, then $P_{ij} = 2\eta \dot{E}_{ij}$.

To find the time, we first find

$$\dot{E}_{ij} = 2\dot{E}_{ij}{}^{\mathrm{M}} \cong 8 \times 10^{-14}/\mathrm{s}$$

Hence, $2\eta\dot{E}_{ij} = 1.6 \times 10^8$ Pa. Using Equation 12-42 with $P_{ij} = P_{ij}^f$, we obtain

$$80 \times 10^6 = 1.6 \times 10^8 + \left[(50 \times 10^6) - (1.6 \times 10^8)\right] \exp\left(-\frac{2.55 \times 10^{10} t}{10^{21}}\right)$$

Performing the arithmetic and taking the logarithm of both sides, we obtain

$$t = 1.25 \times 10^{10}\ \mathrm{s} \cong 396\ \mathrm{a}$$

Application to the Lithosphere The lithosphere is sometimes thought of as a Maxwell body of varying parameters.

- In its upper part where the temperature is below ~ 500° C (see Chapter 13 and 14), η is practically infinite, and the relaxation time exceeds 100 Ma. This is the "elastic lithosphere."
- In its lower part the higher temperature causes a decrease in viscosity (Equation 12-16) and thus in relaxation time. As long as the latter exceeds 1 to 10 Ma, the body moves more or less coherently over geological periods. This is the "viscous lithosphere"; below the continents it is sometimes called the "tectosphere." The seismic velocities (α and β) are essentially the same in the "elastic" and "viscous" lithospheres; therefore, these two parts are together known as the "seismic lithosphere."
- At a depth at which the relaxation time falls below 1...10 Ma, the material may not move with the same velocity as the surface does. This is the asthenosphere. Because of the exponential dependence of the viscosity on the temperature (see Equation 12-16), the transition from the lithosphere to the asthenosphere is relatively rapid (about 10 km). It is generally thought that the drop in seismic velocities and the decrease in viscosity take place over the same depth range.

PROBLEMS

12-7 Is the energy used in deforming a Maxwell body entirely recoverable or entirely lost? Why?

12-8 Two bicycles are identical except for the material of the frames, yet one (A) is much easier to pedal than the other (B). Where does the extra energy needed for pedaling B go? What does this tell you about the constitutive equation of the frame material?

12-9 Repeat the computations in Application 12-1, but suppose that $P_{ij}^f = 1.2$ kbar and that the strain rate is $5\dot{E}_{ij}{}^{\mathrm{M}}$.

13

Temperatures in the Earth

A. Introduction

Direct observations of temperatures within the earth are limited by the maximum depth of mines or oil wells—about 9 km below the surface. Observations that rely on geology (degrees of metamorphism, nodules, and kimberlites, for example) extend this range to over 100 km. But these observations are scattered, and many of them are made in somewhat exceptional areas. We must combine them with laboratory measurements and with knowledge gained from fluid mechanics and other fields to arrive at a picture of the temperatures within the earth. In part, this is a purely academic problem, but it also has important practical consequences. Whether or not oil is present in a given formation, for example, depends in part on the thermal history of this formation; similarly, the presence and richness of mineral deposits depend in a complex way on the conditions of pressure and temperature during geological history.

The problem is made more difficult by the time required for heat transmission. Whereas elastic waves travel through or around the earth in a few minutes or tens of minutes, heat is transmitted through the earth over millions of years. Our knowledge of the temperatures deep within the earth is thus inferential, and much of what follows is speculative (see Verhoogen 1980, p. III).

We now examine the mechanisms of heat transfer and the sources of heat in the earth. In the next chapter, we will see how these factors combine to explain convection.

B. Mechanisms of Heat Transfer

Until the advent of plate tectonic theory, heat conduction was practically the only mechanism of heat transfer to be considered. In this mechanism the thermal agitation of atoms is gradually transmitted to their neighbors. In comparison with elastic wave transmission, heat transmission is very slow.

Heat Conduction

Heat is transported from crustal depths to the surface only by conduction, except in areas where water motion in a vertical direction is important and in volcanoes. The fact that the continents are moving relative to the mantle is not relevant when we consider only the heat transport through the continents. The theory behind the conduction of heat, which was formulated by Fourier (see Chapter 4), may be briefly outlined as follows.

Figure 13-1 shows an infinite two-dimensional wall of finite thickness δx. On the left side the temperature is T_o, on the right side it is T_1. The amount of heat passing through a unit surface of the wall per unit time is the thermal flux, or heat flux, $\mathbf{q}$. It is easy to see that

$$q = -\left(\frac{T_1 - T_o}{\delta x}\right)k$$

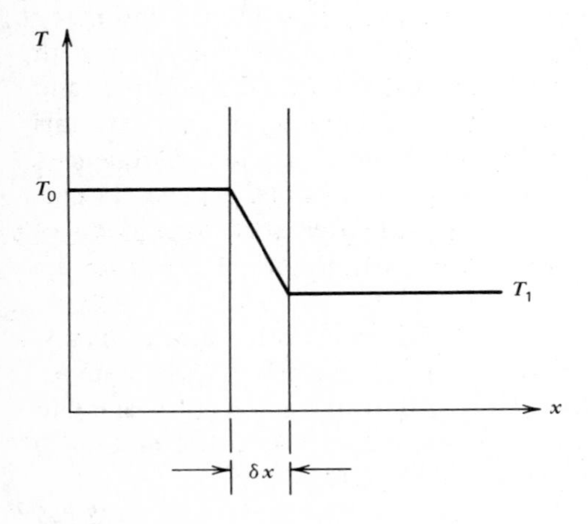

FIG. 13-1 *An infinite wall of thickness δx with temperature T_o on the left, T_1 on the right.*

- The heat goes from high to low temperatures, hence the minus sign.
- The same temperature difference gives rise to different heat fluxes in different materials, hence the proportionality constant k, called the thermal conductivity.

At the limit as $\delta x \to 0$, we obtain

$$q = -k\frac{\partial T}{\partial x} \tag{13-1}$$

In more than one dimension, this becomes

$$\mathbf{q} = -k\,\nabla \mathbf{T} \tag{13-2}$$

Since $\mathbf{q}$ is a flux of heat—that is, of energy—it has the dimensions of power per unit surface (W/m^2), whereas ∇T has the dimensions of °K/m. Thus k has the dimensions of W/(m°K) (see Appendix A).

Figure 13-2 shows volume δV subject only to a heat flux $\mathbf{q}_o$ perpendicular to its left face, and giving off heat with flux $\mathbf{q}_1$ perpendicular to its right face. It will gain heat if $q_o > q_1$ and lose heat if $q_o < q_1$. Hence, its rate of change of temperature is

$$\frac{\partial T}{\partial t} \cong -\frac{1}{\rho c}\left(\frac{q_1 - q_o}{\delta x}\right) \tag{13-3}$$

The reason for the proportionality factor is that the rate at which the material heats up is inversely proportional to its heat capacity, expressed in J/(kg °K), and to its density. As $\delta x \to 0$,

$$\frac{\partial T}{\partial t} = -\frac{1}{\rho c}\frac{\partial q}{\partial x}$$

If the heat flux is two-dimensional, the gain of heat in the y-direction must be added to the gain in the x-direction.

$$\frac{\partial T}{\partial t} = -\frac{1}{\rho c}\left(\frac{\partial q_x}{\partial x} + \frac{\partial q_y}{\partial y}\right) \tag{13-4}$$

because only the x-component of $\mathbf{q}$—that is, q_x (Figure 13-3)—contributes to the change in flux in the x-direction; and the same is true for q_y. But $\mathbf{q} = -k\,\nabla\mathbf{T}$ means (in two dimensions)

$$q_x = -k\frac{\partial T}{\partial x}, \qquad q_y = -k\frac{\partial T}{\partial y} \tag{13-5}$$

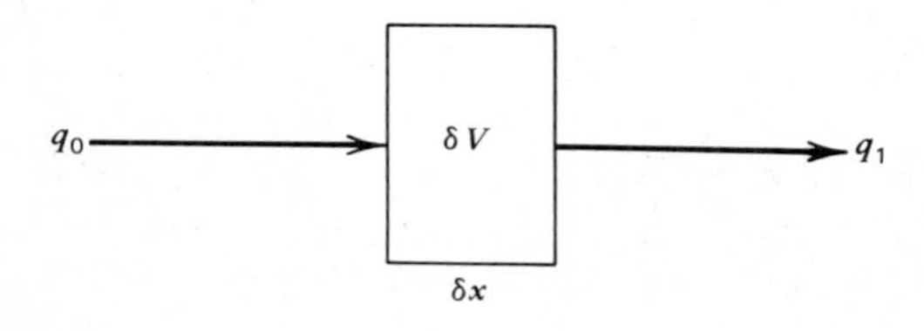

FIG. 13-2 *A volume δV receiving a flux $\mathbf{q}_o$ on its left side and losing $\mathbf{q}_1$ on its right side.*

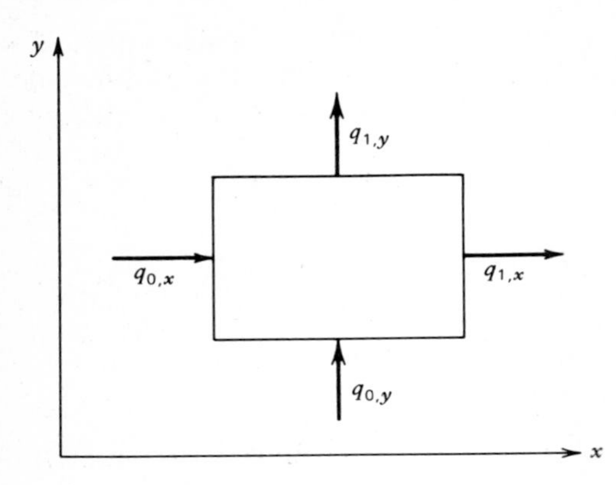

FIG. 13-3 A volume receiving a flux $\mathbf{q}_o$ of components $q_{o,x}$, $q_{o,y}$ and losing a flux $\mathbf{q}_1$ (q_1, x, q_1, y). (In two dimensions.)

Hence, if k is constant,

$$\frac{\partial T}{\partial t} = \frac{k}{\rho c}\left[\frac{\partial}{\partial x}\left(\frac{\partial T}{\partial x}\right) + \frac{\partial}{\partial y}\left(\frac{\partial T}{\partial y}\right)\right]$$

or

$$\frac{\partial T}{\partial t} = \frac{k}{\rho c}\left[\frac{\partial^2 T}{\partial x^2} + \frac{\partial^2 T}{\partial y^2}\right] \tag{13-6}$$

and in three dimensions for constant k,

$$\frac{\partial T}{\partial t} = \frac{k}{\rho c}\nabla^2 T = \kappa \nabla^2 T \tag{13-7}$$

If k varies, we must take its variation into account; $\kappa = k/\rho c$ is the thermal diffusivity. This equation helps to explain why the interior of a house with heavy and thick walls experiences little temperature fluctuation: the large heat capacity and the density of the walls greatly reduce the temperature changes. Such walls are said to have a large "thermal inertia." As we shall see, this is also the case in the earth, where ordinary (i.e., mechanical) inertia is negligible (see Chapter 12, Section C) but thermal inertia is large.

In deriving Equation 13-3, we have assumed that there are no heat sources within δV (Figure 13-2). If this volume contained a source of intensity S, the rate of change of temperature inside the volume would depend not only on the difference between q_1 and q_o, but also on S. We would thus obtain, by exactly the same reasoning,

$$\frac{\partial T}{\partial t} \cong -\frac{1}{\rho c}\frac{q_1 - q_o}{\delta x} + \frac{S}{\rho c} \tag{13-8}$$

Correspondingly, Equation 13-7 would become

$$\frac{\partial T}{\partial t} = \kappa \nabla^2 T + \frac{S}{\rho c} \tag{13-9}$$

where S is measured in W/m³. This term is discussed further in Section 13-C.

Another assumption that we made in deriving Equation 13-7 is that the material does not move. This assumption was made for many decades in investigating the earth, and many thermal histories and geotherms were computed on this basis. Because it is now recognized that motions occur, these results are no longer relevant, but they still appear to exert a shadowy influence.

It is known that k, ρ, and c all vary with temperature and pressure. As a first approximation, though, it appears that the various effects tend to cancel each other, at least in the first few hundred kilometers below the surface.

Two Cases of Special Interest

Periodic Temperature Fluctuations at the Surface In attempting to determine the heat flow in a given area, one must not let the influence of past events distort the results. The temperature 5 cm below the surface is of course greatly dependent on the time of the year, and even on the time of day. The question thus arises, How do the daily and yearly temperature fluctuations at the surface influence the temperatures at depth? In this case, since there is no motion, there is no heat advection (see below). We may further assume that the temperature varies only vertically, and that there are no heat sources. The heat equation thus becomes

$$\frac{\partial T}{\partial t} = \kappa \frac{\partial^2 T}{\partial z^2} \tag{13-10}$$

where the z-axis points downward. We wish to find $T = T(t, z)$ for a surface variation

$$T = T(t, 0) = A \cos 2\pi\nu t \tag{13-11}$$

(The variation of temperature at the surface is given, and the surface is a boundary. This variation is a "boundary condition," in the sense that this term is used in mathematics; geologists use this expression to mean "minimum value" or "maximum value." These two meanings must be carefully distinguished.)

Surmising that the temperature variation at depth has the same period as the one at the surface, perhaps with a phase angle, and is exponentially attenuated with depth, we try

$$T = Ce^{-az} \cos(2\pi\nu t - az) \tag{13-12}$$

Then, computing $\partial T/\partial t$ and $\partial^2 T/\partial z^2$ we find from Equation 13-10

$$-2\pi\nu e^{-az} \sin(2\pi\nu t - az) = -2\kappa a^2 e^{-az} \sin(2\pi\nu t - az)$$

Hence,

$$a = \sqrt{\frac{\pi\nu}{\kappa}} \tag{13-13}$$

This is the attenuation per meter (γ in Chapter 3, Section B). The wavelength is

$$\Lambda = \frac{2\pi}{a} = 2\sqrt{\frac{\pi\kappa}{\nu}} \tag{13-14}$$

Thus, the attenuation per wavelength ($1/Q$ in Chapter 3, Section B) is

$$a\Lambda = 2\pi \tag{13-15}$$

We see that, here again, the attenuation per unit length is proportional to the frequency, and thus the attenuation per wavelength is constant (just like $1/Q$). The latter is very large in this case: $e^{-2\pi} = 2 \times 10^{-3}$. Thus, at one wavelength the temperature variations are entirely negligible. It is easy to estimate Λ for the various periodic fluctuations (ν must be in Hz); using $\kappa \cong 10^{-6}$ m²/s, we find

$$\Lambda_{\text{day}} \cong 1 \text{ m}, \quad \Lambda_{\text{a}} \cong 20 \text{ m}, \quad \Lambda_{10{,}000 \text{ a}} = 2 \text{ km}$$

It is thus easy to make measurements below the zone exposed to daily and yearly fluctuations, but it becomes much more difficult to do so for very long period fluctuations, such as ice ages (which, of course, may have lasted longer than 10 ka and were not strictly periodic).

Sudden Temperature Change at Depth The problem here can be stated as follows. If at a depth D the temperature is suddenly raised by 1 °K, how long will it take for this rise in temperature to travel to the surface? Or more precisely, what is the time τ_c (characteristic time) necessary for the temperature at the surface to rise by 0.5° K? (Since the phenomenon is linear, a rise of A °K will yield a rise of 0.5 A °K in the same time.)

The definition of the problem given above is not exact, because the surface is normally at a fixed temperature; furthermore, the exact solution is very complex. Here we will note only that, as a first approximation (see also Equation 11-4),

$$\tau_c = \frac{D^2}{\kappa} \tag{13-16}$$

The second power of D may be explained in physical terms by the fact that if the rise in temperature occurs at a depth, of, say, $3D$, the heat capacity of the overlying layer will become three times as large; in addition, the thermal gradient will become three times as small. The thermal flux at the surface is thus reduced by a factor of $3 \times 3 = 9$. Using Equation 13-16, we find that for $D \cong 36$ km, with $\kappa \cong 10^{-6}$ m²/s,

$$\tau_c \cong 36 \text{ Ma} \tag{13-17}$$

This is a convenient order of magnitude to remember: for a continental crust of normal thickness, the characteristic time in millions of years is approximately equal to the thickness in kilometers. From this result and the D^2 dependency, it is easy to compute τ_c for any crustal thickness.

PROBLEMS

13-1 What are the characteristic times for events 10 and 100 km deep?

13-2 What is the maximum depth of an event that now yields a surface heat flow anomaly, if you know that it occurred 20 Ma ago?

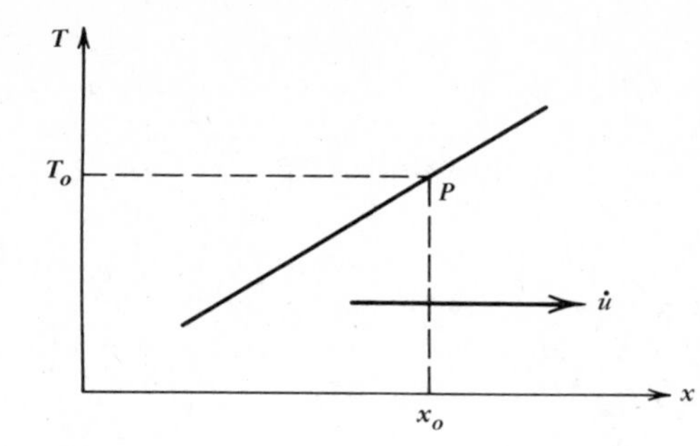

FIG. 13-4 *A fixed point P in a liquid flowing at a velocity $\dot{u}$; the temperature of the liquid increases with x.*

13-3 You are trying to use as little energy as possible in raising the temperature of long rectangular parallelipipedons, which you can cut in half any way you wish. Which way is best, and why? How much energy can you save (in percent)? (*Hint*: How to cut potatoes for rapid cooking.)

Heat Advection

Now that the inside of the earth is no longer considered static, the importance of the resulting heat transfer has been recognized. This mechanism, caused by the motion of the fluid, is called heat advection.

Figure 13-4 shows a fixed point P at x_o surrounded by a liquid flowing in the x-direction. Its temperature is T_o and the temperature of the liquid increases from left to right ($\partial T/\partial x > 0$). The liquid flows at a velocity $\dot{u}$ from left to right. As time passes, neglecting thermal conduction, the temperature T_o will decrease at a rate proportional to $\partial T/\partial x$ and to $\dot{u}$. Hence,

$$\frac{\partial T}{\partial t} = -\dot{u}\frac{\partial T}{\partial x} \tag{13-18}$$

and in three dimensions, neglecting thermal conductivity,

$$\frac{\partial T}{\partial t} = -\left[\dot{u}\frac{\partial T}{\partial x} + \dot{v}\frac{\partial T}{\partial y} + \dot{w}\frac{\partial T}{\partial z}\right] \tag{13-19}$$

If advection, conduction, and heat sources are important, we combine Equations 13-9 and 13-19 to find

$$\frac{\partial T}{\partial t} = \kappa \nabla^2 T - \left[\dot{u}\frac{\partial T}{\partial x} + \dot{v}\frac{\partial T}{\partial y} + \dot{w}\frac{\partial T}{\partial z}\right] + \frac{S}{\rho c} \tag{13-20}$$

Relative Importance of Conduction and Advection

To get an idea of the relative importance of advection and conduction, make the following order-of-magnitude computations. Assume that $\partial T/\partial x$ is on the order of 1 °K/km $= 10^{-3}$ °K/m. Then, $\nabla^2 T$ is on the order of $(10^{-3})^2$ °K/m$^2 = 10^{-6}$ °K/m^2. Because κ is on the order of 10^{-6} m^2/s, $\kappa \nabla^2 T$ is on the order of 10^{-12} °K/s. But if $\dot{u}$ is on the order of 3 cm/a $= 10^{-9}$ m/s, then $u(\partial T/\partial x)$ is also on the order of 10^{-12} °K/s. We conclude that for $u < 1$ cm/a, heat conduction is likely to be dominant, and for $u > 10$ cm/a, heat advection is likely to dominate. This computation is very approximate: if $\nabla T < 1$ °K/km, heat advection will dominate for slower velocities. Still, it gives a rough estimate of the orders of magnitude involved.

We can obtain the same result more directly by computing the Peclet number, Pe (see Chapter 11, Section C), which measures the approximate ratio of heat advected to heat conducted. It is computed as follows. The order of magnitude of the heat advected is

$$\dot{u}\frac{\partial T}{\partial x} = V\frac{°\mathrm{K}}{L}$$

where V is a characteristic velocity and L is a characteristic length. Similarly, the heat conducted is

$$\kappa \nabla^2 T = \kappa \frac{°\mathrm{K}}{L^2}$$

Hence,

$$\mathrm{Pe} = \frac{V\,°\mathrm{K}}{L} \times \frac{L^2}{\kappa\,°\mathrm{K}} = \frac{VL}{\kappa} \tag{13-21}$$

For the example given above,

$$\mathrm{Pe} = \frac{10^{-9} \times 10^3}{10^{-6}} = 1 \qquad (\text{for } L = 1 \text{ km})$$

This shows that for either V or L greater than in the example, $\mathrm{Pe} > 1$, and advection dominates. Care should be taken, however, not to assume that this is a strict dividing line: rather, it is a gradual change. The meaning of the Peclet number is that for $\mathrm{Pe} \lesssim 0.3$, heat diffusion is globally dominant, whereas heat advection dominates for $\mathrm{Pe} \gtrsim 3$.

Radiative Heat Transfer

At high temperatures and in a medium transparent to infrared radiation, radiative heat transfer becomes very important because of the fact that it increases with T^3 (for T in °K). This is the way in which the sun heats the earth. Clark (1957) suggested that this mode of heat transport is also important in the earth. It appears, though, that the opacity of earth materials is such that this mechanism plays only a minor role.

C. Heat Sources

Strictly speaking, for the earth as a whole the only heat sources are radioactive decay and the sun's heat. However, when only a part p of the earth is considered, heat produced outside of p may appear as a heat source. Because this is the usual case, we will include as heat sources such phenomena as isentropic (or adiabatic) heating, shear heating, and residual heat.

Radioactive Heating

In a series of papers starting in 1863, Kelvin used the heat conduction equation (Equation 13-7) to compute the age of the earth. He knew the approximate value of the heat flow, and he assumed that the earth had started from a molten state and gradually cooled down. Since the surface remained close to 0° C during the whole process, it was possible to determine the thermal history of the earth, and in particular to compute the surface heat flow versus time. Kelvin initially found that to account for the present surface heat flow, the age of the earth since solidification could in no case be greater than 400 Ma. He gradually revised his estimate down to less than 40 Ma (1899). At the time, geologists were uncertain about the age of the earth but agreed that the proposed age was insufficient to account for the amount of erosion. In the controversy that followed, the solution was not obvious. We now realize that there are two nonexclusive reasons for Kelvin's error.

1. He did not take into account the possibility of heat advection.
2. He did not know of the existence of radioactivity.

These factors change the results for several reasons. Heat transfer by conduction is extremely slow: Equations 13-16 and 13-17 show, for instance, that the characteristic time for a layer 350 km thick is ~ 3.5 Ga (billions of years). Under these conditions, most of the heat in the mantle is not available to account for the present heat flow. But the motion of the material and the resulting heat advection, mean that material deep within the mantle may be brought close to the surface in ~ 100 Ma, thus making a much larger reservoir of heat available. Also, the present heat flow can be explained by radioactivity, which generates much of it close to the surface (see below). As it happens, the discovery of radioactivity came very shortly after Kelvin's last computation, and radioactivity was then taken to be the only cause of error; the possibility of heat advection was almost completely disregarded for more than 60 years.

Concentration of Radioactive Elements near the Surface In 1906, a few years after the discovery of radioactivity, R. J. Strutt showed that the radioactive elements are concentrated near the surface in the continents. The method that he used was simple. In continents, only conduction need be considered, and furthermore, the temperature changes (almost) only vertically. Finally, the situation is essentially time-independent (i.e., steady-state). Equation 13-9 thus becomes

$$\frac{\partial T}{\partial t} = 0 = \kappa \frac{\partial^2 T}{\partial z^2} + \frac{S}{\rho c} \tag{13-22}$$

Since T is thus a function of z only, and since $\kappa = k/\rho c$, this equation may be written

$$\frac{d^2T}{dz^2} = -\frac{S}{k}$$

Hence,

$$\frac{dT}{dz} = -\frac{Sz}{k} + C \tag{13-23}$$

and

$$T = -\frac{Sz^2}{2k} + Cz + D \tag{13-24}$$

where C and D must be determined by the boundary conditions. At the surface, $z = T = 0$ (in °C); hence $D = 0$. Taking the thermal gradient at the surface to be

$$\left.\frac{dT}{dz}\right]_{z=0} = C = 4.3 \times 10^{-2}\ °\text{C/m}$$

determines C. Although he had only 28 specimens of rock at his disposal, Strutt measured S and was able to conclude that its value is $\sim 1\ \mu\text{W/m}^3$. (The value now used is 0.92 $\mu\text{W/m}^3$.) For $k = 1.7$ W/(m °C)—as against presently used values ranging from 2 to 4 W/(m °C)—Equation 13-24 yields

$$T = (-2.904 \times 10^{-7} z^2) + (4.3 \times 10^{-2} z) \tag{13-25}$$

for z in meters. We can find the depth, z_M, at which the maximum temperature, T_M, occurs by setting dT/dz at zero. If there is no radioactivity below z_M, T will remain constant there. This also satisfies the equality of heat flow at z_M—that is, null. Using the values of S, k, and C given above, we obtain $z_M \cong 73$ km. Inserting this value into Equation 13-25 gives us $T_M \cong 1570$ °C.

To summarize: the present-day heat flow *can* be produced by a layer of rock 73 km thick that has the radioactivity observed near the surface; below this depth, there is no radioactive matter, and the temperature thus remains constant. (Note that the reasoning here shows only that a radioactive layer 73 km thick can account for the observations. It does *not* show that this is the only solution, or the correct one.) Although the details of Strutt's conclusion have been modified, his line of reasoning remains one of the main arguments for the concentration of radioactivity near the surface.

Heat Production by the Various Elements As is well known, radioactive elements decay through complex radioactive reactions to give daughter products. The time necessary for half of the atoms of a given mass to decay is called the half-life (denoted λ_2 here), which varies from element to element. The only elements of importance in heat production are U^{238} and U^{235}, Th^{232}, and K^{40}. Additionally, K exists in a stable (nonradioactive) form. The half-lives and heat production per kg of these elements (for K, the heat production for a kg of

TABLE 13-1 ***Main Radioactive Elements, Half-Lives in Years, Heat Production in W/kg of Element at Present (S_o) and at 4.5 Ga B.P. ($S_{4.5}$), Present Concentration in Regions of the Crust, and Average Concentration in the Whole Earth***

Element	λ_2	S_o	$S_{4.5}$	Oceanic Crust	Continental Shields	Young Continents	Entire Earth
U^{238}	4.468×10^9	94.35×10^{-6}	189.6×10^{-6}	0.42×10^{-6}	0.99×10^{-6}	1.31×10^{-6}	18×10^{-9}
U^{235}	7.038×10^8	4.05×10^{-6}	340.5×10^{-6}	3.0×10^{-9}	7.2×10^{-9}	9.5×10^{-9}	130×10^{-12}
Th^{232}	1.401×10^{10}	26.6×10^{-6}	33.2×10^{-6}	1.68×10^{-6}	4.00×10^{-6}	5.28×10^{-6}	65×10^{-9}
K^{40}*	1.250×10^9	3.5×10^{-9}	41.2×10^{-9}	6.9×10^{-3}	1.63×10^{-2}	2.15×10^{-2}	170×10^{-6} to 800×10^{-6}

*For K, all figures except λ_2 refer to the combination of stable plus radioactive isotopes—that is, ordinary K.

ordinary K) are shown in Table 13-1, in which S_o stands for the present-heat production per kg, and $S_{4.5}$ stands for that 4.5 Ga ago (i.e., at the presumed time of the formation of the earth). In addition, the table gives the present concentrations in various regions of the earth's crust and an educated guess as to the average concentration in the whole earth (see below). The table shows that the low heat production of K^{40} is approximately compensated for by the elements's much greater abundance.

There is much disagreement over how these concentrations vary with depth in the mantle. It is generally thought that uranium and thorium have been largely expelled from the mantle because of their large ionic radii. Because the K/U ratio in the crust is roughly one-eighth of its abundance in meteorites, some geochemists believe that the core has been enriched in potassium in comparison to the crust and mantle, so that its average concentration in the earth may be anywhere between 170×10^{-6} and 800×10^{-6}. (Much the same argument has been made for sulfur.) The net result of this uncertainty is that it is not known whether the earth is presently cooling down or heating up.

Original Heat

By "original heat" or "primitive heat" is meant the heat content of the earth very soon after its formation (Verhoogen 1980, pp. 15–18). It is now generally thought that the earth condensed from a cooling nebular gas, a process which took place below $\sim 1500°$ C ($\sim 1750°$ K). The main source of original heat, however, was the release of gravitational energy by bodies of all sizes falling on the accreting earth. This energy was so large (on the order of 10^{32} J) that a very efficient mechanism of radiating most of it out into space must have been operating. (The total present heat flux of the earth is only on the order of 4×10^{13} W.) The original heat is generally considered to be unimportant at present, but this opinion hinges on the duration of the accretionary process and on the efficiency of the radiating mechanism, both of which are very uncertain.

In addition, it is not known whether iron aggregated first, forming the core, after which the silicates aggregated (heterogeneous accretion) or whether the whole earth condensed first (homogeneous accretion), after which the core formed. The corresponding thermal histories would differ greatly (see below).

Gravitational Energy

"Gravitational energy" includes only the energy released by changes of density distribution after the formation of the earth. The largest source thereof was the formation of the core, provided that the earth accreted homogeneously. This energy also turns out to have been extremely large (10^{31} J) (Verhoogen 1980, p. 20), and to get rid of it without melting the whole earth has proven difficult. It is unlikely that such melting occurred, but the total amount, 10^{31} J, is greater than the present total heat flow at the earth's surface multiplied by the age of the earth. In any case, core formation if it occurred, must have greatly raised the average temperature of the earth.

There is some geochemical evidence suggesting that most of the original heat was dissipated in the very early history of the earth (Verhoogen 1980, p. 18). However, the relative importance of radioactivity and gravitational energy (plus original heat, if any) to the present heat flow remains uncertain. Until about 1980, radioactivity had generally been thought to contribute about three-quarters, but some recent estimates have reduced it to one-quarter or less of the present heat flow (Peltier and Jarvis 1982).

Isentropic Heating

Isentropic heating is not a heat source, but it plays a primary role in the increase of temperature with depth in the earth. A body undergoing a perfectly reversible compression without exchanging heat with the surrounding medium—that is, one at constant entropy—is said to be isentropically heated. The isentropic gradient IG is the rate of increase of the temperature with depth under these conditions; it is often called the adiabatic gradient, but adiabaticity does not imply that the phenomenon is reversible. The isentropic gradient is given by the formula

$$\mathrm{IG} = \left(\frac{\partial T}{\partial z}\right)_S = \frac{\alpha g T}{c_p} \tag{13-26}$$

where the subscript S indicates isentropic conditions, α is the volumetric coefficient of thermal expansion (or expansivity) at constant pressure, and c_p is the heat capacity at constant pressure.

This equation is derived as follows. If the entropy, S, is expressed as a function of T and P, then

$$dS = \left(\frac{\partial S}{\partial T}\right)_P dT + \left(\frac{\partial S}{\partial P}\right)_T dP$$

where the subscripts indicate which of the thermodynamic variables is to be kept constant. Hence, at S constant ($dS = 0$),

$$\left(\frac{dT}{dP}\right)_S = -\frac{(\partial S/\partial P)_T}{(\partial S/\partial T)_P} \tag{13-27}$$

We first will evaluate $(\partial S/\partial T)_P$. By definition, $c_p = (\partial H/\partial T)_P$, where H is the enthalpy. But $dH = V\,dP + T\,dS$, where $V = 1/\rho$ is the specific volume. Hence, at constant P,

$$dH = T\,dS$$

Taking the derivative with respect to T, we obtain

$$\left(\frac{\partial H}{\partial T}\right)_P = T\left(\frac{\partial S}{\partial T}\right)_P$$

which yields

$$c_p = T\left(\frac{\partial S}{\partial T}\right)_P \quad \text{or} \quad \left(\frac{\partial S}{\partial T}\right)_P = \frac{c_p}{T}$$

Next, we compute $(\partial S/\partial P)_T$. To do so, we first note that if

$$dz = K\,dx + L\,dy$$

then in addition

$$dz = \left(\frac{\partial z}{\partial x}\right)_y dx + \left(\frac{\partial z}{\partial y}\right)_x dy$$

Thus,

$$K = \left(\frac{\partial z}{\partial x}\right)_y \quad \text{and} \quad L = \left(\frac{\partial z}{\partial y}\right)_x$$

Hence,

$$\left(\frac{\partial K}{\partial y}\right)_x = \left(\frac{\partial L}{\partial x}\right)_y = \frac{\partial^2 z}{\partial x\,\partial y}$$

In thermodynamics the following relation obtains:

$$dG = V\,dP - S\,dT$$

where G is the Gibbs free energy. Setting $dz = dG$, $K = V$, $dx = dP$, $L = -S$, and $dy = dT$ gives us

$$\left(\frac{\partial V}{\partial T}\right)_P = -\left(\frac{\partial S}{\partial P}\right)_T$$

By the definition of α, we have

$$\left(\frac{\partial V}{\partial T}\right)_P = \alpha V = \frac{\alpha}{\rho}$$

Hence,

$$\left(\frac{\partial S}{\partial P}\right)_T = -\frac{\alpha}{\rho}$$

Introduction of the values of $(\partial S/\partial T)_P$ and $(\partial S/\partial P)_T$ into Equation 13-27 yields, at constant S $(dS = 0)$,

$$\left(\frac{dT}{dP}\right)_S = \frac{\alpha T}{\rho c_p}$$

But in the earth the pressure is close to hydrostatic; thus $P = g\rho z$. And so (for constant g and ρ),

$$dP = g\rho\,dz$$

Thus,

$$\left(\frac{dT}{dP}\right)_S = \left(\frac{dT}{g\rho\,dz}\right)_S = \frac{\alpha T}{\rho c_p}$$

which yields Equation 13-26.

In a liquid cooling slowly by convection, the gradient is close to isentropic. If it were higher (i.e., superisentropic), a particle isentropically displaced downward would have a lower temperature than its surroundings, and thus a greater density, and hence would continue falling. The system would thus be unstable. If the gradient were subisentropic, the system could not convect. (See Chapter 14 for further discussion.)

Equation 13-26 gives a very approximate way of computing the temperature in the earth's mantle. If we assume that T is isentropic (adiabatic) everywhere and varies only with z, we obtain

$$\frac{dT}{T} = \frac{\alpha g}{c_p} dz$$

Assuming that the factor $\alpha g / c_p$ remains constant with depth—for instance, because the variations with temperature and pressure of all the terms approximately cancel each other out—we obtain

$$T = T_o \exp\left[\frac{\alpha g}{c_p}(z - z_o)\right] \tag{13-28}$$

where z_o is the depth at which the temperature starts increasing isentropically and T_o is the temperature at depth z_o, in °K.

Using $z_o = 100$ km, $T_o = 1400\,°K$, $\alpha = 2 \times 10^{-5}/°K$, $g \approx 10$ m/s^2, and $c_p = 800$ J/(kg °K) we find at 2800 km (i.e., near the core-mantle boundary), $T_{2800} \approx 2750\,°K$. If $T_o = 1500\,°K$, then $T_{2800} \approx 2950\,°K$ (dashed line in Figure 13-11). The value of α used above appears to be a minimum; with $\alpha = 2.5 \times 10^{-5}/°K$, T_{2800} becomes $\sim 3250\,°K$ for $T_o = 1500\,°K$. This computation is, however, very approximate, because α and c_p in Equation 13-27 vary with z. The variation of g with z is very minor, because ρ increases with depth in the mantle, whereas ρ in the core is substantially larger than in the mantle. (At 2000 km, $g \approx 10$ m/s^2.) On the one hand, as shown below, the fact that convection occurs in the mantle shows that the temperatures computed above should be increased; on the other hand, the decrease of α with depth means that these temperatures should be decreased. These two effects may approximately cancel each other out. In any case, the temperatures computed at the core-mantle boundary are well within the range of recent estimates—that is, 2450 °K (Brown and Shankland 1981) to 3500 °K (Jeanloz and Richter 1979)—and close to the ~ 2900 °K figure given by Stacey (1977, p. 196), who assumes a slightly superadiabatic gradient.

Shear Heating (Viscous Dissipation)

In a flowing viscous liquid the energy necessary to make the liquid flow despite the viscosity, is transformed into heat. This is not a source, strictly speaking, but in a restricted volume it appears as such, for it provides a way by which mechanical energy, generally diffuse throughout the system, can become concentrated in a rather small volume. For an incompressible fluid, it can be shown (Chandrasekhar 1961, p. 14) that its value per unit volume and per unit time—that is, its power/m^3—is

$$\text{Sh} = P_{xx}\dot{E}_{xx} + P_{yy}\dot{E}_{yy} + \cdots + 2P_{xy}\dot{E}_{xy} + \cdots$$

a formula naturally quite similar to the one that gives the elastic energy per unit volume (Equation 3-11). Since $P_{ij} = 2\eta\dot{E}_{ij}$, it follows that

$$\mathrm{Sh} = 2\eta\left(\dot{E}_{xx}^{\,2} + \dot{E}_{yy}^{\,2} + \cdots + 2\dot{E}_{xy}^{\,2} + \cdots\right) \tag{13-29}$$

Thus, in the previously discussed case (Figure 12-1) of a very simple laminar viscous flow in which all $\dot{E}_{ij}$s vanish except $\dot{E}_{xy}$, the shear heating power is proportional to the *square* of the shear strain rate. In the case in which two bodies separated by a fluid of thickness d move with respect to one another, $\dot{E}_{xy}$ is proportional to their relative velocity (which can often be estimated fairly accurately) and also to the inverse of d (which is often very poorly known). In a model, therefore, if d is assumed to be 100 m instead of its real value of 1 km, the shear strain power per unit volume would be 100 times too large, and the temperatures computed as a result of such an assumption would have little relation with reality. This difficulty is sometimes encountered in attempts to compute temperatures near subduction zones, major faults, and so on.

D. Heat Flow Observations

Equation 13-1 shows that in order to determine the heat flow, it is necessary to measure both the temperature gradient (in fact, the rate of change of temperature with z) and the conductivity. It is further necessary to guard against the effect of slow temperature fluctuations at the surface (see Section 13-B). On the continents, this means making measurements in wells over 100 m deep, at least. On the oceanic bottom the water temperature presumably has remained constant over long periods, but normally it is necessary to make measurements in relatively soft sediments, which in turn necessitates fairly difficult *in situ* measurements of the thermal conductivity. An additional difficulty in oceanic areas less than 50 Ma old is that a significant part of the heat may be advected away by water circulation through the oceanic crust. Sclater et al. (1980) estimate that this process adds roughly 50% to the total measured oceanic heat flow.

There has been much discussion about the significance and analysis of heat flow data, principally for the following reasons.

- The data are scattered and need careful corrections.
- Heat conduction is a very slow process; hence the present heat flow gives no indication of the present temperatures at depths of more than a few kilometers except under steady-state conditions. Such conditions can reasonably be assumed to have existed over many continents for long periods of time, but it must be remembered that the characteristic time of a 35-km layer is ~ 35 Ma, and that this time increases according to the square of the thickness.
- The heat transport mechanism differs below continents and oceans, at least at relatively shallow depths (somewhat less than 100 km).

The most recent results (Sclater et al. 1980) indicate that the average surface heat flow over the continents (including continental shelves) is ~ 57 mW/m^2 and that observed over oceans is ~ 66 mW/m^2, to which somewhat less than 50% should be added for heat advected through water circulation, giving ~ 99 mW/m^2.

Over continents, the observed heat flow decreases approximately exponentially with age from ~ 70 mW/m^2 down to ~ 44 mW/m^2 for ages greater than 1700 Ma (Figure 13-5). It is difficult to know how much of this heat flow results from radioactivity in the continental crust and how much results from heat flow from the mantle. Assuming a crust 40 km thick, a mantle heat flow of 21...34 mW/m^2 below the continents appears probable (see Section 13-E). It is important, however, to realize that part of the subcontinental mantle (the "tectosphere") may move with the continents (see Chapter 12, Section E). This region may extend to perhaps 400 km below the surface, although a depth half as large appears more probable. In this zone, heat can be transmitted only by conduction. Possibly, the tectosphere exists mostly below shields and represents a consequence of the higher solidus temperature resulting from the absence of water there.

The heat flow through the ocean floor decreases approximately according to the inverse of the square root of the age up to an age of ~ 120 Ma (Figure 13-6), but for ages less than ~ 50 Ma, the observed heat flow is reduced because of water circulation through the crust. This decrease with $t^{-1/2}$ is a consequence of the flow pattern produced by convection (see Chapter 14). As we know (see

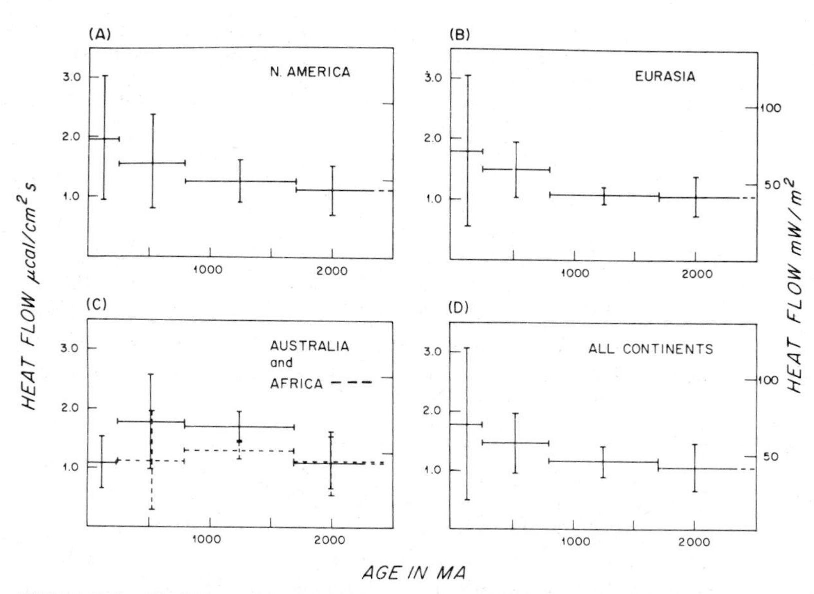

FIG. 13-5 Variation of heat flow with age over the continents. The vertical bars indicate the standard deviations. (From Sclater et al. 1980. Copyright © by the American Geophysical Union.)

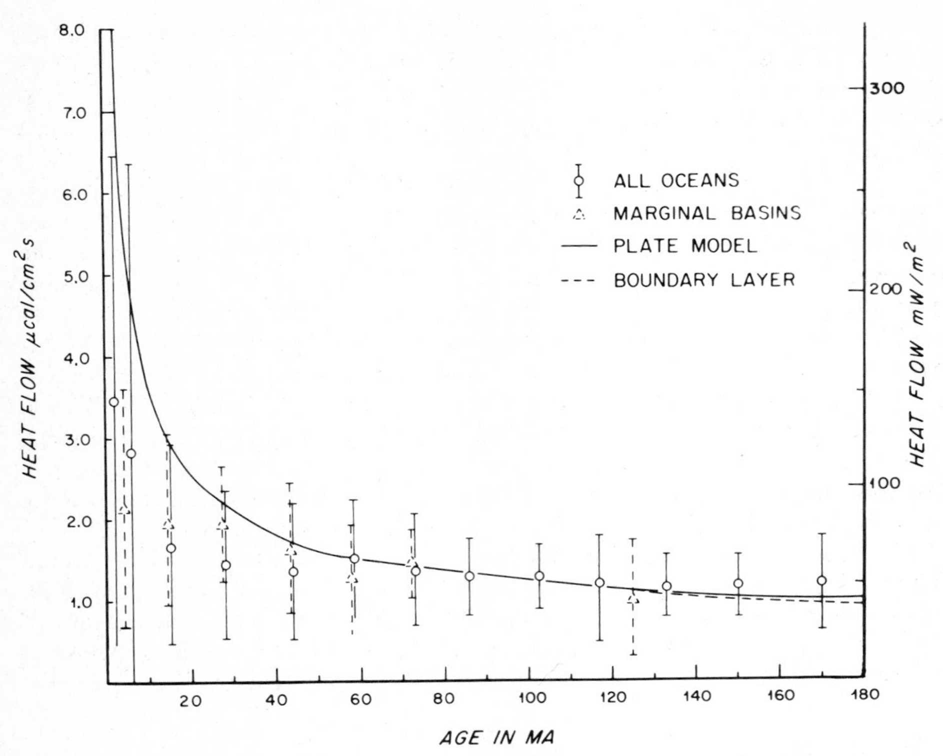

FIG. 13-6 *Oceanic heat flow as a function of the age. For ages less than 50* Ma, *the heat flow is less than the theoretical one because of water circulation. For ages greater than ~120* Ma, *the heat flow is greater than would be expected from the model. (From Sclater et al. 1980. Copyright © by the American Geophysical Union.)*

Figure 13-7), hot material wells up at midoceanic ridges and forms the lithosphere, which gradually cools as it moves away from the ridge. At ages greater than ~120 Ma, the heat flow (49 ± 18 mW/m^2) is higher than would be expected from this model. The reasons for this phenomenon are not entirely clear, but some additional heat from the mantle is necessary to explain it. This model of convection is known as the boundary layer model; it is discussed further in Chapter 14, Section D.

As the model suggests, most of the oceanic heat flow (about 85%) comes from the gradual cooling and thickening of the lithosphere. The average heat flow from the oceanic mantle is probably between 25 and 38 mW/m^2 (although much higher values have been proposed). It is important to note that this is the heat flow through the base of the crust (roughly 10 km thick). The heat flow through the base of the lithosphere increases with age; although it is essentially insignifi-

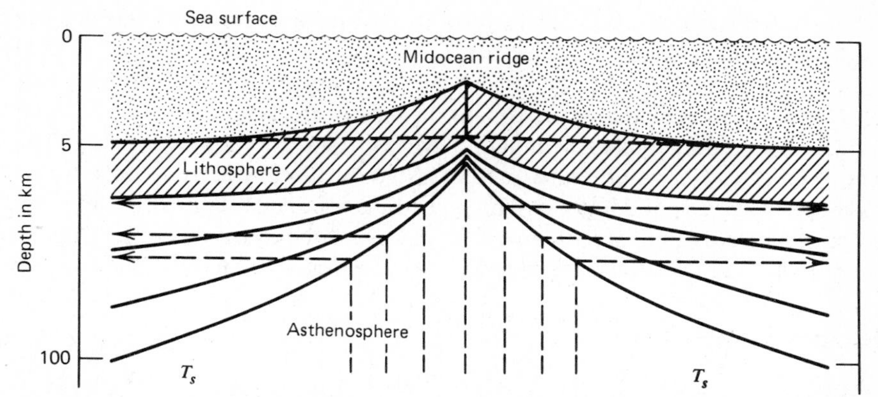

FIG. 13-7 *Schematic model for a thermal boundary layer, showing the material flow (dashed lines) and the concept of a thickening lithosphere. The solidus temperature, T_s, separates solid and partially molten states; the solid lithosphere above is cooler than T_s, and the asthenosphere below is hotter. The dashed lines are (schematic) flow lines. The solid lines indicate isothermal surfaces within the cooling lithosphere. (After Sclater et al. 1980. Copyright © by the American Geophysical Union.)*

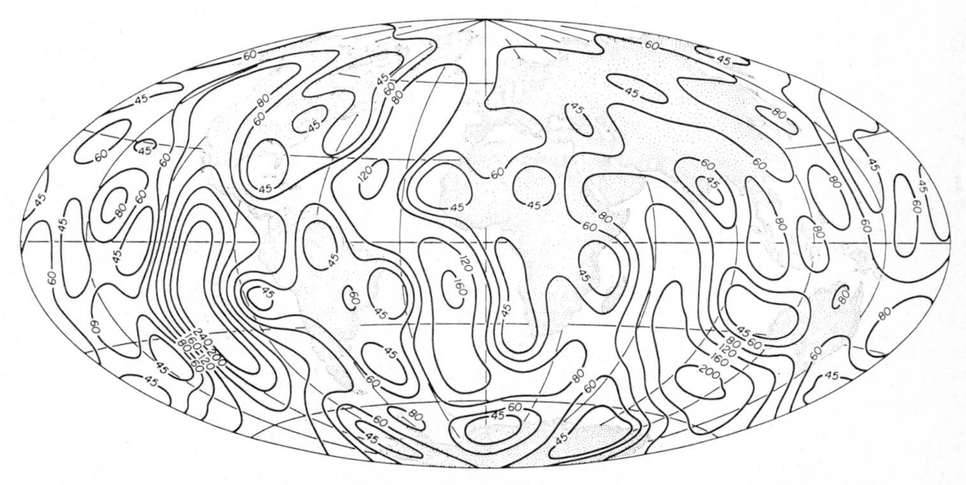

FIG. 13-8 *Smoothed representation of the global heat flow, based both on data and on the relationship between age and heat flow (Chapman and Pollack 1983. By courtesy of David Chapman.)*

cant for ages less than 50 Ma, near 120 Ma it may account for about half of the observed surface heat flow.

The relations between heat flow and age have been used by Chapman and Pollack (1975) to predict the approximate heat flow in areas of the world from which there are no data, including much of Africa, Asia, and South America. The resulting "data" (observations plus predictors) were filtered to pass only wavelengths longer than about 3500 km. The map thus obtained (Figure 13-8) clearly shows the generally high heat flow near the ridges and the low flow on the shields. However, it should not be taken to give the exact heat flow at any specific point.

Reduced Heat Flow

In the northeastern United States (northern New York state and farther east), the surface heat flow is linearly related to the radioactivity of surface crystalline rocks (Birch et al. 1968).

$$q = q_r + DS_o \tag{13-30}$$

(see Figure 13-9), where q_r is the reduced heat flow, or intercept; D is the slope of

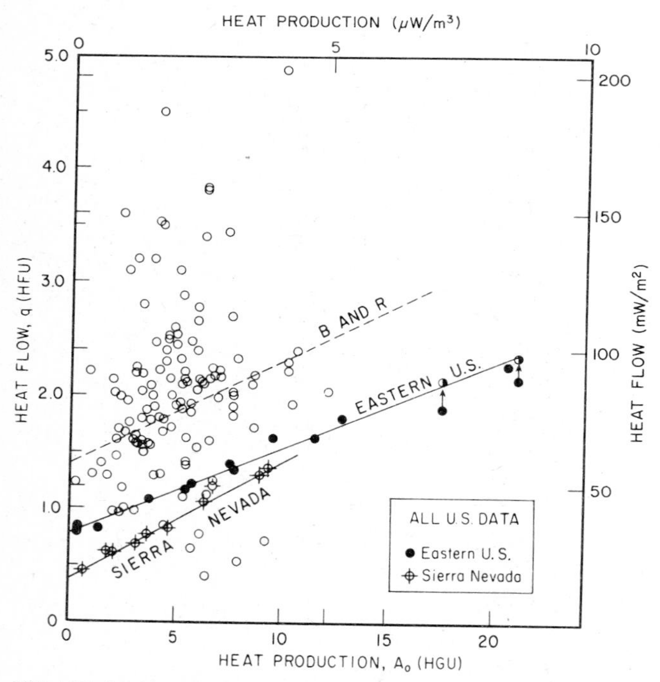

FIG. 13-9 *Heat flow versus radioactive heat production of surface crystalline rocks S_o in the eastern United States, in the Sierra Nevada, and in the rest of the United States (mostly in the Basin and Range). For q in mW/m², S in μW/m³, and D_o in km, the approximate equations of the three lines are, respectively: q = 30 + 8S, q = 15 + 10S, and q = 60 + 9S. (After Lachenbruch and Sass 1977. Copyright © by the American Geophysical Union.)*

the line; and S_o is the radioactivity of rocks near the surface (in W/m^3 or similar unit). The same line appears to characterize the relation in the stable area of the United States and Canada east of the Great Plains, although measurements are sparse (fewer than 20). The relation also holds in the Sierra Nevada, although q_r and D are different.

Equation 13-30 applies to rocks of very different ages, even though they have been eroded to various depths. Thus, for both q and S as functions of the depth z, we have (Lachenbruch and Sass 1977)

$$q(z) = q_r + [D \times S(z)] \tag{13-31}$$

But for a steady-state situation with no motion, Equation 13-8 yields (with z rather than x as variable)

$$\frac{dq}{dz} = S(z) \tag{13-32}$$

Taking the derivative of Equation 13-31 and using Equation 13-32, we obtain

$$D\frac{dS}{dz} = S \tag{13-33}$$

In Equations 13-32 and 13-33, z points in the direction of the outer normal—that is, upward. The more usual convention is to let z point downward. We then obtain

$$D\frac{dS}{dz} = -S \tag{13-34}$$

or

$$S = S_o e^{-z/D} \tag{13-35}$$

If this distribution extends to depth z_M, we obtain, by substituting Equation 13-35 into Equation 13-32 and integrating,

$$q = \left[q_M - \left(D \times S_o e^{-z_M/D}\right)\right] + (D \times S_o) \tag{13-37}$$

Comparing Equations 13-37 and 13-30, we see that the expression in brackets is q_r. Moreover, from Figure 13-9, $D = 7 \ldots 10$ km. If the exponentially fractionated layer (see Equation 13-35) extends through most of the crust, then $z_M \cong 30$ km. Hence, $D \times S_o e^{-z_M/D} \ll q_M$, and thus $q_M \cong q_r$: that is, the reduced heat flow is approximately equal to the mantle heat flow.

Moreover, just as in the case first investigated by Strutt, the steady-state heat conduction equation becomes [with $S = -S_o e^{-z/D}$]

$$\frac{d^2T}{dz^2} = -\frac{S_o}{k} e^{-z/D} \tag{13-38}$$

which can be integrated to yield

$$T = \left[q_r + D^2 S_o\left(1 - e^{-z/D}\right)\right] k^{-1} \tag{13-39}$$

The resulting geotherms for different values of S_o and $k = 2.5$ W/(m s °K) are shown in Figure 13-10.

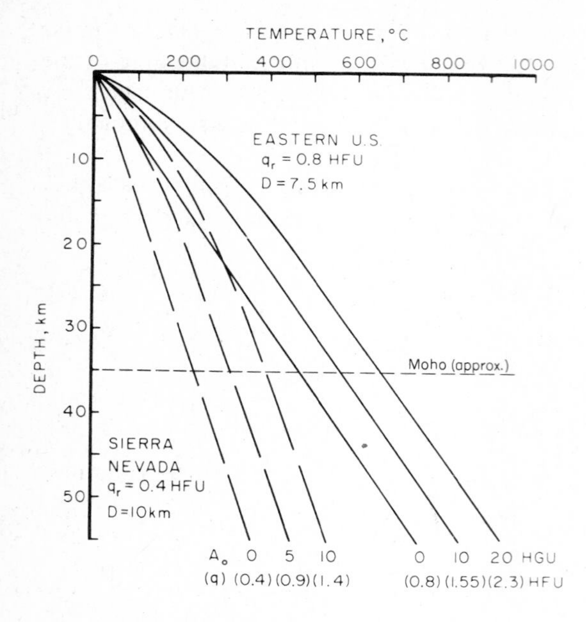

FIG. 13-10 *Geotherms in the eastern United States and in the Sierra Nevada, computed from Equation 13-39 for various values of the surface radioactivity (1 HFU ≅ 43 mW/ m², 1 HGU ≅ 0.43 μW/ m³). (After Lachenbruch and Sass 1977. Copyright © by the American Geophysical Union.)*

If the linear relation between q and S_o that exists in the eastern United States and the Sierra Nevada were general, it would be possible to compute the temperatures in the crust in a continent, provided there was no transport of heat by groundwater. As Figure 13-9 shows, however, this is not the case everywhere in the United States (no data are available from elsewhere). The geotherm in continents thus remains poorly known.

E. Temperatures in the Mantle

The preceding discussion shows that the variation of temperature with depth—the geotherm—in a continent is very different from that below an ocean (see Figure 13-11), and that it also varies with the age of the oceanic crust and below different parts of the same continent. Eventually, though, at a depth of about 200 km (or, perhaps, 400 km), the geotherms converge to the same temperature, possibly near 1300° C, although the subduction zones remain cooler than average, and the uncertainty remains large.

Additional information about the mantle temperature has been obtained from the analysis of kimberlite modules. The minerals in these modules can yield estimates of both the temperature and the pressure at depth. These conditions are widely taken to be representative, but some geophysicists argue that the very reason for the expulsion of kimberlite is that the conditions at depth were abnormal in that area at that time. In any case, temperatures at ~ 100 km are almost certainly between 900° and 1400° C under both continents and oceans

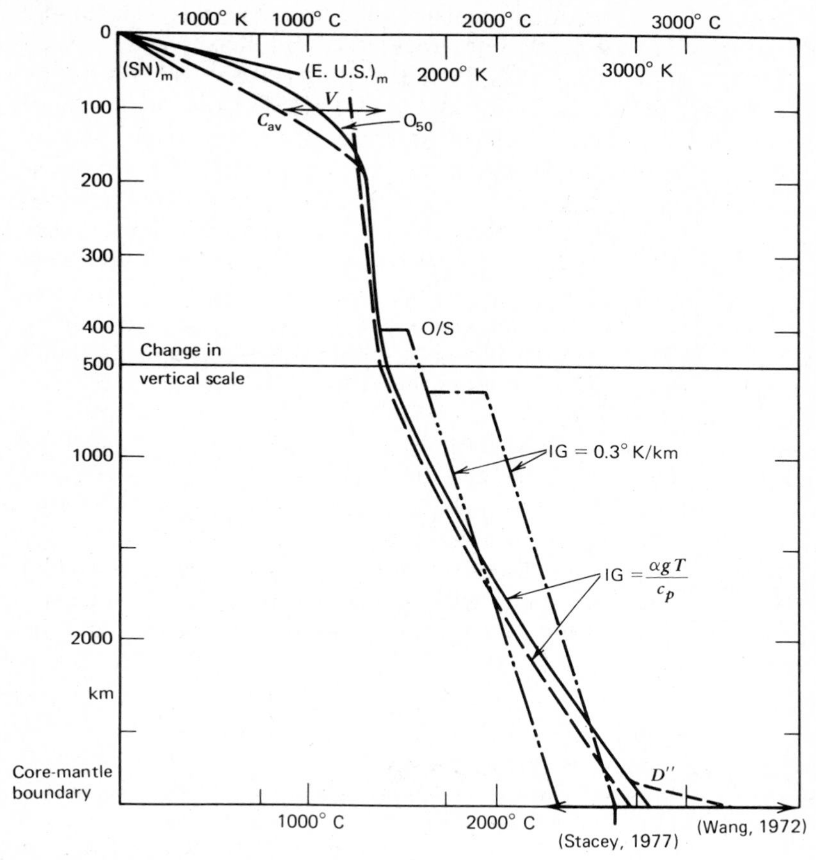

FIG. 13-11 *Approximate geotherms.*

$(SN)_m$:	*minimum temperature below the Sierra Nevada (see Figure 13-10).*
$(E. U.S.)_M$:	*maximum temperature below the eastern United States (see Figure 13-10).*
V:	*range of temperatures given by Verhoogen 1980, p. 32.*
C_{av}:	*average continental geotherm.*
O_{50}:	*oceanic geotherm below a 50-*Ma*-old crust.*
O/S:	*olivine-to-spinel change and corresponding temperature rise.*

At a depth of 500 km, *the vertical scale changes.*

——:	*geotherm computed assuming 1300°*C ≅ *1575°*K *at 200* km *and $IG = \alpha gT/c_p$.*
———:	*the same, with 1225°*C ≅ *1500°*K *at 100* km.
—··—:	*geotherm with IG = 0.3°*K/km *and no temperature rise at 650* km.
—·—:	*the same, but with a hypothetical temperature rise at 650* km.
– – D'':	*possible temperature rise in the D'' layer.*

(Verhoogen 1980, p. 32). The lower temperatures are more probable under water-saturated conditions, the higher ones under dry conditions.

Below this region (see Chapter 14) the temperature (Figure 13-11) increases slightly superisentropically, except for the 400- and 650-km discontinuities, until the core is approached. As noted previously (Chapter 3, Section B), the 400-km discontinuity is due to the phase change from olivine to spinel. The thermodynamic constants of this change are well known: the heat of reaction is $\sim 160 \times 10^3$ J/kg, and c_p is $\sim 10^3$ J/(kg °C). The resulting increase in temperature is thus $160 \times 10^3/10^3 = 160°$ C (or °K). Since there is no agreement on the nature of the 650-km discontinuity, no change in temperature can be ascribed to it. Approximately 200 to 300 km above the core-mantle boundary a new "boundary layer," sometimes called the D″ layer, is approached. Its origin is basically similar to that of the lithosphere, in that the core is losing heat to the mantle (much like the mantle is losing heat to the oceans). The material next to the core heats up gradually (cools down gradually for the lithosphere), when it is hot enough, its density is decreased enough that it rises (the lithosphere is subducted). Because of this motion, heating does occurs not in one place but in a moving layer. Seismic data (decrease of *P*-wave velocity) suggest that the temperature in this layer increases as much as $\sim 1200°$ C (Verhoogen 1980, p. 45), although much smaller estimates are more common. An alternative explanation, based on *SH* waves, is that the composition of the *D″* layer differs from those of both the core and the mantle (Lay and Helmberger 1983).

F. Temperatures in the Core

Many of the difficulties and controversies having to do with the temperatures in the earth would vanish if the composition of the core were known. The density of the core requires that it be mostly iron, but shock wave data require that some lighter elements be present, to reduce the density by about 10%. This also reduces the melting point below that of pure iron, which is $7500° \pm 2000°$ C at the relevant pressure. Geochemists have adduced arguments for Si, O, or S as the main secondary constituent.

There are arguments for each element. Sulfur is currently favored, in part because its cosmic abundance is much greater than its abundance in the crust, and especially because of its great affinity for iron. The lowest melting point (eutectic) of iron in equilibrium with a Fe-S melt is almost completely unknown at the pressures prevailing at the core-mantle boundary. Many theories have been advanced, but none has escaped criticism. The presence of K^{40} in the core is also a contested matter. The best that can be said is that temperatures between 2450° K and 5300° K at the core-mantle boundary have been estimated by various authors. (For reference, the adiabatic temperatures computed in Section C, plus a 500° K increase in layer D″, yield temperatures between $\sim 3300°$ and $\sim 3800°$ K.) The temperatures inside the core are, of course, even less well known than those in the mantle. The only reference point is that the inner core is solid; this might indicate a lower temperature, different phases, different composition, a pressure effect, or a combination of all four. Relevant data are almost nonexistent, apart from the fact that the outer core convects. If this convection is thermal (see Chapter 14) the gradient is superisentropic (superadiabatic); if it stems from differentiation, a subisentropic gradient might exist. All told, the temperature at the inner core boundary is probably between 3400° and 5700° K. Presumably, the temperature varies little in the inner core.

14

Convection in the Earth

A. Introduction

It is somewhat remarkable that convection, which has been used for thousands of years, is still rather poorly understood, and that the first scientific investigation of it was made only in this century. More remarkable still is the fact that the few investigations into convection in the earth published in the 1930s (Pekeris 1935; Hales 1936) were not followed up even by their authors, whose scientific activity continued for over 40 years thereafter.

"Convection" has a wider meaning that the one commonly used in geophysics. One speaks of forced convection when a liquid is mechanically forced to move, and of free convection when the only forces are thermally induced. Mixed convection is also possible. Free convection is thus synonymous with thermally induced gravitational instability, and it will be referred to here as thermal convection. Not fitting into this classification scheme is chemical convection, or compositional convection, which results from chemical differentiation such as occurs (possibly) in the earth's core (see Chapter 11, Section C) or in magma chambers. Chemical convection is believed by a few geophysicists to be the driving force of plate tectonics—an idea that has not been generally accepted, however.

A final point deserves special mention. Many numerical models have been computed showing convection in rectangular two-dimensional enclosures. This geometry is only a convenience and a very rough approximation. It is generally agreed that convection is not a purely two-dimensional phenomenon, but the importance of the flow in the third dimension remains unknown. It is also highly unlikely that convection takes place in a series of closed cells, but the real form of the convection pattern in the earth is not known.

B. Thermal Convection

The first experiments on liquid convection (Bénard 1901) were plagued by a variety of extraneous factors; they resulted, however, in the first physico-mathematical investigation of the phenomenon (Rayleigh 1916). This initial investigation dealt with a liquid heated from below to temperature T_b, while its surface was kept at a fixed temperature, T_s. Although idealized, this is the situation that occurs, for example, when water is heated on a stove. Under these circumstances the water will not quite be uniformly heated. As a result, it will be hotter and thus lighter at some points, where it will rise. This sets the volume in motion. Coming to the surface, the hot liquid spreads out and cools, forming a boundary layer in which the temperature increases sharply from the surface temperature. When the boundary layer is cooler than the surrounding liquid, it sinks and spreads along the bottom. There it is gradually heated, forming a bottom boundary layer that rises when it becomes hotter than its surroundings.

The importance of boundary layers is obvious from common experience. In the winter, window panes prevent the cold outside air from blowing into a house (advection of heat). In addition, the inside air near the panes forms a boundary layer where the temperature rapidly changes from that of the panes to that of the room, and a similar situation obtains for the outside air. The relatively high thermal conductivity of the glass is thus not important: what restricts the thermal exchange is the existence of these boundary layers. Double windows are useful because they double the number of boundary layers, not because of any thermal property of the glass.

Returning now to the convecting liquid, and considering the situation purely qualitatively, we see that if T_b differs very little (whatever this may mean) from T_s, the liquid will not move: heat will be transported by conduction alone. It is also intuitively clear that this situation is easier to achieve in a high-viscosity liquid than in a low-viscosity one.

Rayleigh (1916) showed that convection occurs if a quantity now called the Rayleigh number, Ra, exceeds a critical value, Ra_c. Ra is given by the formula

$$\mathrm{Ra} = \frac{\alpha g D^4 s}{\kappa \nu} \tag{14-1}$$

where α is the thermal expansivity, s is the "superisentropic" (superadiabatic) gradient (i.e., the amount by which the actual gradient exceeds the isentropic, or adiabatic, gradient), g is the value of gravity, D is the depth of the layer, κ is the thermal diffusivity, and ν is the kinematic viscosity $(=\eta/\rho)$.

Rayleigh's analysis was later refined and extended to take into account such questions as whether the liquid slips freely at the upper and lower surfaces (free-slip conditions) or whether it cannot move horizontally there (no-slip conditions). Depending on these conditions, Ra_c varies from ~ 658 to ~ 1710. (Ra is nondimensional, just like Re). When Ra is just above Ra_c—that is, at marginal stability—the liquid forms convection cells whose horizontal dimensions are about $\sqrt{2}$ that of the depth of the layer. The cells are thus roughly equidimensional.

The situation in the earth's mantle differs in several ways from those investigated above.

1. The heat probably comes mostly from more or less uniformly distributed sources. An unknown proportion, generally thought to be about 25%, comes from the core, but some researchers believe that this proportion is approximately 75% (Peltier 1983). Although the mixed (distributed sources plus heat at the bottom) case has not been investigated in detail, the case of distributed sources has been solved.
2. The geometry is spherical rather than planar. It is known (Chandrasekhar 1961, pp. 227–250) that the Rayleigh number applicable to points 1 and 2 may be written

$$\mathrm{Ra} = \frac{\alpha g S R^5}{3k\kappa\nu} \tag{14-2}$$

   where α, κ, ν, g, and k have been previously defined, R is the outer radius of the sphere or spherical shell, and S is the power of the source (W/m^3). Ra_c ranges from ~ 3000 for a sphere to a maximum of $\sim 30{,}000$ for a spherical shell with the same geometry as the mantle and no-slip boundaries (Chandrasekhar 1961, pp. 249–250). With the values used previously (Chapter 13, Section C)—that is, $\alpha = 2\times10^{-5}/°\mathrm{K}$, $g \approx 10$ m/s^2, $S = 3.4\times10^{-8}$ W/m^3, $R \cong 6000$ km, $k = 3$ W/m°C, $\kappa = 1.2\times10^{-6}$ m^2/s, and $\nu = 2\times10^{17}$ m^2/s (which corresponds to $\eta \cong 10^{21}$ Pa s $= 10^{22}$ P, see Verhoogen 1980, p. 105)—we obtain $\mathrm{Ra} \approx 3\times10^{10} \cong 10^6 \ldots 10^7\ \mathrm{Ra}_c$. The earth must thus be vigorously convecting, and it is probable that at such a highly "supercritical" Rayleigh number the convection pattern varies with time. It also seems likely that the time needed to effect a significant change in the pattern of circulation must be at least on the order of one complete (but fictitious) turn of a particle around a convection cell with a diameter equal to the depth of the mantle. At 5 cm/a average velocity, this suggests that the motion of the plates was appreciably different ~ 150 Ma ago from what it is at present. As an order-or-magnitude estimate, this is not unreasonable.
3. The viscosity varies with temperature. Although the oceanic lithosphere is probably closer to being elastic than to being viscous, it can bear only a certain

shear stress, beyond which it fractures. On a global scale it might thus perhaps be said to be thixotropic, with a high viscosity where the plates move horizontally and the stresses are low and a low viscosity near subduction zones, where the stresses are high. A high viscosity of 10^{23} Pa s and a low one of 10^{21} or less have been suggested (Kopitzke 1979), but the existence of such a rheology remains speculative. Some numerical experiments (Houston and De Bremaecker 1975; Kopitzke, 1979) suggest that it may cause cells that are much wider than they are deep, but definitive confirmation is lacking.

Prandtl Number

The problem of convection in the earth is rendered more difficult by the fact that two very different time scales are involved. On the one hand, the viscosity of the earth is so large that a change in circulation at one place is (practically) instantaneously felt everywhere; on the other hand, the heat conductivity is so low that it takes billions of years for heat effects to be transmitted over a few hundred kilometers. The ratio of these two time scales is called the Prandtl number.

$$\mathrm{Pr} = \frac{\nu}{\kappa} \tag{14-3}$$

which is also nondimensional. In the earth it is on the order of 10^{23}. What this means physically can be explained as follows (in somewhat anthropromorphic terms). If a collision between plates occurs anywhere on earth, the other plates immediately "know about it" mechanically, but their temperatures only "get the message" many hundreds of millions of years later. The net result is that the temperature field of the earth cannot adjust to the velocity field. But convection in the earth is thermally driven. The large Prandtl number thus suggests that the driving force "does not know what is going on" and contributes to a time-dependent convection pattern. Whether this is true or not, the fact remains that there are no data on convection when the Rayleigh number is highly supercritical and $\mathrm{Pr} \gg \mathrm{Ra}$. The preceding discussion is thus speculative.

C. Single-Cell or Multicell Convection

It is generally agreed that thermal convection takes place in the mantle, but there is no complete agreement beyond this. The main question at issue is whether the mantle is convecting as one cell throughout its depth ("mantlewide convection") or as a series of superposed cells. The chief argument for superposed convection cells is that subducting slabs apparently do not penetrate deeper than ~ 700 km and are often in compression near their maximum depth. This suggests that they may be pushing against some barrier, such as the lower boundary of the upper convection cell. The cause of the phenomenon (if it exists) might be the much higher (but unknown) viscosity of oxides below ~ 700 km. This viscosity might be high enough to cause the lower mantle to convect too slowly for single-cell convection. No data clearly support this hypothesis. Alternatively, the lower mantle might have a slightly higher density than the upper mantle, presumably as a result of (continuing?) differentiation. A density contrast of 1% (i.e., ~ 40 kg/m^3) would suffice as a barrier (Richter 1979), but it remains uncertain whether such a contrast exists, or whether it could arise from a primitive and vigorously convecting mantle. If it does exist, it would cause two more boundary layers to come into being: the lower one of the upper cell and the upper one of the lower cell. The temperatures of the lower mantle and core would correspondingly be higher than otherwise. As previously noted, data on these temperatures are essentially absent.

Present opinion appears slightly to favor single-cell (mantlewide) convection. For convection to exist, the average thermal gradient must be superisentropic (superadiabatic). It has also been argued that only a slight superisentropic condition is possible, because of both the vigor of the convection and the decrease of η with T. A contrary view is that convection can only take place if the viscosity is low enough, which means that the material is close to (or at) the solidus (Stacey 1977, p. 195). There is also some support for the approximate constancy (within about one order of magnitude) of the viscosity along an adiabat. This constancy of viscosity with depth is compatible with the rise in elevation following the last glaciation (Peltier 1983).

As to the effects of phase changes, it is now agreed that they are relatively minor and do not present an obstacle to convection. Indeed, as first suggested by Vening Meinesz (Heiskanen and Vening Meinesz 1958, p. 406), the change from olivine to spinel should increase the velocity of convection, and local effects on the temperature field may result (see Chapter 13, Section E). The nature and effect of the discontinuity near 650–700 km is unknown.

D. Convection and Bathymetry

As we saw in Chapter 13, Section D, the heat flow *decreases* approximately as the inverse of the square root of t, the age of the ocean floor. This relationship is valid for $t < 120$ Ma. It is a consequence of sea-floor spreading. Because the lithosphere cools and thus contracts as it moves away from the ridge, it is logical to expect that the depth d of the ocean will *increase* approximately according to $t^{1/2}$. Plots of d versus $t^{1/2}$ are shown in Figure 14-1 for the North Pacific and the North Atlantic. As can be seen, the plots depart from the $d = At^{1/2}$ law for ~ 70 Ma. The reason for the difference between this 70 Ma and the 120 Ma resulting from heat-flow observations (Figure 13-6) is that bathymetry is an "integrating" indicator of the influence of the temperature at depth: that is, it depends on the temperature between ~ 120 km and the surface, whereas heat flow depends on

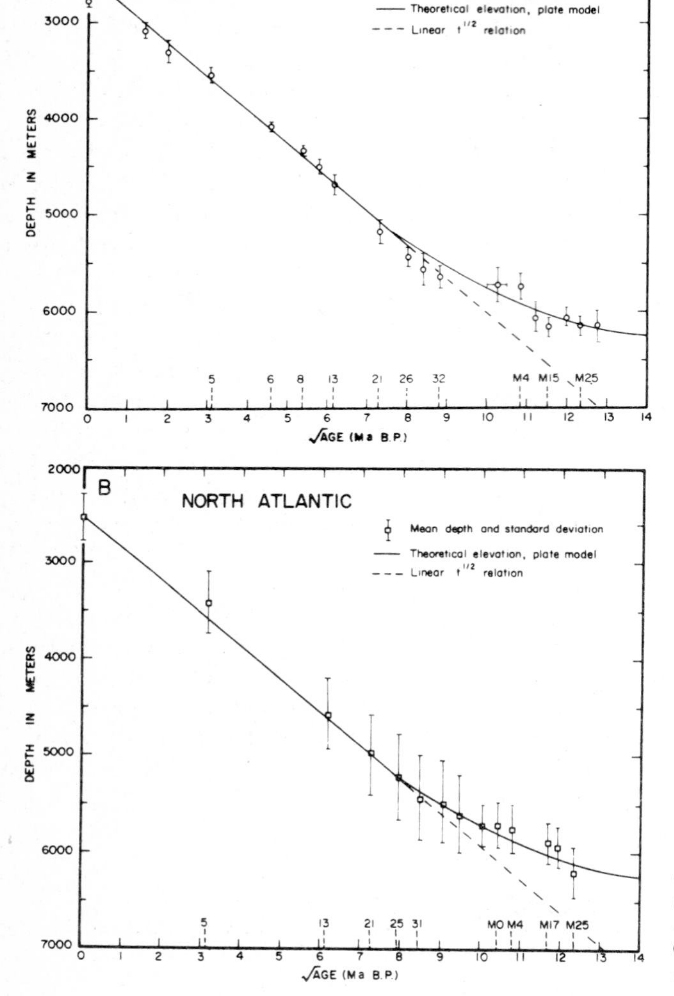

FIG. 14-1 *Mean depths and standard deviations versus the square root of the age for the North Pacific and the North Atlantic. The ages of some magnetic anomalies are also shown. The fit of the bathymetry is obviously very good for the plate model (solid line), but not so good for the boundary layer model (dashed line). (From Parsons and Sclater 1977. Copyright © by the American Geophysical Union.)*

the near-surface temperature only. The bathymetry is thus a more sensitive general indicator.

Two models have been proposed to account for the relation between age and bathymetry: the boundary layer model and the plate model. In the boundary layer model (Parker and Oldenburg 1973), the only heat source is the heat of solidification of the magma injected at the ridge (Figure 13-7). The bottom of the lithosphere is at the "melting temperature"—that is, the solidus, taken to be 1200° C. The depth of the lithosphere gradually increases as the lithosphere moves away from the ridge. Because the lithosphere receives no heat from anywhere and continuously looses heat at the ocean bottom, it thickens indefinitely as $t^{1/2}$; the bathymetry follows the same law. Figure 14-1 shows that this is not the case for $t \gtrsim 70$ Ma.

A more successful model, originally proposed by McKenzie (1967), postulates that the "plate" has a constant thickness d (originally 50 km, now often ~ 120) and a constant bottom temperature T_b (originally 550° C, now often ~ 1300), and that it is created at the ridge at this same temperature and moves away from the ridge at a constant velocity. (It may be unfortunate to use the term "plate" in this connection, since it has nothing to do with the "plate" of "plate tectonics.") The physical properties of the "plate" vary with depth, especially near the ridge, where part of it is essentially rigid and part of it is viscous. The bathymetry resulting from this model is in very good agreement with observational evidence (see Figure 14-1).

A much more difficult question arises from this result. Boundary layer theory shows that the thickness of the boundary layer should vary according to $t^{1/2}$ in a simple convection model. How can the necessary additional heat be brought up? Or, to put it otherwise: How can the constant temperature T_b be maintained at depth d? There is no agreement on this subject, although a variety of proposals have been made. The essential difficulties in testing the validity of these proposals are that laboratory experiments at very high Rayleigh numbers and higher Prandtl numbers appear impossible, that the numerical computations made so far are two-dimensional, and that no analytic solution has been obtained. Some authors (Schubert et al. 1976) have proposed that this additional heat is produced by high shear-heating at the bottom of a rigid lithosphere moving on a viscous stationary asthenosphere, but the dynamics of this situation have been questioned. Others have cited distributed radioactivity (Forsyth 1977) or a smaller-scale circulation in the third dimension (Richter and Parsons 1975). Finally, some two-dimensional numerical models suggest that additional heat is not needed and that the peculiar nonlinear and temperature-dependent rheology of the plates adequately explains the observations (Kopitzke 1979).

In any case, the existence of the age-depth relation is not in doubt. This relation also shows that under the oceans the pressure at a depth d departs very little from the lithostatic pressure—for if such were not the case, the height of each column (above d) would not be a function only of the temperature.

E. Convection and Plate Tectonics

Thermal convection occurs in the earth's mantle. The energy comes partly from radioactivity and partly from the core, but the proportion is uncertain. Heating is thus partly a volumetric process, but cooling takes place at the surface only. This difference makes it likely that rising columns of material are broader than descending columns, although the exact effects of variable viscosity are not known with certainty. It is generally thought that the most vigorously rising columns are under oceanic ridges; however, there may be other regions under oceanic plates or continents where material rises less vigorously.

The question is often asked, "What is the source of the energy that moves the plates?" This question can be answered by means of a simple analogy: The cars on a roller coaster are first dragged up by a motor, then they descend under the action of gravity. If there were no motor, the roller coaster would soon stop. Now imagine that the cars are much more numerous—in fact, that they completely fill the tracks. The motor is still needed to make the cars move, but now all the cars move at the same speed. We can look at the whole system and say, "It is the motor that makes it move." Or we can look at a car on a downhill course, notice that it goes much more slowly than it would if it were alone, and say, "This car is moving the others." Each statement is correct. A similar situation occurs in plate tectonics, only it is much more complex. To pursue our example, say that you have heard that a fancy roller coaster has been built on which 12 trains of cars ride at the same time. These trains are connected by cables, springs, and dashpots that are constrained to stay right in the middle of the tracks and at a constant height above them. Greatly intrigued, you take a tour bus to go look at this ingenious new machine. When you get there, unfortunately, a thick fog covers the area. Climbing a tower, you look down to see the tracks emerging from the fog at about a dozen places and trains of cars moving on them. You start taking notes, but the bus driver is disturbed by the fog and decides to leave after only a few minutes. When you arrive home, everybody asks you to explain why some of the trains move faster than others... .

This analogy is not so far-fetched as it may seem. Only the motions at the surface of the earth are known—the circulation in the mantle is not (just as the fog covers most of the tracks). The coupling between the plates can be roughly evaluated at the surface, but no one knows how the motions couple at depth or how much energy this requires (just as the springs and dashpots between the trains are unknown). Finally, we know the pattern of plate motions at present but not much about it in the past (Schult and Gordon 1984). (So you leave after spending only a few minutes.)

Let us now examine the convecting mantle from a physical point of view. Looking at the whole system as at a heat engine, we know that it requires a source and a sink. The source is the radioactivity and the heat from the core, and the sink is the earth's surface. Looking at various parts of the system, we can distinguish a number of sources. When we say that source S_1 is more important than S_2, this means that convection would (instantaneously) slow down more rapidly if S_1 vanished than if S_2 did (just as a roller-coaster car on a steep slope contributes more energy than one on a gentle slope). The sources usually distinguished include the following three.

1. The slope down from the ridge crest.
2. The thickening of the lithosphere with age and the consequent increase in the average density of the upper 100 km.
3. The pull of the heavy, cold slab.

The first two forces are, strictly speaking, distributed from the ridge to the subduction zone; most of the thickening and the largest change in elevation, however, occur relatively close to the ridge. Therefore, these two forces are often known collectively as the "ridge push."

The convective drag (either contributing to or resisting the motion) at the base of the plate is another force that can be quantified, because it can be assumed to be proportional to the plate area.

The pull of the slab is very large, but most of the pull is used up locally. This is easy to understand if you remember that, say, a solid steel sphere falling through viscous oil acquires only a limited terminal velocity despite the large difference in density between oil and steel. Alternatively, you might remember the roller-coaster analogy: it is as if a car on a downward slope had to use most of its increase in potential energy to overcome its friction on the tracks. The terminal velocity, or free-fall velocity, cannot be determined theoretically in the present case without additional assumptions of very uncertain validity.

The only part of the resistive force that can be localized and taken into account is the drag of the subducting lithosphere under the continent. The largest part of the resistive force is probably provided by the viscosity of the mantle and depends on the circulation pattern. For example, it clearly is difficult to get a stream of highly viscous material to turn over a relatively short distance.

Despite the difficulties mentioned above, the relative importance of slab pull, ridge push, and drag has been the subject of numerous investigations based on three main methods. The relative importance is measured per effective unit length of ridge or trench, but the effective length of two ridges say, that are pushing a plate in opposite directions is the difference between the length of the ridges.

1. The first method (Richardson et al. 1979) computes the forces on the periphery of the plates under different assumptions. These forces must balance out, since each plate is in equilibrium. This constitutes a first constraint. The direction of the resultant principal stresses in the plates can then be computed at any point and compared with observations of the focal mechanisms of intraplate earthquakes. The latter are rather scarce, but enough of them occur to provide a basis for comparison. The difficulty with the method stems from the fact that earthquake mechanisms reflect the state of stress in only a relatively small volume—which might not be representative, because of local conditions. In any case, this method yields a slab pull approximately twice as large as the ridge push and a small drag.
2. The second method (Richardson 1983) computes the velocity of each plate on the basis of the forces, assuming a highly viscous lithosphere moving over a less viscous asthenosphere. In this case, the ridge push is at least twice as large as the slab pull, and the drag forces are small.
3. Finally, multiple linear regression has been used (Carlson 1981) to establish a statistical relation between the forces acting on the plates and the plate velocities. Here the result is that the ridge push provides only 25%, or even less, of the driving force (Carlson et al. 1983).

It is clear that the conclusions reached through these different methods are incompatible. Whether there is enough information available to solve the problem at present remains uncertain: the largest part of the system i.e., the mantle, is inaccessible and the forces there are unknown, furthermore the present situation may be a purely transient one.

Although there is thus little doubt that thermal convection is the driving mechanism of plate tectonics, the implications of this process on the temperature and chemistry of the earth's interior are subjects still fully open for research.

References

Abel, N. 1881. Solution de quelques problèmes à l'aide d'intégrales définies. *In*: *Oeuvres complètes*, 1: 11–27. Christiana: Grøndahl and Son.

Adams, L. H., and E. D. Williamson, 1923. The composition of the Earth's interior. *Smithsonian Inst. Ann. Rept.*: 241–260.

Aki, K., and P. G. Richards. 1980. *Quantitative seismology*, 1. San Francisco: Freeman.

Artyushkov, E. V. 1973. Stresses in the lithosphere caused by crustal thickness inhomogeneities. *J. Geophys. Res.* 78: 7675–7708.

Barazangi, M., and B. Isacks. 1971. Lateral variation of seismic-wave attenuation in the Upper Mantle above the inclined earthquake zone of the Tonga Island arc: deep anomaly in the Upper Mantle. *J. Geophys. Res.* 76: 8493–8516.

Barenblatt, G. I., V. I. Keilis-Borok, and M. M. Vishik. 1981. Model of clustering of earthquakes. *Nat. Acad. Sci. Proc.* (U.S.A.) 78: 5284–5287.

Bénard, H. 1901. Les tourbillons cellulaires dans une nappe liquide transportant de la chaleur par convection en régime permanent. *Ann. Chimie et Physique*, *s. 7*, 23: 62–144.

Birch, F. 1952. Elasticity and constitution of the Earth's interior. *J. Geophys. Res.* 57: 227–286.

______. 1966. Compressibility, elastic constants. *In*: S. P. Clark (ed.), *Handbook of Physical Constants. Geol. Soc. Am.*, *Mem.*, 97.

Birch, F., R. F. Roy, and E. R. Decker. 1968. Heat flow and thermal history in New England and New York. *In*: E. A. Zen, W. S. White, J. B. Hadley, and J. B.

Thompson (eds.), *Studies of Appalachian Geology, Northern and Maritime*, 437–451. New York: Interscience.
Blakemore, R. P., and R. B. Frankel. 1981. Magnetic navigation in bacteria. *Scientific Amer.* 245 (6): 58–65.
Bloom, A. L. 1960. Optical pumping. *Scientific Amer.* 203: 72–80.
Bomford, G. 1952. *Geodesy*. Oxford: Clarendon Press.
Bowin, D., G. M. Purdy, C. Johnston, G. Shor, L. Lawver, H. M. S. Hartono, and T. Jezek. 1980. Arc-continent collision in Banda Sea region. *Am. Assoc. Petroleum Geol., Bull.*, 64: 868–915.
Brekhovskikh, L. M. 1960. *Waves in layered media* (English trans.). New York: Academic Press.
Bronowski, J. 1965. *Science and human values* (rev. ed.). New York: Harper and Row.
Brown, J. M., and T. J. Shankland. 1981. Thermodynamic parameters in the Earth as determined from seismic profiles. *Geophys. J. R. Astr. Soc.* 66: 579–596.
Bucher, W. H. 1956. Role of gravity in orogenesis. *Geol. Soc. Am. Bull.* 67: 1295–1318.
Bullard, E. C. 1936. Gravity measurements in East Africa. *Roy. Soc. London, Phil. Trans., A*, 25: 445–531.
Bullen, K. E. 1963. *An Introduction to the Theory of Seismology*. Cambridge: Cambridge Univ. Press.
Byerly, P. 1942. *Seismology*. New York: Prentice-Hall.
Cagniard, L. 1939. *Réflexion et réfraction des ondes séismiques progressives*. Paris: Gauthier-Villars.
______. 1962. *Reflection and Refraction of Progressive Seismic Waves* (English trans.). New York: McGraw-Hill.
Carlson, R. L. 1981. Boundary forces and plate velocities. *Geophys. Res. L.* 8: 958–961.
Carlson, R. L., T. W. C. Hilde, and S. Uyeda. 1983. The driving mechanism of plate tectonics: relation to age of lithosphere at trenches. *Geophys. Res. L.* 10: 297–300.
Carrigan, C. R., and D. Gubbins. 1979. The source of the Earth's magnetic field. *Scientific Amer.* 240 (2): 118–130.
Chandrasekhar, S. 1961. *Hydrodynamic and Hydromagnetic Stability*. Oxford: Clarendon Press.
Chapman, D. S., and H. N. Pollack. 1975. Global heat flow: a new look. *Earth Planet. Sci. L.* 28: 23–32.
Chapman, M. E., and M. Talwani. 1979. Comparison of gravimetric geoids with Geos 3 altimetric geoid. *J. Geophys. Res.* 84: 3803–3816.
Chen, W. P., and P. Molnar. 1983. Focal depths of intracontinental and intraplate earthquakes and their implications for the thermal and mechanical properties of the lithosphere. *J. Geophys. Res.* 88: 4183–4214.
Clark, S. P. Jr. 1957. Radiative transfer in the Earth's mantle. *Trans. Am. Geophys. U.* 38: 931–938.
______ (ed.). 1966. *Handbook of Physical Constants. Geol. Soc. Am., Mem.*, 97.
Cochran, J. R., and M. Talwani. 1979. Marine gravimetry. *Rev. Geophys. Space Phys.* 17: 1387–1397.
Cook, F. A., L. D. Brown, S. Kaufman, J. E. Oliver, and T. A. Petersen. 1981. COCORP seismic profiling of the Appalachian orogen beneath the Coastal Plain of Georgia. *Geol. Soc. Am., Bull.*, 92: 738–748.
Coons, R. L., G. P. Woollard, and G. Hershey. 1967. Structural significance and analysis of mid-continent gravity high. *Am. Assoc. Petroleum Geol., Bull.*, 51: 2381–2399.
Corbató, C. E. 1965. A least-squares procedure for gravity interpretation. *Geophysics* 30: 228–233.

Cox, A. 1968. Lengths of geomagnetic polarity reversals. *J. Geophys. Res.* 73: 3247–3260.
Dobrin, M. B. 1976. *Introduction to Geophysical Prospecting*. 3rd ed. New York: McGraw-Hill.
Dorman, J. 1969. Seismic surface-wave data on the Upper Mantle. *In*: P. J. Hart (ed.), *The Earth's Crust and Upper Mantle*, 257–265. *Am. Geophys. U., Monogr.*, 13.
Dorman, L. M., and B. T. R. Lewis. 1970. Experimental isostasy. 1. Theory of the determination of the Earth's isostatic response to a concentrated load. *J. Geophys. Res.* 75: 3357–3365.
Dziewonski, A. M., and D. L. Anderson. 1981. Preliminary reference Earth model. *Phys. Earth Planet. Int.* 25: 297–356.
Dziewonski, A. M., A. L. Hales, and E. R. Lapwood. 1975. Parametrically simple Earth models consistent with geophysical data. *Phys. Earth Planet. Int.* 10: 12–48.
EIB. 1979. Seismic gaps: some questions and answers. *Earthquake Inf. Bull.* 11 (6): 221–223.
Ellers, F. S. 1982. Advanced offshore oil platforms. *Scientific Amer.* 246 (4): 39–49.
Elsasser, W. M. 1958. The Earth as a dynamo. *Scientific Amer.* 198 (5): 44–48.
Ewing, W. M., W. S. Jardetzky, and F. Press. 1957. *Elastic Waves in Layered Media*. New York: McGraw-Hill.
Ewing, (W.) M., and F. Press. 1955. Geophysical contrasts between continents and ocean basins, *In*: A. Poldervart (ed.), *The Crust of the Earth*, 1–18. *Geol. Soc. Am., Spec. Paper* 62.
Fitch, T. J., and C. H. Scholz. 1971. Mechanism of underthrusting in Southwest Japan: a model of convergent plate interaction. *J. Geophys. Res.* 76: 7260–7292.
Forsyth, D. 1977. The evolution of the upper mantle beneath mid-ocean ridges. *Tectonophysics* 38: 89–118.
______. 1979. Lithospheric flexure. *Rev. Geophys. Space Phys.* 17: 1109–1114.
Frank, S. 1982. Ascending droplets in the Earth's core. *Phys. Earth Planet. Int.* 27: 249–254.
Furumoto, M., and I. Nakanishi. 1983. Source times and scaling relations of large earthquakes. *J. Geophys. Res.* 88: 2191–2198.
Galileo, G. 1638. *Dialogues Concerning Two New Sciences*. Trans. by H. Grew and A. De Salvio, 1914. New York: MacMillan.
Gamow, G. 1962. *Gravity*. Garden City, N.Y.: Doubleday.
Garland, G. D. 1979. *Introduction to Geophysics* (2nd ed.). Philadelphia: W. B. Saunders.
Gelfand, I. M., Sh. A. Guberman, V. I. Keilis-Borok, L. Knopoff, F. Press, E. Ya. Ranzman, I. M. Rotwain, and A. M. Sadovsky. 1976. Pattern recognition applied to earthquake epicenters in California, *Phys. Earth Planet. Int.* 11: 227–283.
Gilbert, F., and A. M. Dziewonski. 1975. An application of normal mode theory to the retrieval of structural parameters and source mechanisms from seismic spectra. *Roy. Soc. London Phil. Trans., A*, 278: 187–269.
Gilbert G. (or W.). 1600. De magnete, magneticisque corporibus, et de magno magnete tellure. London: Petrus Short. 240 pp. (*G*: Latin initial. *W*: English initial.)
Goree, W. S., and M. Fuller. 1976. Magnetometers using RF-driven Squids and their application in rock magnetism and paleomagnetism. *Rev. Geophysics Space Phys.* 14: 591–608.
Grow, J. A., and C. O. Bowin. 1975. Evidence for high density crust and mantle beneath the Chile trench due to the descending lithosphere. *J. Geophys. Res.* 80: 1449–1458.
Grow, J. A., R. E. Mattik, and J. S. Schlee. 1979. Multichannel depth sections and interval

velocities over outer continental shelf and upper continental slope between Cape Hatteras and Cape Cod. *In*: J. S. Watkins, L. Montadert, and P. W. Dickerson (eds.), *Geological and Geophysical Investigations of Continental Margins*, 65–83. *Am. Assoc. Petroleum Geol., Mem.*, 29.

Guillemin, E. A. 1949. *The Mathematics of Circuit Analysis*. New York: Wiley.

Gutenberg, B. 1939. The elastic constants in the interior of the Earth, *In*: B. Gutenberg (ed.), *Internal Constitution of the Earth* (1st ed.), 345–360. New York: McGraw-Hill.

______. 1959. *Physics of the Earth's Interior*. New York: Academic Press.

Hale, L. D., and G. A. Thompson. 1982. The seismic reflection character of the continental Mohorovicic discontinuity. *J. Geophys. Res.* 87: 4625–4635.

Hales, A. L. 1936. Convection currents in the Earth. *Roy. Astr. Soc. Geophys. Suppl.* 3: 372–379.

Hannon, W. J. 1964. An application of the Haskell-Thomson matrix method to the synthesis of surface wave motion due to dilatational waves. *Seism. Soc. Am., Bull.*, 54: 2067–2079.

Harris, L. D., and K. C. Bayer. 1979. Sequential development of the Appalachian orogen above a master decollement—a hypothesis. *Geology* 7: 568–572.

Haskell, N. A. 1960. Crustal reflection of plane SH waves. *J. Geophys. Res.* 65: 4147–4150.

______. 1962. Crustal reflection of P and SV waves. *J. Geophys. Res.* 67: 4751–4767.

Heirtzler, J. R., G. O. Dickson, E. M. Herron, W. C. Pitman III, and X. Le Pichon. 1968. Marine magnetic anomalies, geomagnetic field reversals and motions of the ocean floor and continents. *J. Geophys. Res.* 73: 2119–2136.

Heiskanen, W. A., and F. A. Vening Meinesz. 1958. *The Earth and its Gravity Field*. New York: McGraw-Hill.

Houston, M. H. Jr., and J. Cl. De Bremaecker. 1975. Numerical models of convection in the upper mantle. *J. Geophys. Res.* 80: 742–751.

Hubbert, M. K. 1937. Theory of scale models as applied to the study of geologic structures. *Geol. Soc. Am., Bull.* 48: 1459–1519.

Inglis, D. R. 1955. Theory of the Earth's magnetism. *Rev. Modern Phys.* 27: 212–248.

Isacks, B. L., J. Oliver, and L. R. Sykes. 1968. Seismology and the new global tectonics. *J. Geophys. Res.* 73: 5855–5899.

Jackson, J. D. 1962. *Classical Electrodynamics*. New York: Wiley.

Jacobs, J. A. 1975. *The Earth's Core*. London: Academic Press.

Jaeger, J. C. 1969. *Elasticity, Fracture and Flow*. London: Methuen.

Jeanloz, R. and F. M. Richter, 1979. Convection, composition and the thermal state of the lower mantle. *J. Geophys. Res.* 84: 5497–5504.

Jeffreys, H. 1926. On compressional waves in two superposed layers. *Camb. Phil. Soc., Proc.*, 23: 472–481.

______. 1959. *The Earth* (4th ed.). Cambridge: Cambridge Univ. Press.

Kanamori, H. 1978. Quantification of earthquakes. *Nature* 271: 411–414.

______. 1981. The nature of seismicity patterns before large earthquakes. *In*: *Earthquake Prediction, An International Review*. *Am. Geophys. U.*, M. Ewing s., 4: 1–19.

Kanamori, H., and J. W. Given. 1981. Use of long-period surface waves for rapid determination of earthquake-source parameters. *Phys. Earth Planet. Int.* 27: 8–31.

Kellogg, O. D. 1953. *Foundations of Potential Theory*. New York: Dover.

Kelvin (William Thomson, Baron). 1863. On the secular cooling of the Earth. *Phil. Mag.*, s. 4, 25: 1–14.

______. 1899. The age of the Earth as an abode fitted for life. *Phil. Mag.* 47: 66–90.

Kopitzke, U. 1979. Finite element convection models: comparison of shallow and deep

mantle convection, and temperatures in the mantle. *J. Geophysics* 46: 97–121.
Kovach, R. L. 1978. Seismic surface waves and crustal and upper mantle structures. *Rev. Geophys. Space Phys.* 16: 1–13.
Kuhn, T. S. 1970. *The Structure of Scientific Revolutions* (2nd ed.). Chicago: Univ. of Chicago Press.
Kunz, K. S. 1957. *Numerical Analysis*. New York: McGraw-Hill.
LaBrecque, J. L., D. V. Kent, and S. C. Cande. 1977. Revised magnetic polarity time-scale for late Cretaceous and Cenozoic time. *Geology* 5: 330–335.
Lachenbruch, A. H., and J. H. Sass. 1977. Heat flow in the United States and the thermal regime of the crust. *In*: J. G. Heacock (ed.), *The Earth's Crust*, 626–675. *Am. Geophys. U., Geophys. Monogr.*, 20.
Lambert, J. H. 1894. *Anmerkungen und Zusätze zur Entwerfung der Land- und Himmelscharten*. Leipzig: Wilhelm Engelman.
Lanczos, C. 1956. *Applied Analysis*. Englewood Cliffs, N.J.: Prentice-Hall.
Larson, R. L., and T. W. Hilde. 1975. A revised time scale of magnetic reversals for the late Cretaceous and early Jurassic. *J. Geophys. Res.* 80: 2586–2594.
Lay, T., and D. V. Helmberger. 1983. A shear velocity discontinuity in the lower mantle. *Geophys. Res. L.* 10: 63–66.
Lay, T., and H. Kanamori. 1981. An asperity model of large earthquake sequences. *In*: *Earthquake Prediction—An International Review. Am. Geophys. U.*, M. Ewing s., 4: 579–592.
Lee, Y. W. 1960. *Statistical Theory of Communication*. New York: Wiley.
Le Mouël, J. L., J. Ducruix, and C. Ha Duyen. 1983. On the recent variation of the apparent westward drift rate. *Geophys. Res. L.* 10: 369–372.
Le Pichon, X., J. Francheteau, and J. Bonnin. 1976. *Plate Tectonics*, New York: Elsevier.
Le Pichon, X., and P. Huchon. 1984. Geoid, Pangea and convection. *Earth Planet. Sci. L.* 67: 123–135.
Lerch, F. J., S. M. Klosko, R. E. Laubscher, and C. A. Wagner. 1979. Gravity model improvement using Geos 3 (GEM 9 and 10). *J. Geophys. Res.* 84: 3897–3916.
Lichtman, A. J., and V. I. Keilis-Borok. 1981. Pattern recognition applied to presidential elections in the United States, 1860–1980: Role of integral social, economic, and political traits. *Nat. Acad. Sci.* (U.S.A.) *Proc.* 78: 7230–7234.
Lisitzin, E. 1965. The mean sea level of the world ocean. *Commen. Phys.-Math.* (Helsinki) 30 (7): 1–35.
Liu, C. S., D. T. Sandwell, and J. R. Curray. 1982. The negative gravity field over the 85° E ridge. *J. Geophys. Res.* 87: 7673–7686.
Love, A. E. H. 1967. *Some Problems of Geodynamics*. New York: Dover.
Macdonald, K. C., and B. P. Luyendyk. 1981. The crest of the East Pacific Rise. *Scientific Amer.* 244 (5): 100–116.
Macelwane, J. B. 1939. Evidence on the interior of the Earth derived from seismic sources. *In*: B. Gutenberg (ed.), *Internal Constitution of the Earth* (1st ed.), 219–290. New York: McGraw-Hill.
MacMillan, W. D. 1958. *The Theory of the Potential*. New York: Dover.
Markowitz, W. 1973. SI, the international system of units. *Geophysical Surveys* 1: 217–241.
Marsh, J. G., R. E. Cheney, T. V. Martin, and J. J. McCarthy. 1982. Computation of a precise mean sea surface in the eastern North Pacific using SEASAT altimetry. *EOS, Trans. Am. Geophys. U.* 63 (9): 178–179 and cover.
McElhinny, M. W. 1973. *Paleomagnetism and Plate Tectonics*. Cambridge: Cambridge Univ. Press.

McElhinny, M. W., and W. E. Senanayake. 1982. Variations in the geomagnetic dipole. 1: The past 50,000 years. *J. Geomag. Geoelectr*. 34: 39–51.
McKenzie, D. P. 1967. Some remarks on heat flow and gravity anomalies. *J. Geophys. Res*. 72: 6261–6273.
McKenzie, D., and C. Bowin. 1976. The relationship between bathymetry and gravity in the Atlantic Ocean. *J. Geophys. Res*. 81: 1903–1915.
Minster, J. B., and D. L. Anderson. 1981. A model of dislocation-controlled rheology for the mantle. *Roy. Soc. London Phil. Trans., A*., 299: 319–356.
More, L. T. 1934. *Isaac Newton, A Biography*. New York: Scribners.
Murnaghan, F. D. 1951. *Finite Deformation of an Elastic Solid*. New York: Wiley.
Muskat, M. 1933. The theory of refraction shooting. *Physics* 4: 14–28.
Nettleton, L. L. 1940. *Geophysical Prospecting for Oil*. New York: McGraw-Hill.
______. 1962. Gravity and magnetics for geologists and seismologists. *Am. Assoc. Pet. Geol., Bull*., 46: 1815–1838.
______. 1971. Elementary gravity and magnetics for geologists and seismologists. *Soc Expl. Geophys., Monogr*., 1.
______. 1976. *Gravity and Magnetics in Oil Prospecting*. New York: McGraw-Hill.
Newton, I. 1695(?). Portsmouth Coll., sec. 1, div. 10, number 41. As quoted in L. T. More 1934, p. 290.
O'Connell, R. J., and B. Budiansky. 1977. Viscoelastic properties of fluid-saturated cracked solids. *J. Geophys. Res*. 82: 5719–5735.
Officer, C. B. 174. *Introduction to Theoretical Geophysics*. New York: Springer-Verlag.
Oliver, J. 1978. Exploration of the continental basement by seismic reflection profiling. *Nature* 275: 485–488.
Olson, P. 1981. A simple physical model for the terrestrial dynamo. *J. Geophys. Res*. 86: 10875–10882.
Opdyke, N. D., B. Glass, J. D. Hays, and J. Foster. 1966. Paleomagnetic study of Antarctic deep-sea cores. *Science* 154: 349–357.
Parker, E. N. 1983. Magnetic fields in the cosmos. *Scientific Amer*. 249 (2): 44–54.
Parker, R. L., and D. W. Oldenburg. 1973. Thermal model of ocean ridges, *Nature Phys. Sci*. 242: 137–139.
Parsons, B., and J. G. Sclater. 1977. An analysis of the variation of ocean floor bathymetry and heat flow with age. *J. Geophys. Res*. 82: 803–827.
Peddie, N. W. 1983. International geomagnetic reference field—its evolution and the difference in total field intensity between new and old models for 1965–1980. *Geophysics* 48: 1691–1696.
Peirce, B. O. 1929. *A Short Table of Integrals* (3rd ed.). Boston: Ginn.
Pekeris, C. L. 1935. Thermal convection in the interior of the Earth. *Roy. Astr. Soc. Geophys. Suppl*. 3: 343–367.
Peltier, W. R. 1983. Constraint on deep mantle viscosity from Lageos acceleration data. *Nature* 304: 434–436.
Peltier, W. R. and G. T. Jarvis. 1982. Whole mantle convection and the thermal evolution of the Earth. *Physics Earth Planet. Inter*. 29: 281–304.
Plouff, D., 1976. Gravity and magnetic fields of polygonal prisms and application to magnetic terrain corrections. *Geophysics* 41: 727–741.
Press, F., and R. Siever, 1982. *Earth* (3rd ed.). San Francisco: Freeman.
Ramsey, A. S. 1961. *An Introduction to the Theory of Newtonian Attraction*. Cambridge: Cambridge Univ. Press.
Rayleigh (J. W. Strutt, 3rd Baron). 1887. On waves propagated along the plane surface of

an elastic solid. *London Math. Soc., Proc.* 17: 4–11.

______. 1894 (repr. 1945). *The Theory of Sound*, 2 vols. New York: Dover.

______. 1916. On convective currents in a layer of fluid when the higher temperature is on the underside. *Phil. Mag.*, s.6, 32: 529–546.

Reid, H. F. 1910. *The Mechanics of the Earthquake, The California Earthquake of April 18, 1906*. Report of the State Investigation Commission, 2. Washington, D.C.: Carnegie Inst.

Reiner, M. 1960. *Lectures on Theoretical Rheology*. Amsterdam: North-Holland.

Richardson, R. M. 1983. Ridge push: how important as a driving force (Abstr.). Symposium on oceanic lithosphere, Texas A&M Univ. April 28–29, 1983.

Richardson, R. M., S. C. Solomon, and N. H. Sleep. 1979. Tectonic stresses in the plates. *Rev. Geophysics Space Phys.* 17: 981–1019.

Richter, C. F. 1958. *Elementary Seismology*. San Francisco: Freeman.

Richter, F. M. 1979. Focal mechanisms and seismic energy release of deep and intermediate earthquakes in the Tonga-Kermadec region and their bearing on the depth extent of mantle flow. *J. Geophys. Res.* 84: 6783–6795.

Richter, F. M., and B. Parsons. 1975. On the interaction of two scales of convection in the mantle. *J. Geophys. Res.* 80: 2529–2541.

Rikitake, T. 1976. *Earthquake Prediction*. New York: Elsevier.

Sacks, I. S., S. Suyehiro, D. W. Evertson, and Y. Yamagishi. 1971. Sacks-Evertson strainmeter, its installation in Japan and some preliminary results concerning strain steps. *Pap. Meteorol. Geophys.* 22: 195–208.

Sacks, I. S., S. Suyehiro, A. T. Linde, and J. A. Snoke. 1982. Stress redistribution and slow earthquakes. *Tectonophysics* 81: 311–318.

Schubert, G., C. Froideveaux, and D. A. Yuen. 1976. Oceanic lithosphere and asthenosphere: thermal and mechanical structure. *J. Geophys. Res.* 81: 3525–3540.

Schult, F. R. and R. G. Gordon. 1984. Root mean square velocities of the continents with respect to the hot spots since the Early Jurassic. *J. Geophys. Res.* 89: 1789–1800.

Sclater, J. G., S. Hellinger, and C. Tapscott. 1977. The paleobathymetry of the Atlantic Ocean from the Jurassic to the present. *J. Geology* 85: 509–552.

Sclater, J. G., C. Jaupart, and D. Galson. 1980. The heat flow through oceanic and continental crust and the heat loss of the Earth. *Rev. Geophys. Space Phys.* 18: 269–311.

Simon, R. B. 1981. *Earthquake Interpretations*. Los Altos, Cal.: William Kaufman.

Smithson, S. B. 1978. Modeling continental crust: Structural and chemical constraints. *Geophys. Res. L.* 5: 749–752.

Smythe, W. R. 1950. *Static and Dynamic Electricity*. New York: McGraw-Hill.

Stacey, F. D. 1977. *Physics of the Earth* (2nd ed.). New York: Wiley.

Stark, M., and D. W. Forsyth. 1983. The geoid, small-scale convection, and differential travel time anomalies of shear waves in the central Indian Ocean. *J. Geophys. Res.* 88: 2273–2288.

Stauder, W. 1968. Tensional character of earthquake foci beneath the Aleutian trench with relation to sea-floor spreading. *J. Geophys. Res.* 73: 7693–7701.

Strutt, J. W. *See*: Rayleigh.

Strutt, R. J. 1906. On the distribution of radium in the Earth's crust and on the Earth's internal heat. *Roy. Soc. London, Proc., A*, 77: 472–485.

Sykes, L. R. 1967. Mechanism of earthquakes and nature of faulting on the mid-oceanic ridges. *J. Geophys. Res.* 72: 2131–2153.

Talwani, M., and J. M. Heirtzler. 1964. Computation of magnetic anomalies caused by two-dimensional structures of arbitrary shape. *In*: G. A. Parks (ed.), *Computers in the Mineral Industries*, 464–479. Stanford Univ. Publ., 9, 1.

Talwani, M., P. Stoffa, P. Buhl, C. Windisch, and J. B. Diebold. 1982. Seismic multichannel towed arrays in the exploration of the continental crust. *Tectonophysics* 81: 273–300.

Talwani, M., J. L. Worzel, and M. Landisman. 1959. Rapid gravity computations for two-dimensional bodies with application to the Mendocino submarine fracture zone. *J. Geophys. Res.* 64: 49–59.

Tarantola, A., and B. Valette. 1982*a*. Generalized nonlinear inverse problems solved using the least squares criterion. *Rev. Geophys. Space Phys.* 20: 219–232.

______. 1982*b*. Inverse problems = quest for information. *J. Geophys.* 50: 159–170.

Tarling, D. H. 1971. *Principles and Applications of Paleomagnetism*. London: Chapman and Hall.

Telford, W. M., L. P. Geldart, R. E. Sheriff, and D. A. Keys. 1976. *Applied Geophysics*. Cambridge: Cambridge Univ. Press.

Thompson, G. A., and M. Talwani. 1964. Crustal structure from Pacific Basin to Central Nevada. *J. Geophys. Res.* 69: 4813–4837.

Thomson, W. *See*: Kelvin.

Turcotte, D. L., and G. Schubert. 1982. *Geodynamics*. New York: Wiley.

Turner, F. J., and L. E. Weiss. 1963. *Structural Analysis of Metamorphic Tectonites*. New York: McGraw-Hill.

Vaniček, P., R. O. Castle, and E. I. Balazs. 1980. Geodetic leveling and its applications. *Rev. Geophys. Space Phys.* 18: 505–524.

Verhoogen, J. 1980. *Energetics of the Earth*. Washington, D.C.: Nat. Acad. Sci.

Vestine, E. H., C. Laporte, I. Lange, C. Cooper, and W. C. Hendrix. 1947. *Description of the Earth's Main Magnetic Field and Its Secular Change, 1905–1945*. Carnegie Inst. Wash. Publ. 578.

Vine, F. J. 1966. Spreading of the ocean floor: new evidence. *Science* 154: 1405–1415.

Wadati, K. 1928. Shallow and deep earthquakes. *Geophys. Mag.* (Tokyo) 1: 161–202.

Wegener, A. 1912. Die Entstehung der Kontinente. *Peterm. Geogr. Mitt.* 58: 185–194, 253–256, and 305–309.

Williams, H., F. A. Cook, D. S. Albaugh, L. D. Brown, S. Kaufman, J. E. Oliver, L. D. Harris, K. C. Bayer, J. C. Reed, B. Bryant, and R. D. Hatcher. 1980. Comments and replies on 'Thin-skinned tectonics in the crystalline southern Appalachians: COCORP seismic reflection profiling of the Blue Ridge and Piedmont' and 'Sequential development of the Appalachian orogen above a master décollement. A hypothesis'. *Geology* 8: 211–216.

Zimmerman, J. E., and W. H. Campbell. 1975. Test of cryogenic SQUID for geomagnetic field measurements. *Geophysics* 40: 269–284.

Zumberge, M. A. and J. E. Faller. 1983. Results from an absolute gravity survey in the United States. *J. Geophys. Res.* 88: 7495–7502.

APPENDIX A

Mechanical Quantities:

Internationally Recommended Symbols, Names of Units, Abbreviations, Dimensions, and Conversions

Quantity	Symbol	SI Units			cgs and Other Units		Conversion
		Name of Unit	Abbreviation	Dimension	Name of Unit	Abbreviation	
Mass	m	kilogram	kg	kg	gram	g	1 kg = 10^3 g
Length	l	meter	m	m	centimeter	cm	1 m = 10^2 cm
Time	t	second	s	s	year	a	1 a ≈ 3.156×10^7 s
Acceleration	a	—	m s^{-2}	m s^{-2}	milligal gravity unit	mgal g.u.	1 m s^{-2} = 10^5 mgal 1 m s^{-2} = 10^6 g.u.
Energy or work	E	joule	J	kg m^2 s^{-2}	erg calorie kilowatt hour	— cal kWh	1 J = 10^7 ergs 1 J ≈ 0.239 cal 1 J = 2.78×10^{-7} kWh
Force	$\mathbf{F}$	newton	N	kg m s^{-2}	dyne	—	1 N = 10^5 dynes
Heat capacity	c	—	J kg^{-1}/°K	m^2 s^{-2}/°K	—	cal g^{-1}/°K	1 J kg^{-1}/°K ≈ 2.39×10^{-4} cal g^{-1}/°K
Heat flow	$\mathbf{q}$	—	W m^{-2}	kg s^{-3}	heat flow unit	HFU = μcal cm^{-2} s^{-1}	1 W m^{-2} ≈ 23.9 HFU

Heat generation	—	—	$W\ m^{-3}$	$kg\ m^{-1}\ s^{-3}$	heat generation unit	HGU = $10^{-7}\ \mu cal\ cm^{-3}\ s^{-1}$	$1\ W\ m^{-3} \approx 2.39 \times 10^{6}$ HGU
Power	P	watt	W	$kg\ m^{2}\ s^{-3}$	erg per second	$erg\ s^{-1}$	$1\ W = 10^{7}\ ergs\ s^{-1}$
Pressure	$P^{(1)}$	pascal	Pa	$kg\ m^{-1}\ s^{-2}$	bar	—	$1\ Pa = 10^{-5}$ bar
					dynes per cm^2	$dynes\ cm^{-2}$	$1\ Pa = 10\ dynes\ cm^{-2}$
Thermal conductivity	$k^{(2)}$	—	$W\ m^{-1}/°K$	$kg\ m\ s^{-3}/°K$	—	$cal\ cm^{-1}\ s^{-1}/°K$	$1\ W\ m^{-1}/°K \approx 2.39 \times 10^{-3}\ cal\ cm^{-1}\ s^{-1}/°K$
Thermal diffusivity	$\kappa^{(3)}$	—	$m^2\ s^{-1}$	$m^2\ s^{-1}$	—	$cm^2\ s^{-1}$	$1\ m^2\ s^{-1} = 10^4\ cm^2\ s^{-1}$
Velocity	v	—	$m\ s^{-1}$	$m\ s^{-1}$	cm/year	$cm\ a^{-1}$	$1\ m\ s^{-1} \approx 3.156 \times 10^{9}\ cm\ a^{-1}$
Viscosity (dynamic)	η	—	Pa s	$kg\ m^{-1}\ s^{-1}$	poise	P	1 Pa s = 10 P
Viscosity (kinematic)	ν	—	—	$m^2\ s^{-1}$	stokes	—	$1\ m^2\ s^{-1} = 10^4$ Stokes

[1] p is recommended.

[2] λ is recommended.

[3] a is recommended.

APPENDIX B

Magnetic Quantities:

Internationally Adopted Symbols, Names of Units, Abbreviations, Dimensions, and Conversions from SI to emu

Quantity	Symbol	SI Units			emu Units			Conversion
		Name	Abbrev	Dimensions	Name	Abbrev	Dimensions	
Dipole moment[1]	$(M)^2$	(1) ampere m^2 (2) tesla m^3	A m^2 T m^3	A m^2 A^{-1} kg m^3 s^{-2}	maxwell cm = gauss cm^3	(1) Mx cm = G cm^3	(1) $g^{1/2}$ $cm^{5/2}$ s^{-1} $(\mu^{1/2})$	(1) 1 A $m^2 = 10^3$ G cm^3 (2) 1 T $m^3 = 10^{10}$ G cm^3
Electric current	I	ampere	A	A	abampere	—	$g^{1/2}$ $cm^{1/2}$ s^{-1} (μ)	1 A = 0.1 abampere
Intensity of magnetization[1]	(**J**)	(1) ampere/m (2) tesla	A m^{-1} T	A m^{-1} A^{-1} kg s^{-2}	gauss	G	$g^{1/2}$ $cm^{-1/2}$ s^{-1} $(\mu^{1/2})$	(1) 1 A $m^{-1} = 10^{-3}$ G (2) 1 T = 10^4 G
Magnetic field strength	**H**	—	A m^{-1}	A m^{-1}	oersted	Oe	$g^{1/2}$ $cm^{-1/2}$ s^{-1} $(\mu^{-1/2})$	1 A $m^{-1} = 4\pi 10^{-3}$ Oe
Magnetic flux	Φ	weber	Wb = V s	A^{-1} kg m^2 s^{-2}	maxwell	Mx = G cm^2	$g^{1/2}$ $cm^{3/2}$ s^{-1} $(\mu^{1/2})$	1 Wb = 10^8 Mx
Magnetic induction	**B**	tesla or weber/m^2	T or Wb m^{-2}	A^{-1} kg s^{-2}	gauss	G	$g^{1/2}$ $cm^{-1/2}$ s^{-1} $(\mu^{1/2})$	1 T = 10^4 G = 10^4 Mx cm^{-2} 1 nT = 1 γ = 10^{-5} G
Magnetic permeability	μ	—	NA^{-2} = Hm^{-1}	A^{-2} kg m s^{-2}	—	—	1 or G/Oe	1 H $m^{-1} = 10^7/4\pi$ G/Oe
Magnetic susceptibility	χ	none (see text)	—	—	—	—	—	1 SI = $1/4\pi$ emu

[1]There are two possibilities for M and **J**. The first (1) defines a dipole in terms of current in a loop; it is becoming more common. The second defines it in terms of a dipole moment. The units are different (see Chapter 12, Section C).

[2]Symbols in parentheses have not been adopted internationally.

APPENDIX C

Selected Data on the Earth

1. Global Data

Equatorial radius	6.37814×10^6 m
Polar radius	6.35675×10^6 m
Flattening	1/298.247
Volume	1.083×10^{21} m^3
Mass	5.973×10^{24} kg
Mean density	5.515×10^3 kg/m^3
Total surface (land surface, 29%; ocean surface, 71%)	5.1×10^{14} m^2
Radius of the core	3.48×10^6 m
Mean core density	11.03×10^3 kg/m^3
Mean mantle density	4.439×10^3 kg/m^3
Mean ocean depth	3800 m
Equatorial gravity	9.7803185 m/s^2
Magnetic moment	7.95×10^{22} A m^2

Average mantle viscosity	10^{21} Pa s
Total heat flow	4×10^{13} W
Mean continental heat flow	57 mW/m^2
Mean sub-oceanic heat flow	99 mW/m^2
Observed mean sub-oceanic heat flow	66 mW/m^2
Typical thermal properties in the mantle	
Heat capacity at constant pressure	1000...1200 J/(kg °K)
Thermal conductivity	2.5...3.0 W/(m °K)
Thermal expansivity	$2...4 \cdot 10^{-5}$/°K
Heat sources	$2.5...5 \times 10^{-8}$ W/m^3
Density $\times 10^{-3}$	
3.3 (top)	4.0 (~ 600 km)
5.0 (1800 km)	5.5 (core boundary)

Distribution of Radioactive Elements: see Table 13-1

Gravitational constant, G	6.67×10^{-11} m^3/(kg s^2)
Magnetic permeability of free space, μ_o	$4\pi \times 10^{-7}$ N/A^2

2. Data on Specific Rock Types at Surface Pressure and Temperature

(For Each Property: Approximate Minimum, Average, and Maximum Values)

Rock type	α in km/s	β in km/s	ρ in 10^3 kg/m^3	χ in 10^{-4} SI unit
Limestone	1.7, 4.2?, 7.0	2.7, 3.1, 3.6	1.93, 2.54, 2.90	0.25, 2.9, 35
Sandstone	1.4, 4.0?, 4.3	—	1.61, 2.32, 2.76	0, 4.0, 200
Shale	2.3, 3.9, 4.7	—	1.77, 2.42, 2.45	0.6, 6.5, 190
Metamorphic	3.5, 5.3, 7.5	2.5, 3.0, 3.2	2.40, 2.74, 3.10	0, 44, 730
Acidic igneous	4.8, 5.3, 5.8	2.9, 3.0, 3.2	2.30, 2.61, 3.11	0.4, 81, 820
Basic igneous	5.1, 5.9, 6.4	2.7, 3.2, 3.5	2.09, 2.79, 3.17	5.5, 330, 1200

APPENDIX D

Notations

$\dot{}$	Dot above a letter,—for example, $\dot{u} = \partial u / \partial t$
$\approx$ or $\cong$	Approximately equal
$\lesssim$	Less than approximately
$\gtrsim$	Greater than approximately
$\hat{}$	Circumflex: unit vector
α	*P*-wave velocity; Coefficient of thermal expansion (expansivity)
β	*S*-wave velocity
γ	Attenuation
γ_i	Gravity anomaly of type *i*
Δ	Epicentral distance
δg_i	Correction of type *i*
δ_{ij}	Kronecker delta, 1 or 0
θ	Dilatation
η	(Dynamic) viscosity
κ	Incompressibility; Thermal diffusivity
Λ	Wavelength
λ	First Lamé constant
λ_2	Half-life
μ	Modulus of rigidity; Magnetic permeability; As a prefix: *micro* (10^{-6})
μ_o	Magnetic permeability of free space
ν	Frequency; Dynamic viscosity
ρ	Density
σ	Poisson's ratio; Surface density
τ	"Average stress" (Equation 12-8)
τ_c	Characteristic time

τ_r	Relaxation time
ϕ	Latitude
ϕ_1, ϕ_n	Phase angle
χ	Magnetic susceptibility
ω	Angular frequency
∇	Gradient, del
∇^2	Laplacian, del square
A	ampere
A	Amplitude of a wave
a	Abbreviation for *annum* (year)
a	Acceleration
A_n	Amplitude of the nth Fourier cosine coefficient
AP	Auxiliary plane
$\mathbf{B}$, B_x, B_y, B_z	Magnetic induction and components
B_n	Amplitude of the nth Fourier sine coefficient
B.P.	Before present
C	Compression (fault plane method)
C_n	Amplitude of the nth Fourier coefficient
c	As a prefix: *centi* (10^{-2})
c	Heat capacity; Phase velocity
D	Dilatation (fault plane method)
d	Magnetic declination (variation)
E	Young's modulus; Energy (released in an earthquake)
$\dot{E}_{ij}$	Component of the strain rate deviator
$\exp(a)$	e^a
e_{xx}, e_{xy}, $e_{ij}\ldots$	Strains
F	Gravitational field (in general)
f	Flattening
FP	Fault plane
G	gauss; As a prefix: *giga* (10^9)
G	Constant of gravitation
g	Gram
g	Gravity
g_{obs}	Observed gravity
g_{ref}	Gravity on reference surface
g.u.	Gravity unit
h	Hour
$\mathbf{H}$	Magnetic field
HFU	Heat flow unit
HGU	Heat generation unit
I	Electric current
i	Magnetic inclination
IG	Isentropic gradient
J	joule
$\mathbf{J}$	Intensity of magnetization
$\hat{\mathbf{j}}$	Direction of **J**
k	As a prefix: *kilo* (10^3)
k	Wavenumber ($=1/\Lambda$); Thermal conductivity
l	Length of a simple pendulum
L	Characteristic length
LVZ	Low-velocity zone
m	Meter; As a prefix: *milli* (10^{-3}); As an index: *minimum*
M	As a prefix: *mega* (10^6); As an index: *maximum*
$\mathbf{M}$	Magnetic moment
M, M_s	Surface wave magnitude
M_o	Seismic moment
m_b	Body wave magnitude
N	newton
N	Distance from the geoid to the reference ellipsoid
n	As a prefix: *nano* (10^{-9})

$\hat{\mathbf{n}}$	Outer normal
Oe	oersted
P	poise
P	Pressure; Power
P_{ij}	Component of the stress deviator
$\mathbf{p}_x$	Traction across x-plane
p_{xx}, p_{yy}, $p_{ij}\ldots$	Components of the stress tensor
$\mathbf{q}$	Heat flow
Q	Quality factor, general
Q_α, Q_β	Quality factors for P- and S-waves
R_A, R_E	Reflection coefficients for amplitude and energy
R_e	Earth's radius (average)
R_{eq}, R_{po}	Equatorial and polar radii
r	Radius
RHS	Right-hand side
s	Superisentropic (superadiabatic) thermal gradient
T	tesla
T	Period
T_o	Time interval; Free period of a pendulum
T_A, T_E	Transmission coefficients for amplitude and energy
U	Group velocity; Gravitational potential
V	volt
VGP	Virtual geomagnetic pole (paleopole)
W	watt
W	Magnetic potential
Z	Acoustic impedance

APPENDIX E

Some Useful Mathematical Concepts

1. Partial Derivative

The elevation (z) of the ground above sea level normally varies with x and y, where, say, the x-axis points east and the y-axis points north. Since z depends on x and y, it is the dependent variable; x and y are the independent variables.

Starting from a point, z will generally vary as we go east or north. The rate of change of z in the x-direction is written $\partial z/\partial x$, where the symbol $\partial/\partial x$ indicates that y is kept constant in this computation. We say that $\partial z/\partial x$ is the partial derivative of z with respect to x. Similarly, $\partial z/\partial y$ is the rate of change of z in the y-direction with x kept constant. It follows that $\partial z/\partial x$ is the slope in the east-west direction and $\partial z/\partial y$ the slope in the north-south direction.

It is not necessary for the dependent variable to be of the same nature (i.e., have the same dimensions) as the independent variables. If the temperature of a surface, for instance, is a function of x and y, $\partial T/\partial x$ is the rate of variation of T in the x-direction and $\partial T/\partial y$ is that in the y-direction.

If the functional relationship between the dependent variable (T or z in the preceding examples) and the independent variable (x and y in the examples) is known, we can compute $\partial T/\partial x$ just as if it were dT/dx and y were a constant.

The same is true for $\partial T/\partial y$. For example, if

$$T = a + bx + cy + dx^2 + ey^2 + fxy$$

then

$$\frac{\partial T}{\partial x} = b + 2\,dx + fy$$

$$\frac{\partial T}{\partial y} = c + 2ey + fx$$

Finally, we can assume that T is a function of x, y, and z. We then have three partial derivatives, $\partial T/\partial x$, $\partial T/\partial y$, and $\partial T/\partial z$.

2. Directional Derivative

In many cases of practical importance the functional relationship between the dependent and the independent variables is unknown. It is then necessary to compute the partial derivatives in an approximate manner. The following simple case can easily be generalized. Let the temperature T of a plane vary with x and y in a manner that is not known analytically; however, the temperature can be measured at any point. We first wish to find an approximate value for the rate of change of T at a point A along an arbitrary direction $\hat{\mathbf{s}}$ (Figure E-1), where $\hat{\mathbf{s}}$ is a vector of unit length. This approximate rate of change is $\delta T/\delta s$; to find it, we take an arbitrary but small length, δs, from A to B, and measure T at A and B—T_A and T_B, respectively. Then,

$$\frac{\partial T}{\partial s} \cong \frac{\delta T}{\delta s} = \frac{T_B - T_A}{\delta s} \tag{E-1}$$

Note that we have taken the final value (T_B) minus the initial value (T_A) and divided by the length (considered positive). The term $\partial T/\partial s$ is the directional derivative in the s-direction, and Equation E-1 enables us to compute it approximately. The approximation improves as δs decreases, provided that T_B and

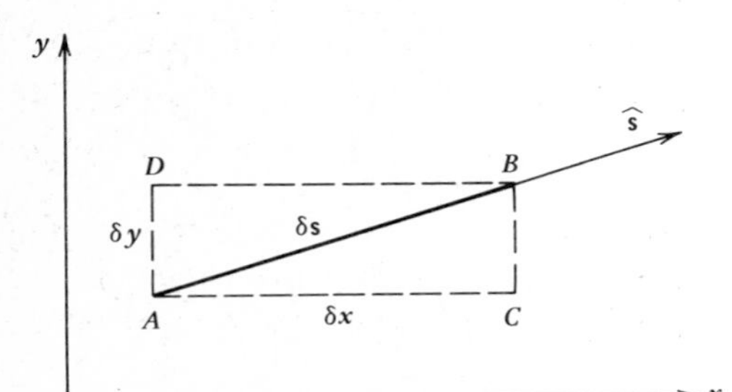

FIG. E-1 *Pertaining to the approximate computation of $\partial T/\partial s$, $\partial T/\partial x$, $\partial T/\partial y$.*

T_A are measured exactly. Using the same reasoning in the x- and y-directions, we obtain

$$\frac{\partial T}{\partial x} \cong \frac{\delta T}{\delta x} = \frac{T_C - T_A}{\delta x} \tag{E-2}$$

$$\frac{\partial T}{\partial y} \cong \frac{\delta T}{\delta y} = \frac{T_D - T_A}{\delta y} \tag{E-3}$$

Finally, if δx and δy are small,

$$T_B - T_C \cong T_D - T_A$$

Thus,

$$T_B - T_A = (T_B - T_C) + (T_C - T_A) \cong (T_D - T_A) + (T_C - T_A)$$

Hence,

$$\frac{\delta T}{\delta s}\,\delta s \cong \frac{\delta T}{\delta y}\,\delta y + \frac{\delta T}{\delta x}\,\delta x$$

and with δs, δx, and $\delta y \to ds, dx, dy$,

$$\frac{\partial T}{\partial s}\,ds = \frac{\partial T}{\partial x}\,dx + \frac{\partial T}{\partial y}\,dy \tag{E-4}$$

3. Gradient

Let us now examine the situation at A along a direction, say, $\hat{\mathbf{t}}$, where T remains constant (Figure E-2). Along this direction, by hypothesis,

$$0 = \frac{\partial T}{\partial t} \cong \frac{\delta T}{\delta t}$$

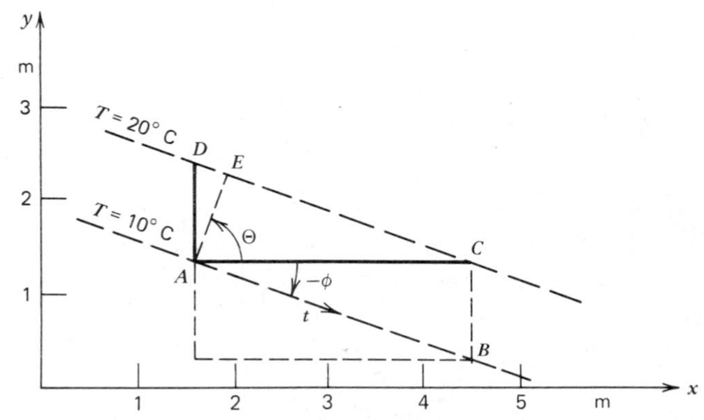

FIG. E-2 *Lines of constant T, direction (AE) of the gradient, and angles involved in the computation.*

Applying Equation E-4 to the t-direction, we obtain, along that direction *only*,

$$0 = \frac{\partial T}{\partial x}\,dx + \frac{\partial T}{\partial y}\,dy$$

Hence,

$$\frac{dy}{dx} = -\frac{\partial T/\partial x}{\partial T/\partial y} = \tan\phi \tag{E-5}$$

where dy/dx is the direction of the tangent to the lines of constant T. The figure shows that in the present case,

$$\frac{\partial T}{\partial x} \cong \frac{\delta T}{\delta x} = \frac{T_C - T_A}{AC} = \frac{10^\circ\,\mathrm{C}}{2.8\ \mathrm{m}} \cong 3.6^\circ\,\mathrm{C/m}$$

$$\frac{\partial T}{\partial y} \cong \frac{\delta T}{\delta y} = \frac{T_D - T_A}{AD} = \frac{10^\circ\,\mathrm{C}}{1.0\ \mathrm{m}} = 10^\circ\,\mathrm{C/m}$$

Hence, from Equation E-5,

$$\tan\phi = \frac{dy}{dx} \cong -\frac{3.6}{10} = -0.36$$

To show that this is correct, we can write

$$-\frac{dy}{dx} \cong -\frac{\delta y}{\delta x} = -\frac{CB}{AC} \cong -\frac{AD}{AC} = -\frac{AD}{10^\circ\,\mathrm{C}} \times \frac{10^\circ\,\mathrm{C}}{AC}$$

But $AD = \delta y$, $AC = \delta x$, and $10^\circ\,\mathrm{C} = \delta T$; hence,

$$-\frac{dy}{dx} = -\frac{1}{\delta T/\delta y} \times \frac{\delta T}{\delta x} = -\frac{\delta T/\delta x}{\delta T/\delta y}$$

The direction of most rapid increase in temperature, say, $\hat{\mathbf{g}}$, is perpendicular to that of constant temperature. By definition, $\hat{\mathbf{g}}$ is the direction of the gradient. The magnitude of the gradient is the rate of change of the temperature in that direction. In Figure E-2, $\hat{\mathbf{g}}$ is in the direction of AE. Since $\phi < 0$, $\Theta - \phi = \pi/2$, and using Equation E-5, we find

$$\tan\Theta = \frac{\sin\Theta}{\cos\Theta} = \frac{\cos\phi}{-\sin\phi} = -\frac{1}{\tan\phi} = \frac{\partial T/\partial y}{\partial T/\partial x}$$

This shows that the x- and y-components of a unit vector in the AE direction (Figure E-3) are proportional to $\partial T/\partial x$ and $\partial T/\partial y$, respectively. Moreover,

$$T_E - T_A \cong \frac{\partial T}{\partial g}\,\delta g$$

where $\delta g = AE$ and $\partial T/\partial g$ is the directional derivative in the g-direction—that is, it is the magnitude of the gradient. We thus see that the x- and y-components of $\nabla \mathbf{T}$ are $\partial T/\partial x$ and $\partial T/\partial y$, respectively. In vector notation,

$$\nabla \mathbf{T} = \hat{\mathbf{i}}\frac{\partial T}{\partial x} + \hat{\mathbf{j}}\frac{\partial T}{\partial y}$$

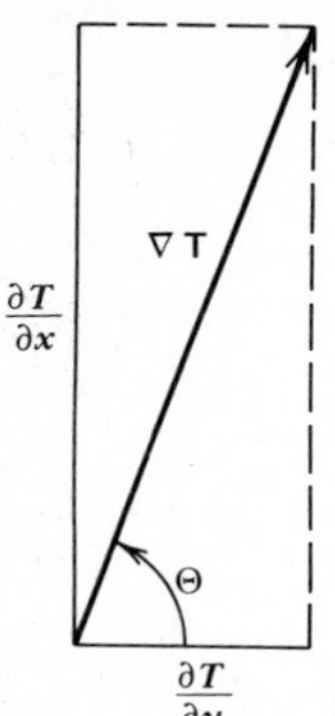

FIG. E-3 *The vector ΔT, its components $\partial T/\partial x$ and $\partial T/y$, and the angle Θ.*

where the symbol ∇ indicates the gradient and is read "del." In three dimensions, the term $\hat{\mathbf{k}}(\partial T/\partial z)$ would be added.

The lines of $T =$ constant were drawn as straight lines in Figure E-2. This is not the case in reality. But as δx and δy become smaller, this approximation becomes better; and at the limit, for $\delta x = dx$ and $\delta y = dy$, the approximation is exact; that is, in the neighborhood of A the line of constant T is a straight line.

APPENDIX F

Greek Alphabet

Name of Letter	Greek Alphabet
Alpha	α
Beta	β
Gamma	γ
Delta	Δ, δ
Epsilon	ϵ
Zeta	ζ
Eta	η
Theta	Θ, θ
Iota	ι
Kappa	κ
Lambda	Λ, λ
Mu	μ

Name of Letter	Greek Alphabet
Nu	ν
Xi	ξ
Omicron	o
Pi	π
Rho	ρ
Sigma	Σ, σ
Tau	τ
Upsilon	υ
Phi	ϕ
Chi	χ
Psi	ψ
Omega	ω

SUBJECT INDEX

Page numbers in *italic* indicate major discussion of entry.